CHEMISTRY
The Central Science

Laboratory Experiments

John H. Nelson
University of Nevada

Kenneth C. Kemp
University of Nevada

Sixth Edition

CHEMISTRY

The Central Science

Theodore L. Brown
University of Illinois

H. Eugene LeMay, Jr.
University of Nevada

Bruce E. Bursten
The Ohio State University

Prentice Hall, Englewood Cliffs, NJ 07632

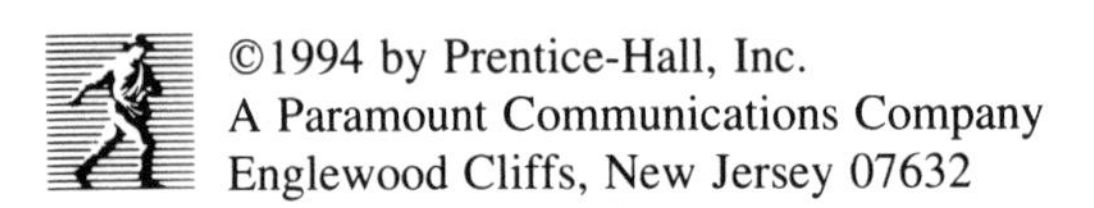

A Paramount Communications Company
Englewood Cliffs, New Jersey 07632

Printed in the United States of America

10 9 8 7 6 5 4 3

ISBN 0-13-338708-9

Prentice-Hall International (UK) Limited, *London*
Prentice-Hall of Australia Pty. Limited, *Sydney*
Prentice-Hall Canada Inc., *Toronto*
Prentice-Hall Hispanoamericana, S.A., *Mexico*
Prentice-Hall of India Private Limited, *New Delhi*
Prentice-Hall of Japan, Inc., *Tokyo*
Simon & Schuster Asia Pte. Ltd., *Singapore*
Editora Prentice-Hall do Brasil, Ltda., *Rio de Janeiro*

Contents

*Approximate time required to complete experiment.

Preface

Most students who take freshman chemistry are not planning for a career in this discipline. As a result, the introductory chemistry course usually serves several functions at various levels. It begins the training process for those who seek to become chemists. It introduces nonscience students to chemistry as an important, useful, and, we hope, interesting and rewarding part of their general education. It also should stimulate those students who are seeking the intellectual challenges and sense of purpose they hope to obtain from a career.

This manual has been written with these objectives in mind to accompany the sixth edition of the text *Chemistry: The Central Science* by Theodore L. Brown, H. Eugene LeMay, Jr., and Bruce E. Bursten. Each of the experiments is self-contained, with sufficient background material to conduct and understand the experiment. Each has a pedagogical objective to exemplify one or more specific principles. Because the experiments are self-contained, they may be undertaken in any order; however, we have found for our General Chemistry course that the sequence of Experiments 1 through 7 provides the firmest background and introduction.

To assist the students, we have included review questions to be answered before the experiments are begun. These are designed both to help the student understand the experiment and as an incentive to read the experiment in advance. As a further incentive, answers to some of these questions are provided in Appendix K.

We have made an effort to minimize the cost of the experiments. We have at the same time striven for a broad representation of the essential principles while keeping in mind that many students gain no other exposure to analytical techniques. Consequently, balances, pH meters, and spectrophotometers are used in some of the experiments. A list of necessary materials is given at the beginning of each experiment.

Each of the experiments contains a detachable report sheet that is easily graded, and most experiments contain unknowns. Very few of the experiments may be "dry-labbed."

In this sixth edition we have carefully edited all experiments for accuracy, safety and cost, and have increased the number of review questions and changed several of the existing ones. We have incorporated many of the valuable suggestions from users of the fifth edition for improving many of the experiments. A few of the suggestions for improving the experiments more properly concern the instructor and have therefore been included in the Instructor's Manual. Several of the experiments have been modified to eliminate the use of toxic chemicals, such as benzene and bromoform, and to reduce the amount of waste produced.

We owe a sincere debt of gratitude to the hundreds of students who have tested these experiments and commented on the directions. We wish to thank all of you who have sent us corrections and suggestions for improving these experiments and hope that you continue to make recommendations for their improvement.

We gratefully acknowledge the valuable suggestions and encouraging comments of H. Eugene LeMay, Jr. We wish to thank Dr. Roy G. Garvey, North Dakota State University and Ted Sakano, Rockland Community College for their criticisms and suggestions for improving this edition. We also wish to express our appreciation to Mary Hornby and Paul Banks for their cooperation and editorial assistance in the preparation of this manual.

JOHN H. NELSON
KENNETH C. KEMP

University of Nevada, Reno

To the Student

You are about to engage in what for most of you will be an unique experience. You are going to collect experimental data on your own and use your reasoning powers to draw logical conclusions about the meaning of these data. Your laboratory periods are short, and in most instances, there will not be enough time to come to the laboratory unaware of what you are to do, collect your experimental data, make conclusions and/or calculations regarding them, clean up, and hand in your results. Thus, you should *read the experimental procedure in advance* so that you can work in the lab most efficiently.

After you've read through the experiment, try to answer the review questions we've included at the end of each experiment. These questions will help you to understand the experiment and give you another reason to read over the experiment in advance. You can check most of your own answers against the answers we've included in Appendix K.

Some of your experiments will also contain an element of *danger*. For this and other reasons, there are laboratory instructors present to assist you. They are your friends. Treat them well and above all don't be afraid to ask them questions. Within reason, they will be glad to help you.

Chemistry is an experimental science. The knowledge that has been accumulated through previous experiments provides the basis for today's chemistry courses. The information now being gathered will form the basis of future courses. There are basically two types of experiments that chemists conduct:

1. Qualitative—to determine the nature of processes, which are often unanticipated and sometimes unpredictable.
2. Quantitative—to determine the amount of a measurable change in mass, volume, or temperature, for example, including the time rate of change on processes for which the qualitative data are already known.

It is much easier to appreciate and comprehend the science of chemistry, if you actually participate in experimentation. Although there are many descriptions of the scientific method, the reasoning process involved is difficult to appreciate without performing experiments. Invariably there are experimental difficulties encountered in the laboratory that require care and patience to overcome. There are four objectives for you, the student, in the laboratory:

1. To develop the skills necessary to obtain and evaluate a reliable original result.

2. To record your results for future use.
3. To be able to draw conclusions regarding your result (with the aid of some coaching and reading in the beginning).
4. To learn to communicate your results critically and knowledgeably.

By attentively reading over the experiments in advance, and by carefully following directions and working safely in the laboratory, you will be able to accomplish all these objectives. Good luck and best wishes for an error-free and accident-free term.

Laboratory Safety and Work Instructions

LABORATORY SAFETY

The laboratory can be but is not necessarily a dangerous place. When intelligent precautions and a proper understanding of techniques are employed, the laboratory is no more dangerous than any other classroom. Most of the precautions are just common-sense practices. These include the following:

1. Wear *approved* eye protection (including splash guards) at all times while in the laboratory. (*No one will be admitted without it*.) Your safety eye protection may be slightly different from that shown, but it must include shatterproof lenses and side shields to provide protection from splashes.

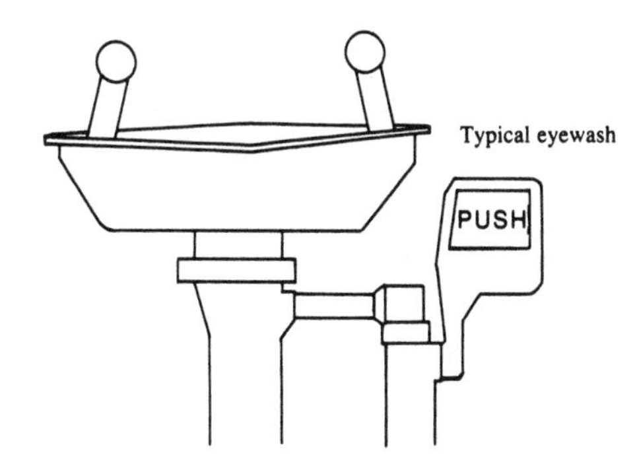

Typical eyewash

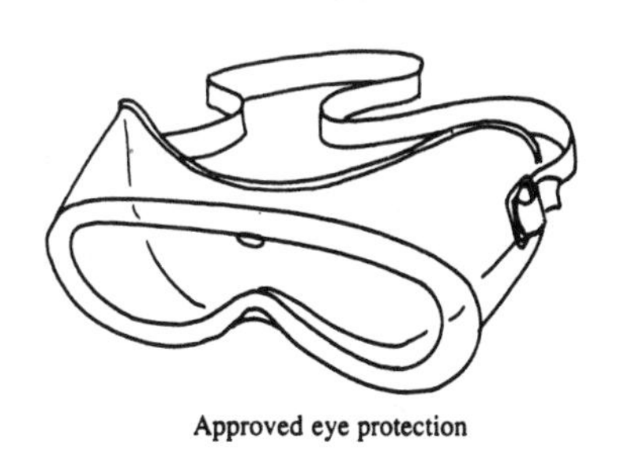
Approved eye protection

   The laboratory has an eyewash fountain available for your use. In the event that a chemical splashes near your eyes, you should use the fountain BEFORE THE MATERIAL RUNS BEHIND YOUR EYEGLASSES AND INTO YOUR EYES. The eyewash has a "panic bar," which enables its easy activation in an emergency.
2. Wear shoes at all times. (*No one will be admitted without them*.)
3. Eating, drinking, and smoking are strictly prohibited in the laboratory at all times.
4. Know where to find and how to use safety and first-aid equipment (see the first page of this book).
5. Consider all chemicals to be hazardous unless you are instructed otherwise. **Dispose of chemicals as instructed by your instructor.** Follow the explicit instructions given in the experiments.

6. If chemicals come into contact with your skin or eyes, wash immediately with copious amounts of water and then consult your laboratory instructor.
7. Never taste anything. Never directly smell the source of any vapor or gas; instead, by means of your cupped hand, bring a small sample to your nose. Chemicals are not to be used to obtain a "high" or clear your sinuses.

8. Perform in the hood any reactions involving skin-irritating or dangerous chemicals, or unpleasant odors. A typical fume exhaust hood is shown.

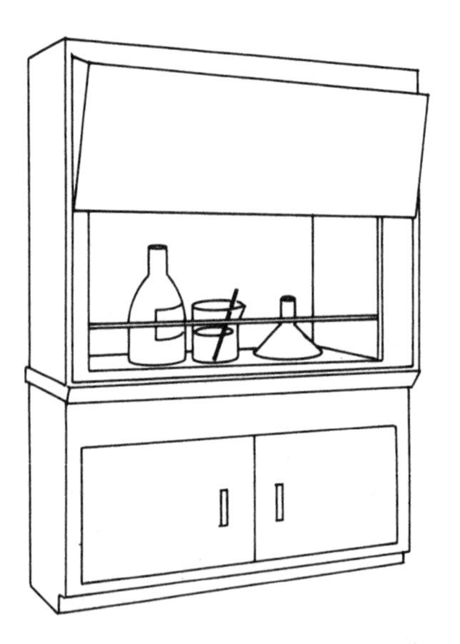

Exhaust hoods have fans to exhaust fumes out of the hood and away from the user. The hood should be used when noxious, hazardous, and flammable materials are being studied. It also has a shatterproof glass window, which may be used as a shield to protect you from minor explosions. Reagents that evolve toxic fumes are stored in the hood. Return these reagents to the hood after their use.

9. Never point a test tube that you are heating at yourself or your neighbor—it may erupt like a geyser.

10. Do not perform *any* unauthorized experiments.

11. Clean up all broken glassware *immediately*.

12. Always pour acids into water, not water into acid, because the heat of solution will cause the water to boil and the acid to spatter. "Do as you oughter, pour acid into water."

13. Avoid rubbing your eyes unless you *know* that your hands are clean.

14. When inserting glass tubing or thermometers into stoppers, *lubricate the tubing and the hole in the stopper with glycerol or water*. Wrap the rod in a towel and grasp it as close to the end being inserted as possible. Slide the glass into the rubber stopper with a twisting motion. Do not push. Finally, remove the excess lubricant by wiping with a towel. Keep your hands as close together as possible in order to reduce leverage.

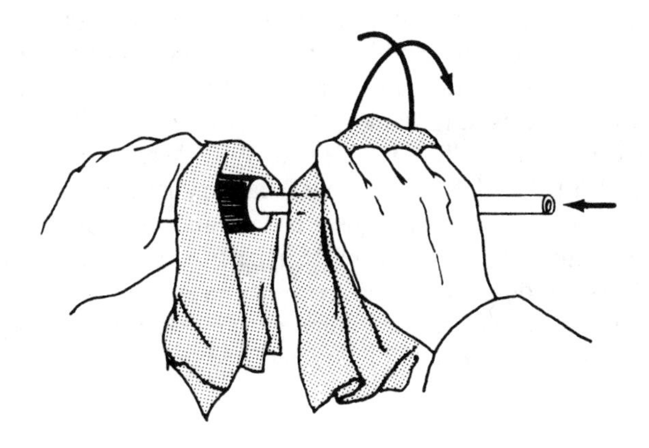

15. NOTIFY THE INSTRUCTOR IMMEDIATELY IN CASE OF AN ACCIDENT.

16. Many common reagents, for example, alcohols, acetone, and especially ether, are highly flammable. *Do not use them anywhere near open flames.*

17. Observe all special precautions mentioned in experiments.

18. Learn the location of fire protection devices.

In the unlikely event that a large chemical fire occurs, carbon dioxide fire extinguishers are available in the lab (usually mounted near one of the exits from the room). An example of a typical carbon dioxide fire extinguisher is shown below.

In order to activate the extinguisher, you must pull the metal safety ring from the handle and then depress the handle. Direct the output from the extinguisher at the base of the flames. The carbon dioxide smothers the flames and cools the flammable material quickly. If you use the fire extinguisher, be sure to turn the extinguisher in at the stockroom so that it can be refilled immediately. If the carbon dioxiode extinguisher does not extinguish the fire, evacuate the laboratory immediately and call the fire department.

One of the most frightening and potentially most serious accidents is the ignition of one's clothing. Therefore, certain types of clothing are hazardous in the laboratory and must *not* be worn. Since *sleeves* are most likely to come closest to flames, ANY CLOTHING THAT HAS BULKY OR LOOSE SLEEVES SHOULD NOT BE WORN IN THE LABORATORY. Ideally, students should wear laboratory coats with tightly fitting sleeves. Long hair also presents a hazard and must be tied back.

If a student's clothing or hair catches fire, his or her neighbors should take prompt action to prevent severe burns. Most laboratories have a water shower for such emergencies. A typical laboratory emergency water shower has the following appearance:

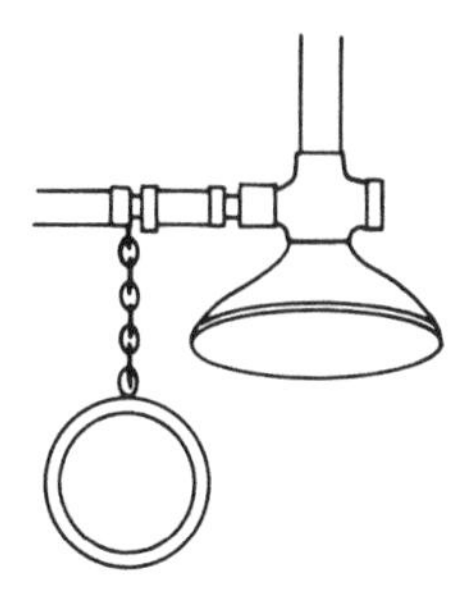

In case someone's clothing or hair is on fire, immediately lead the person to the shower and pull the metal ring. Safety showers generally dump 40 to 50 gallons of water, which should extinguish the flames. These showers cannot be shut off once the metal ring has been pulled. Therefore, the shower cannot be demonstrated. (Showers are checked for proper operation on a regular basis, however.)

19. Whenever possible use hot plates in place of Bunsen burners.

BASIC INSTRUCTIONS FOR LABORATORY WORK

1. Read the assignment *before* coming to the laboratory.
2. Work independently unless instructed to do otherwise.
3. Record your results directly onto your report sheet or notebook. DO NOT RECOPY FROM ANOTHER PIECE OF PAPER.
4. Work conscientiously to avoid accidents.
5. Dispose of excess reagents as instructed by your instructor. NEVER RETURN REAGENTS TO THE REAGENT BOTTLE.
6. Do not place reagent-bottle stoppers on the desk; hold them in your hand. Your laboratory instructor will show you how to do this. Replace the stopper on the same bottle, never on a different one.
7. Leave reagent bottles on the shelf where you found them.
8. Use only the amount of reagent called for; avoid excesses.
9. Whenever instructed to use water in these experiments, use distilled water unless instructed to do otherwise.
10. Keep your area clean.
11. Do not borrow apparatus from other desks. If you need extra equipment, obtain it from the stockroom.
12. When weighing, do not place chemicals directly on the balance.
13. Do not weigh hot or warm objects. Objects should be at room temperature.
14. Do not put hot objects on the desk top. Place them on a wire gauze or heat-resistant pad.

COMMON LABORATORY APPARATUS

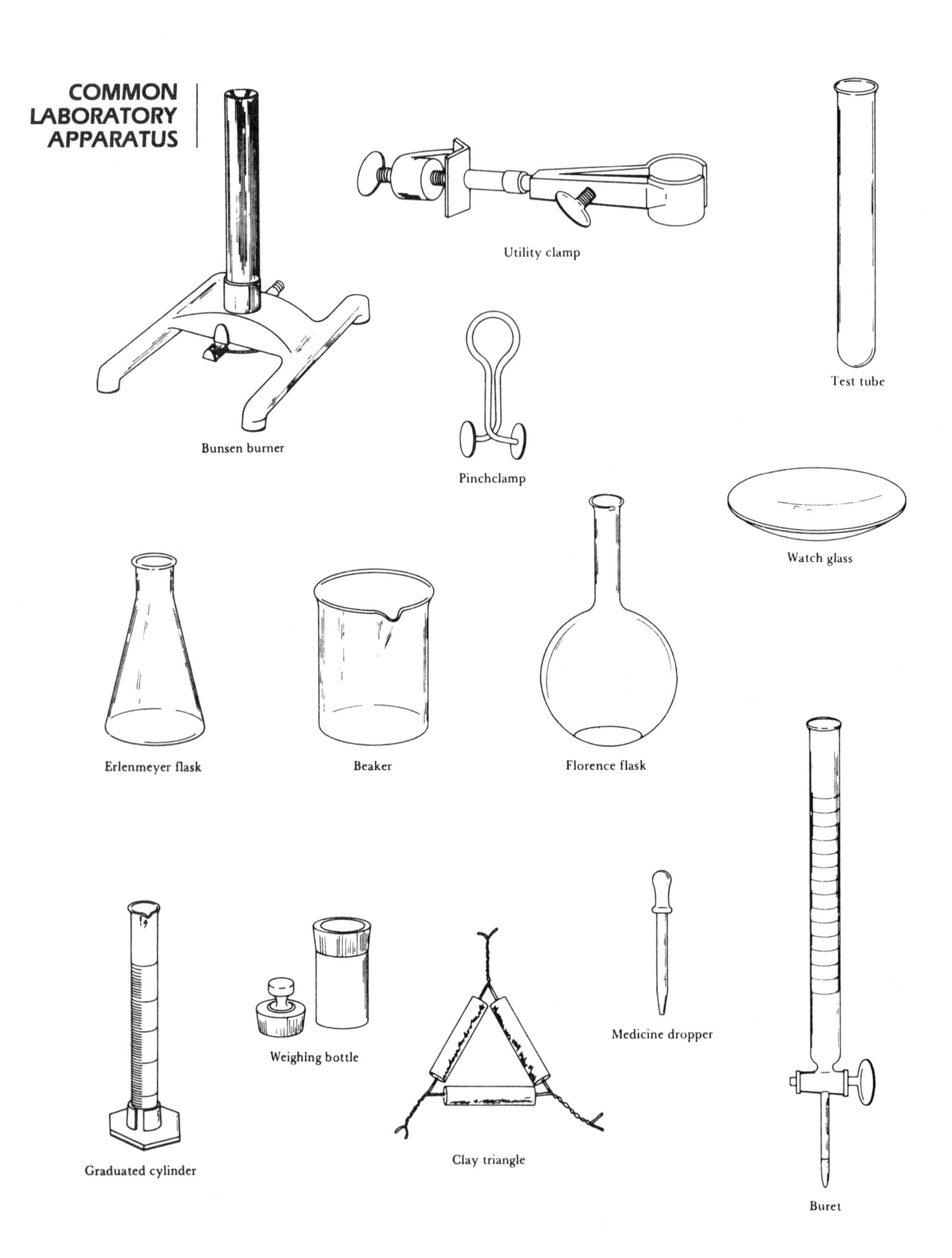

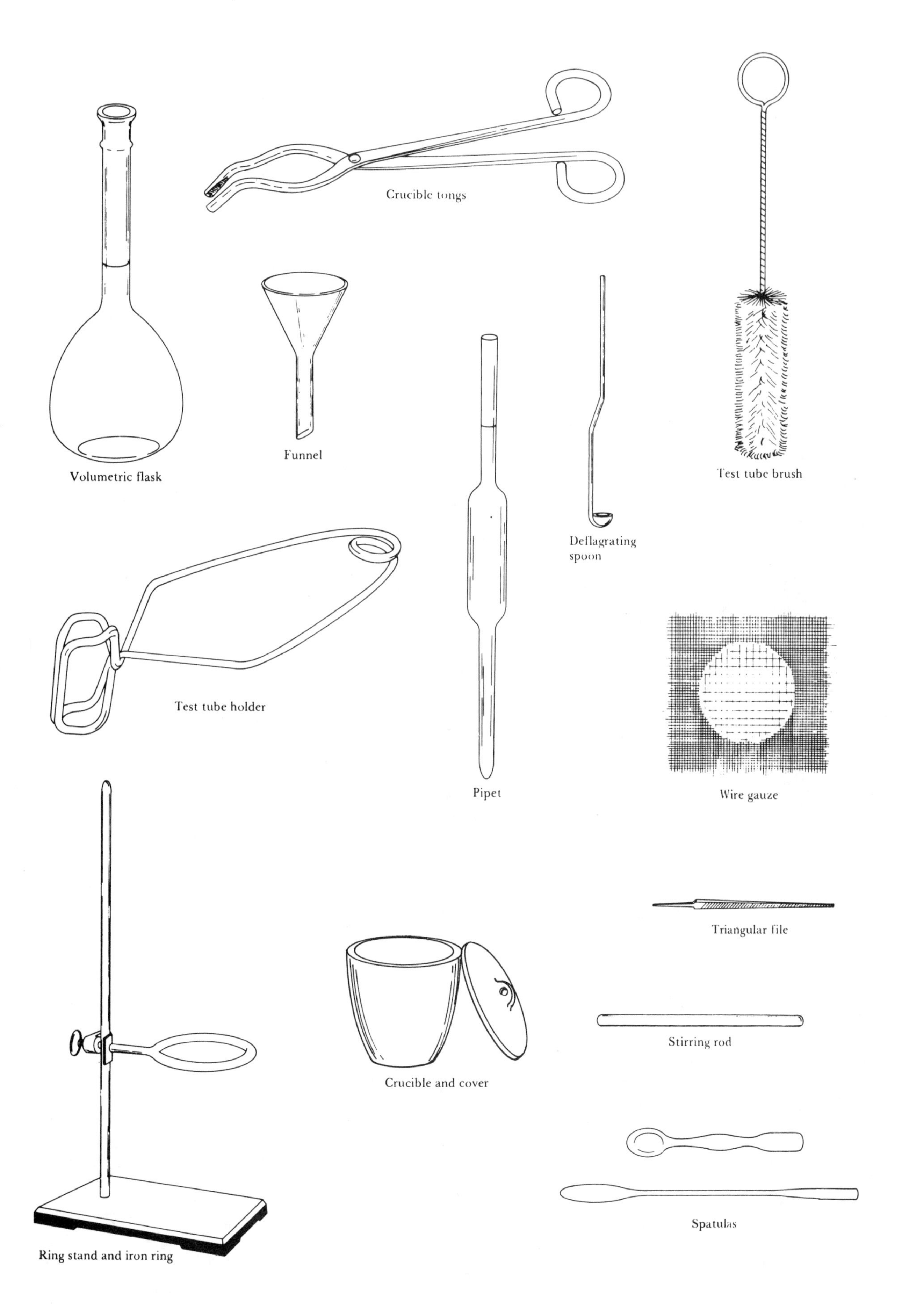
Crucible tongs
Volumetric flask
Funnel
Test tube brush
Deflagrating spoon
Test tube holder
Pipet
Wire gauze
Triangular file
Crucible and cover
Stirring rod
Spatulas
Ring stand and iron ring

CHEMISTRY
The Central Science

Basic Laboratory Techniques

EXPERIMENT 1

OBJECTIVE

To learn the use of common, simple laboratory equipment.

APPARATUS AND CHEMICALS

balance
150-mL beaker
250-mL beaker
Bunsen burner and hose
clamp
50-mL Erlenmeyer flask
125-mL Erlenmeyer flask
50- or 100-mL graduated cylinder
iron ring and ring stand
meterstick
10-mL pipet
rubber bulb
thermometer
wing tip
ice
barometer

DISCUSSION

Chemistry is an experimental science. It depends upon careful observation and the use of good laboratory techniques. In this experiment you will become familiar with some basic operations that will help you throughout this course. Your success as well as your safety in future experiments will depend upon your mastering these fundamental operations.

Because every measurement made in the laboratory is really an approximation, it is important that the numbers you record reflect the accuracy of the device you use to make the measurement. Appendix A of this manual contains a section on significant figures and measurements that you may find helpful in performing this experiment. Our system of weights and measures, the metric system, was originally based mainly upon fundamental properties of one of the world's most abundant substances, water. The system is summarized in Table 1.1. Conversions within the metric system are quite simple once you have committed to memory the meaning of the prefixes given in Table 1.2.

Recently, scientists have started to use a briefer version of the metric system of units in which the basic units for length, mass, and time are the meter, the kilogram, and the second. This system of units, known as the International System of Units, is commonly referred to as the SI system and is preferred in scientific work. A comparison of some common SI, metric, and English units is presented in Table 1.3.

Conversions within the metric system are quite easy if you remember the definitions for the prefixes and use dimensional analysis in problem solving.

In Table 1.1, the prefix *means* the power of 10. For example, 5.4 *centi*meters means 5.4×10^{-2} meter; *centi-* has the same meaning as $\times 10^{-2}$.

Table 1.1 Units of Measurement in the Metric System

Measurement	Unit and definition
Mass or weight	Gram (g) = weight of 1 cubic centimeter (cm^3) of water at 4°C and 760 mm Hg Mass = quantity of material Weight = mass × gravitational force
Length	Meter (m) = 100 cm = 1000 millimeters (mm) = 39.37 in.
Volume	Liter (L) = volume of 1 kilogram (kg) of H_2O at 4°C
Temperature	°C, measures heat intensity: $°C = \frac{5}{9}(°F - 32)$ or $°F = \frac{9}{5}°C + 32$
Heat	1 calorie (cal), amount of heat required to raise 1 g of water 1°C: 1 cal = 4.184 joules (J)
Density	*d*, usually g/mL for liquids and g/L for gases: $d = \frac{\text{mass}}{\text{unit volume}}$
Specific gravity	sp gr, dimensionless: $\text{sp gr} = \frac{\text{density of a substance}}{\text{density of a reference substance}}$

Table 1.2 The Meaning of Prefixes in the Metric System

Prefix	Meaning (power of 10)	Abbreviation
femto-	10^{-15}	f
pico-	10^{-12}	p
nano-	10^{-9}	n
micro-	10^{-6}	μ
milli-	10^{-3}	m
centi-	10^{-2}	c
deci-	10^{-1}	d
kilo-	10^{3}	k
mega-	10^{6}	M
giga-	10^{9}	G

Table 1.3 Comparison of SI, Metric, and English Units

Physical quantity	SI unit	Some common Metric units	Conversion factors
Length	Meter (m)	Meter (m) Centimeter (cm)	1 m = 10^2 cm 1 m = 39.37 in. 1 in. = 2.54 cm
Volume	Cubic meter (m^3)	Liter (L) Milliliter (ml)*	1 L = 10^3 cm^3 1 L = 10^{-3} m^3 1 L = 1.06 qt
Mass	Kilogram (kg)	Gram (g) Milligram (mg)	1 kg = 10^3 g 1 kg = 2.205 lb 1 lb = 453.6 g
Energy	Joule (J)	Calorie (cal)	1 cal = 4.184 J
Temperature	Kelvin (K)	Degree celsius (°C)	0K = −273.15°C $°C = \frac{5}{9}(°F - 32)$ $°F = \frac{5}{9}°C + 32$

*A mL is the same volume as a cubic centimeter: 1 mL = 1 cm^3

EXAMPLE 1.1

Convert 6.7 nanograms to milligrams.

Solution:

$$(6.7 \text{ ng})\left(\frac{10^{-9}\text{g}}{1 \text{ ng}}\right)\left(\frac{1 \text{ mg}}{10^{-3}\text{g}}\right) = 6.7 \times 10^{-6} \text{ mg}$$

Notice that the conversion factors have no effect on the magnitude (only the power of 10) of the mass measurement.

The quantities presented in Table 1.1 are measured with the aid of various pieces of apparatus. A brief description of some measuring devices follows.

Laboratory Balance

A laboratory balance is used to obtain the mass of various objects. There are several different varieties of balances, with various limits on their accuracy. Three of these balances are pictured in Figures 1.1 and 1.2. Most modern laboratories possess single-pan balances with only two knife edges. These are the most accurate balances; generally, they are also the simplest to use and are the most delicate and expensive. A comparison of the basic operation of a single-pan balance and that of a double-pan balance is given in Figure 1.3. The amount of material to be weighed and the accuracy required determine which balance you should use.

Meter Rule

The standard unit of length is the meter (m), which is 39.37 in. in length. A metric rule, or meterstick, is divided into centimeters (1 cm = 0.01 m; 1 m = 100 cm) and millimeters (1 mm = 0.001 m; 1 m = 1000 mm). It follows that 1 in. is 2.54 cm. (Convince yourself of this, since it is a good exercise in dimensional analysis.)

Graduated Cylinders

Graduated cylinders are tall, cylindrical vessels with graduations scribed along the side of the cylinder. Since volumes are measured in these cylinders by measuring the height of a column of liquid, it is critical that the cylinder have a uniform diameter along its entire height. Obviously, a tall cylinder with a

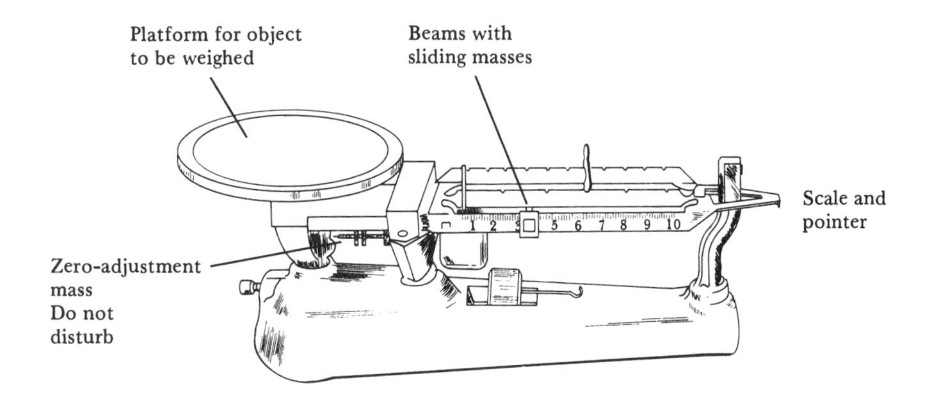

Figure 1.1 A laboratory platform balance (one knife edge) for crude weighing (0.1 g).

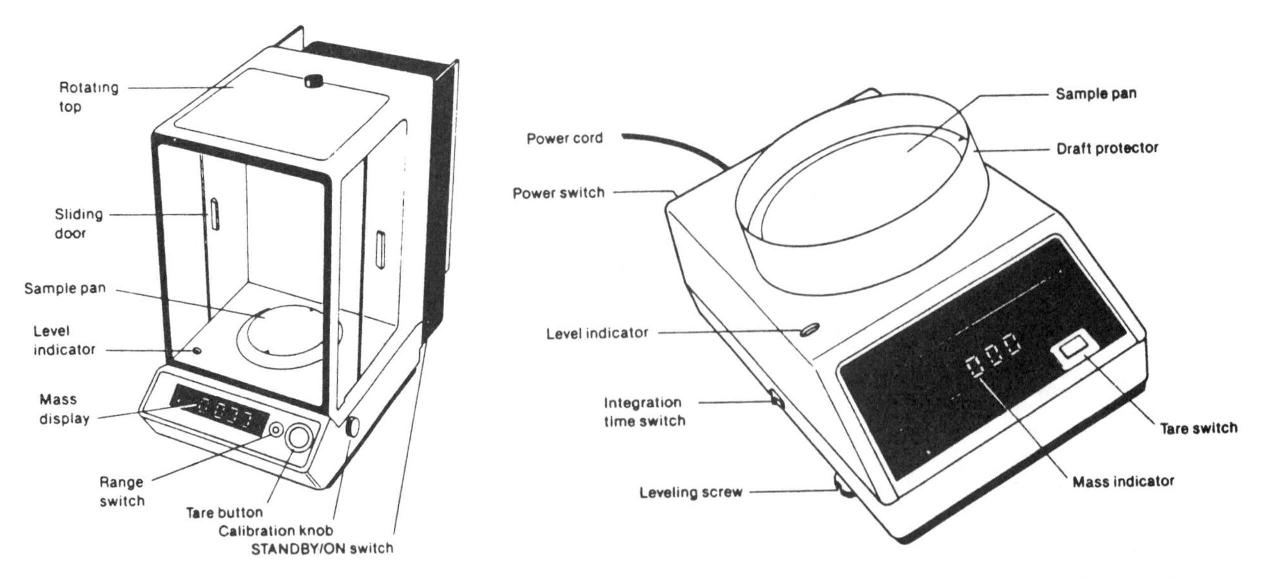

Figure 1.2 Digital electronic balances. The balance gives the mass directly when an object to be weighed is placed on the pan (0.0001 to 1 g depending upon the balance range).

small diameter will be more accurate than a short one with a large diameter. A liter (L) is divided into milliliters (mL) such that 1 mL = 0.001 L, 1 L = 1000 mL, and 1 L = 1.06 qt.

Thermometers

Most thermometers are based upon the principle that liquids expand when heated. Most common thermometers use mercury as the liquid. These thermometers are constructed so that a uniform-diameter capillary tube surmounts

Figure 1.3 A comparison of the essential features of the 2-pan and the 1-pan analytical balances. Substituting the flask for the two equivalent weights maintains a constant sensitivity on the 1-pan type, and also eliminates errors due to unequal lever arms.

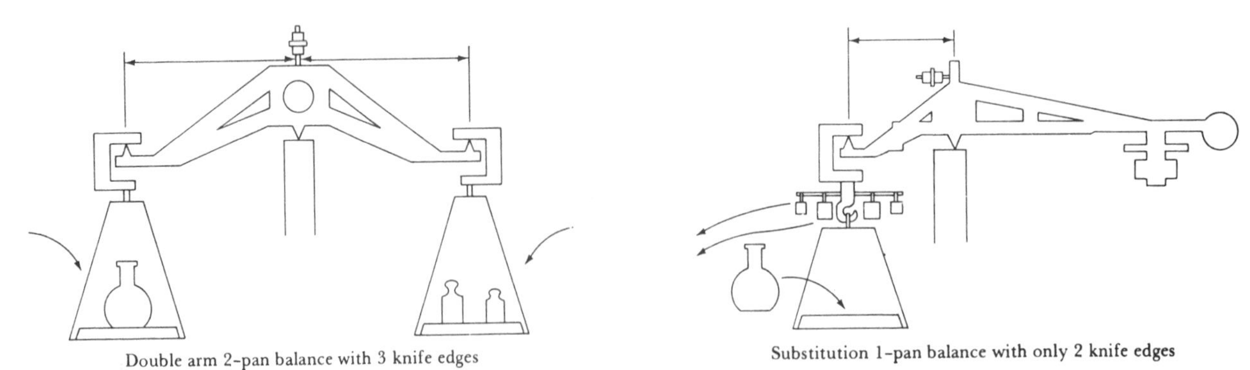

a mercury reservoir. To calibrate a thermometer, one defines two reference points, normally the freezing point of water (0°C, 32°F) and the boiling point of water (100°C, 212°F) at 1 atm of pressure (1 atm = 760 mm Hg).* Once these points are marked on the capillary, its length is then subdivided into uniform divisions called *degrees*. There are 100° between these two points on the Celsius, (°C, or centigrade) scale and 180° between those two points on the Fahrenheit (°F) scale.

Pipets

Pipets are glass vessels that are constructed and calibrated so as to deliver a precisely known volume of liquid at a given temperature. The markings on the pipet illustrated in Figure 1.4 signify that this pipet was calibrated to deliver (TD) 10.0 mL of liquid at 25°C. *Always* use a rubber bulb to fill a pipet. NEVER USE YOUR MOUTH! A TD pipet should not be blown empty.

It is important that you be aware that every measuring device, regardless of what it may be, has limitations in its accuracy. Moreover, to take full advantage of a given measuring instrument you should be familiar with or evaluate its accuracy. Careful examination of the subdivisions on the device will indicate the maximum accuracy you can expect of that particular tool. In this experiment you will determine the accuracy of your 10-mL pipet. The approximate accuracy of some of the equipment you will use in this course is given in Table 1.4.

Not only should you obtain a measurement to the highest degree of accuracy that the device or instrument permits, but you should also record the reading or measurement in a manner that reflects the accuracy of the instrument (see the section on significant figures in Appendix A). For example, a mass obtained from an analytical balance should be observed and recorded to the nearest 0.0001 g. This is illustrated in Table 1.5.

PROCEDURE

A. The Meterstick

Examine the meterstick and observe that one side is ruled in inches, while the other is ruled in centimeters. Measure and record the length and width of your lab book in both units. Mathematically convert the two measurements to show that they are equivalent.

B. The Bunsen Burner and Operations with Glass Tubing

The Bunsen burner is a convenient source of heat in the laboratory. Although there are several varieties, their principle of operation is the same and is similar to that of the common gas stove. The Bunsen burner requires gas and air, which it mixes in various proportions. The amount of air and gas mixed in the chamber is varied by use of the two adjustments illustrated in Figure 1.5. The relative proportions of gas and air determine the temperature of the flame.

* 1 mm Hg is also called 1 torr.

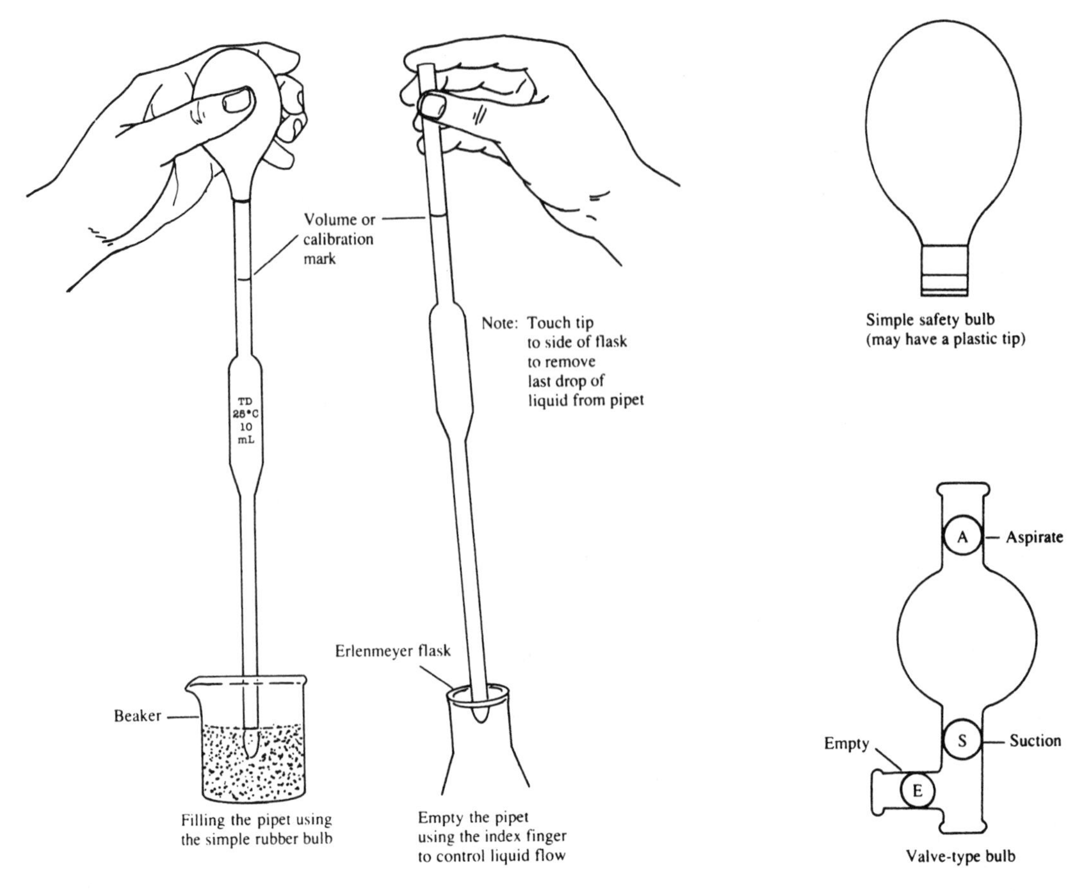

Figure 1.4 A typical volumetric pipet, rubber bulbs, and the pipet filling technique.

Table 1.4 Equipment Accuracy

Equipment	Accuracy
Analytical balance	±0.0001 g (±0.1 mg)
Top-loading balance	±0.001 g (±1 mg)
Meterstick	±0.1 cm (±1 mm)
Graduated cylinder	±0.1 mL
Pipet	±0.02 mL
Buret	±0.02 mL
Thermometer	±0.2°C

Table 1.5 Obtaining Significant Figures

Analytical balance	Top loader
85.9 g (incorrect)	85.9 g (incorrect)
85.93 g (incorrect)	85.93 g (incorrect)
85.932 g (incorrect)	85.932 g (correct)
85.9322 (correct)	

Examine your burner and locate the gas and air flow adjustments (valves) (see Figure 1.5). Determine how each valve operates before connecting the burner to the gas outlet. Close both valves; connect a rubber hose to the gas outlet on the burner and the desk; then open the desk valve about two-thirds of the way. Strike a match or use a gas lighter. Hold the lighted match to the side and just below the top of the barrel of the burner while gradually opening the gas valve on the burner to obtain a flame about 3 or 4 in. high. Gradually open and adjust the air valve until you obtain a pale blue flame with an inner cone as shown in Figure 1.5.

Glass is not a crystalline solid, but rather a supercooled liquid. Thus when it is heated it softens, flows, and can be worked. Crystalline solids melt rather than soften when heated.

Soda-lime glass (or soft glass) is made by heating a mixture of sodium carbonate, Na_2CO_3; calcium carbonate, $CaCO_3$; and silicon dioxide, SiO_2. It softens in the region of 300–400°C. It can easily be worked using a Bunsen burner, but because of its high temperature coefficient of expansion, it must be heated and cooled gradually to avoid undue strain or breakage. *Annealing* by a mild reheating and uniform, slow cooling is often wise. Such glass must not be laid on a cold surface while it is hot, because this introduces strains, and the glass will crack or even shatter.

Borosilicate glass (such as Pyrex or Kimax) does not soften much below 700–800°C and must be worked in an oxygen–natural gas flame or blowtorch.

Figure 1.5 Typical burners.

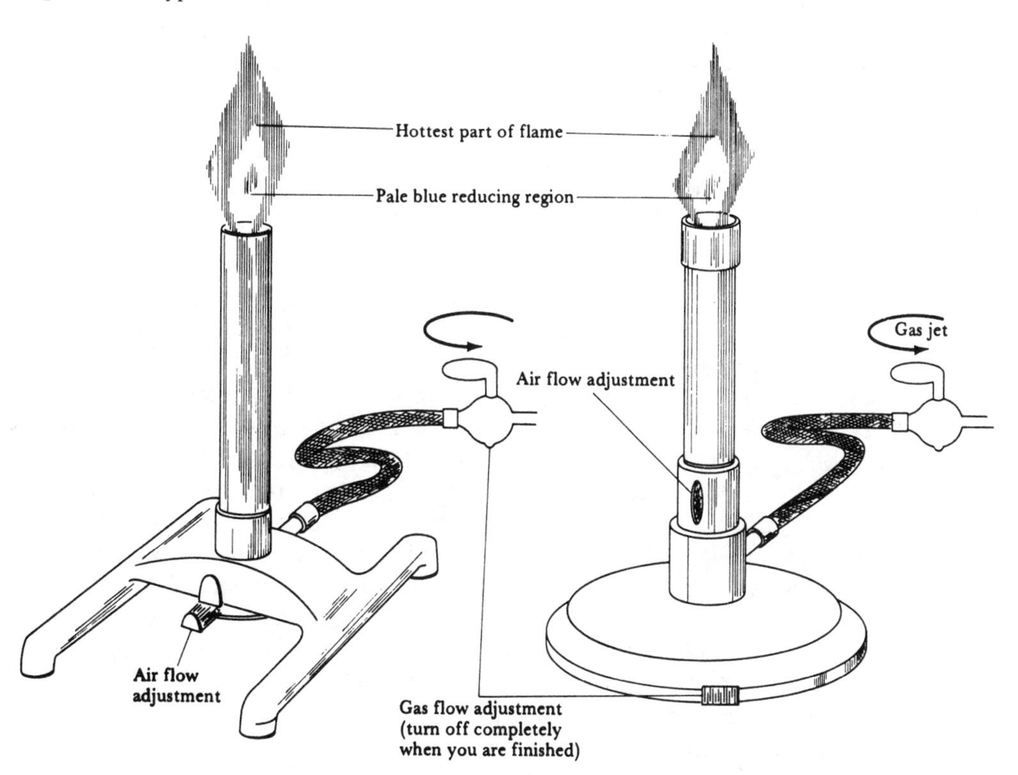

Because it has a low temperature coefficient of expansion, objects made from it can withstand sudden temperature changes. For this reason, among others, most of today's laboratory glassware is made from borosilicate glass. However, care must be exercised with it also. Uneven or rapid heating or cooling will cause the glass to shatter.

Study the illustrations in Figures 1.6 and 1.7 thoroughly and observe your instructor's demonstrations of how to work the glass. *Construct a right-angle bend and a 60° bend and two dropper tips for use in future experiments.* Use a wing tip on your Bunsen burner for the construction of the angle bends.

C. The Graduated Cylinder

Examine the 100-mL graduated cylinder and notice that it is scribed in milliliters. Fill the cylinder approximately half full with water. Notice that the *meniscus* (curved surface of the water) is concave (see Figure 1.8).

The *lowest* point on the curve is always read as the volume, never the upper level. Avoid errors due to parallax; different and erroneous readings are obtained if the eye is not perpendicular to the scale. Read the volume of water to the nearest 0.1 mL. Record this volume. Measure the maximum amount of water that your 125-mL Erlenmeyer flask will hold. Record this volume.

D. Relationship Between Temperature Scales

Both the Fahrenheit (°F) and Celsius (°C) temperature scales are linear scales that are based upon the physical properties of water. On the Celsius scale, water freezes at 0°C and boils at 100°C, whereas on the Fahrenheit scale, water freezes at 32°F and boils at 212°F. Using these two values, make a plot of degrees Fahrenheit versus degrees Celsius on the graph paper provided. Determine the Celsius equivalent of 40°F using your graph. The relationship between these two temperature scales is linear, that is, it is, of the form $y = mx + b$. Consult Appendix B regarding linear relationships and determine the equation that relates degrees Fahrenheit to degrees Celsius; then compute the Celsius equivalent of 40°F using this relationship.

E. The Thermometer and its Calibration

This part of the experiment is performed to check the accuracy of your thermometer. These measurements will show how measured temperatures (read from thermometer) compare with true temperatures (the boiling and freezing points of water). The freezing point of water is 0°C; the boiling point depends upon atmospheric pressure and is calculated as shown in Example 1.2. Place approximately 50 mL of ice in a 250-mL beaker and cover the ice with distilled water. Allow about 15 min for the mixture to come to equilibrium and then measure and record the temperature of the mixture. *Theoretically, this temperature is 0°C.* Now, set up a 250-mL beaker on a wire gauze and iron ring as shown in Figure 1.9. Fill the beaker about half full with distilled water. Adjust your burner to give maximum heating and begin heating the water. (*Time can be saved if the water is heated while other parts of the experiment are being conducted.*) Periodically determine the temperature of the water with the thermometer, but be careful not to touch the walls of the beaker with the thermometer bulb. Record the boiling point (b.p.) of the water. Using the data given in Example 1.2, determine the *true boiling point at the observed*

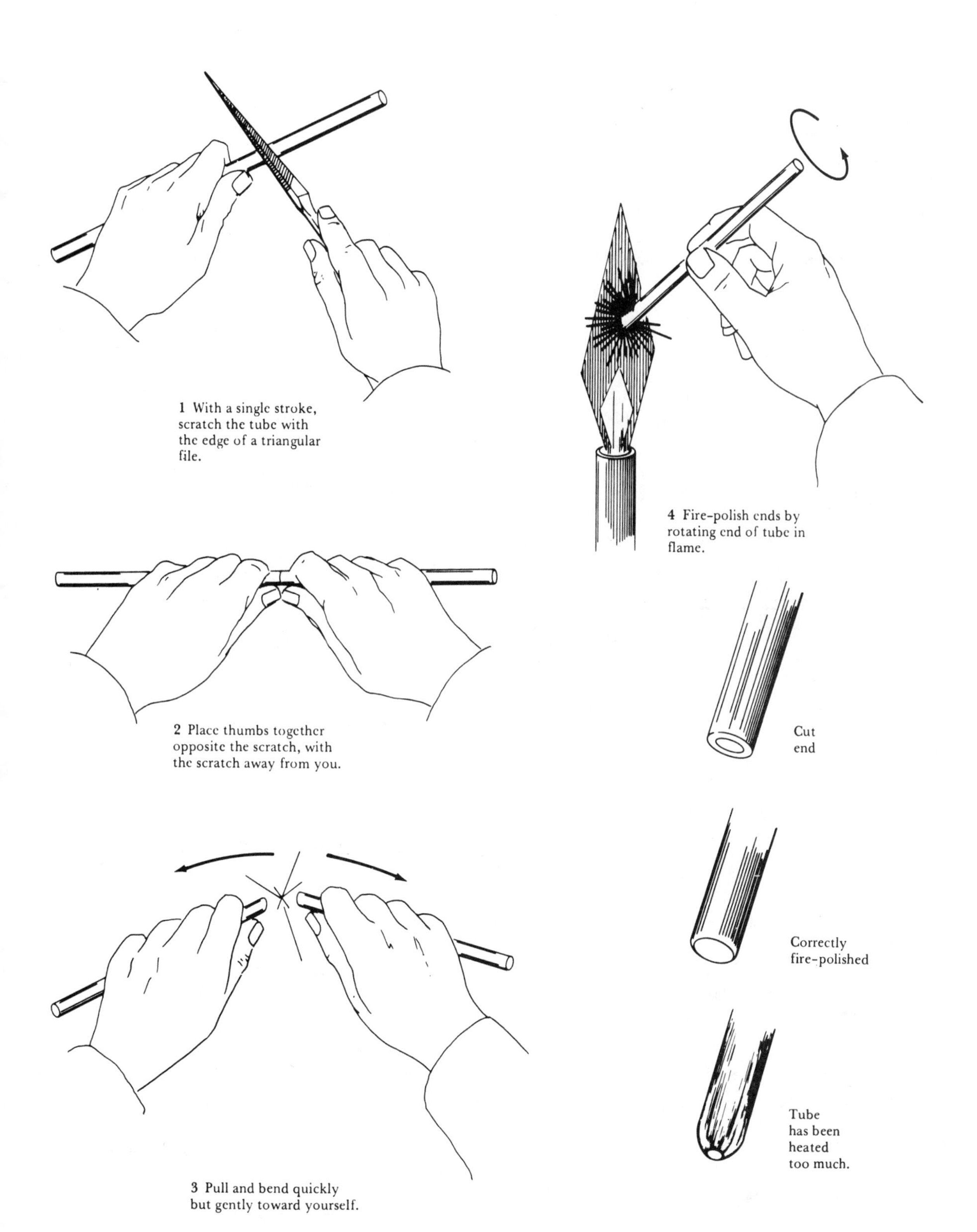

Figure 1.6 Cutting and fire-polishing a glass tube.

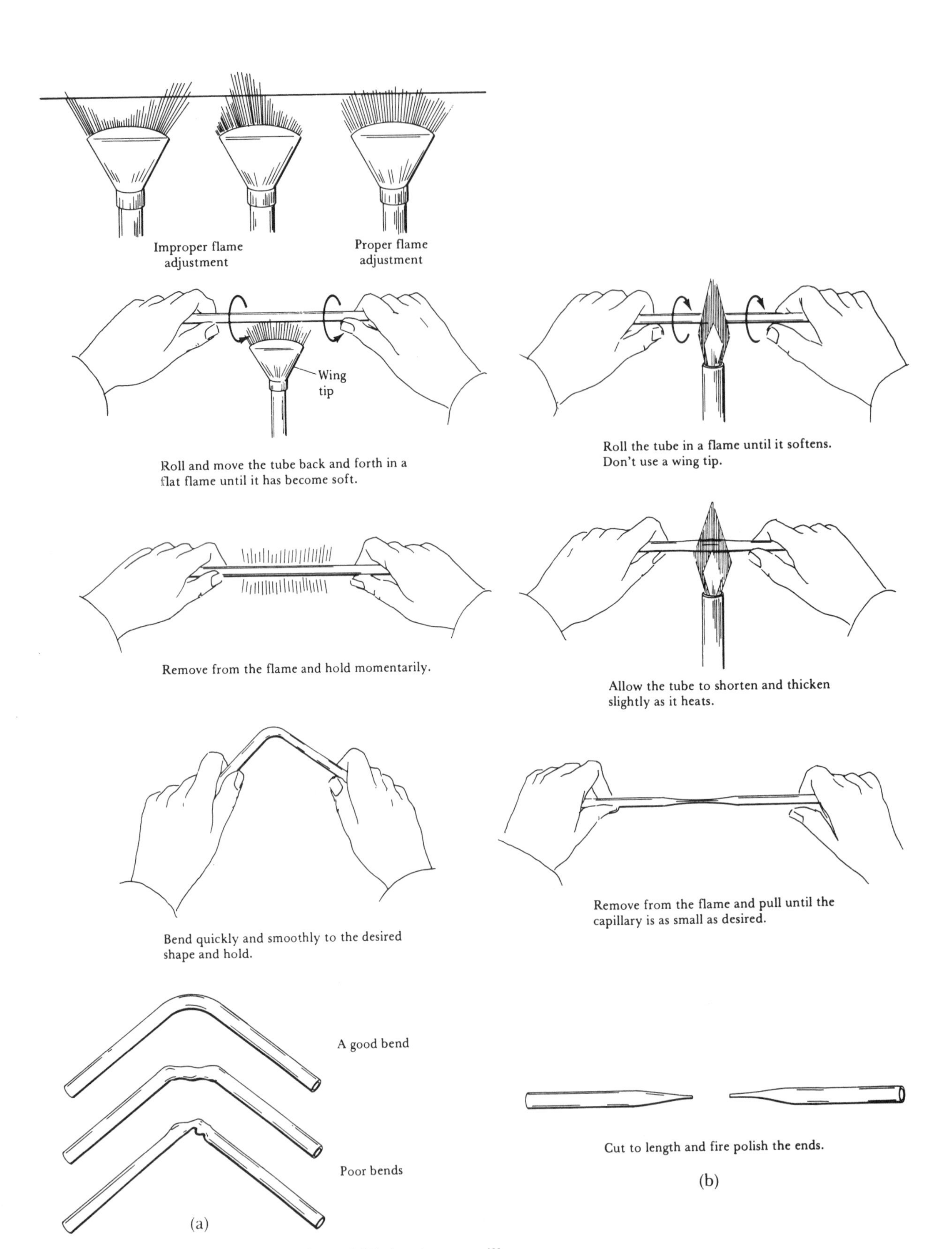

Figure 1.7 (a) Bending a glass tube and (b) drawing a capillary.

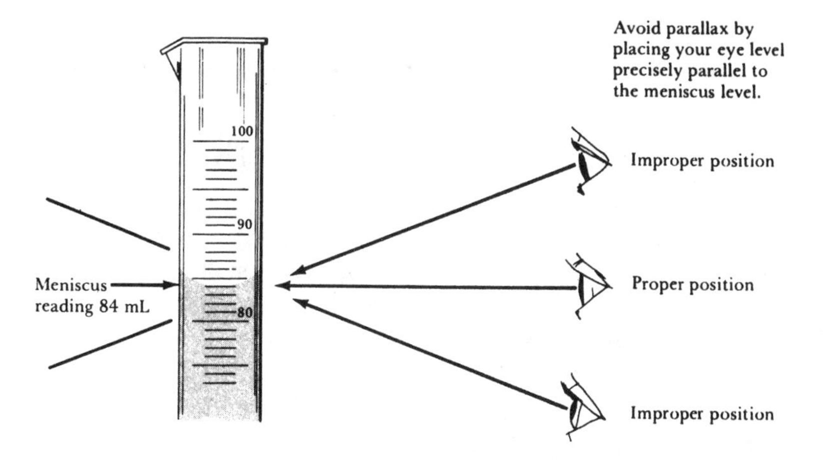

Figure 1.8 Proper eye position for taking volume readings.

atmospheric pressure. Obtain the atmospheric pressure from your laboratory instructor.

EXAMPLE 1.2

Determine the boiling point of water at 659.3 mm Hg.

Solution: Temperature corrections to the boiling point of water are calculated using the following formula:

$$\text{b.p. correction} = (760 \text{ mm Hg} - \text{atmospheric pressure}) \times (0.037°\text{C/mm})$$

The correction at 659.3 mm Hg is therefore

$$\text{b.p. correction} = (760 \text{ mm Hg} - 659.3 \text{ mm Hg}) \times (0.037°\text{C/mm}) = 3.7°\text{C}$$

The true boiling point is thus

$$100.0°\text{C} - 3.7°\text{C} = 96.3°\text{C}$$

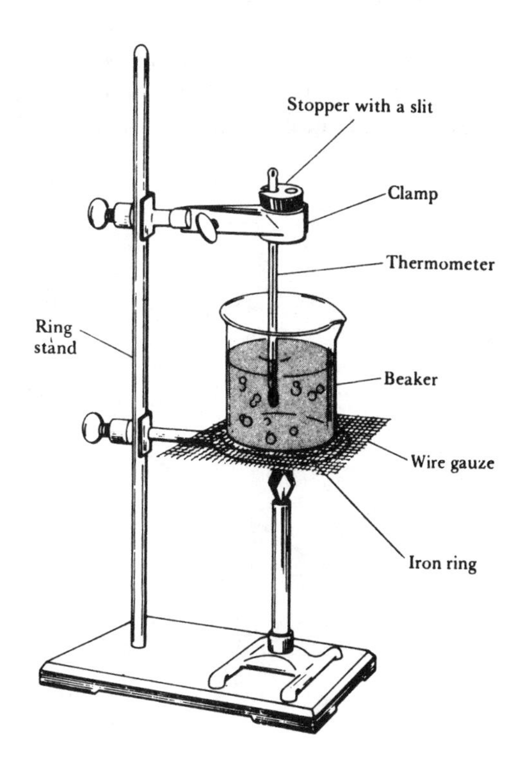

Figure 1.9 Apparatus setup for thermometer calibration.

Using the graph paper provided, construct a thermometer-calibration curve like the one shown in Figure 1.10 by plotting observed temperatures versus true temperatures for the boiling and freezing points of water.

F. Using the Balance to Calibrate Your 10-mL Pipet

Weighing an object on a single-pan balance is a simple matter. Because of the sensitivity and the expense of the balance (some cost more than $2500) you must be careful in its use. Directions for operation of single-pan balances vary with make and model. Your laboratory instructor will explain how to use the balance. Regardless of the balance you use, proper care of the balance requires that you observe the following:

1. Do not drop an object on the pan.
2. Center the object on the pan.
3. Do not place chemicals directly on the pan; use a beaker, watch glass, weighing bottle, or weighing paper.
4. Do not weigh hot or warm objects; objects must be at room temperature.
5. Return all weights to the zero position after weighing.
6. Clean up any chemical spills in the balance area.
7. Inform your instructor if the balance is not operating correctly; do not attempt to repair it yourself.

Weigh a penny and record its mass.

The following method is used to calibrate a pipet or other volumetric glassware. Obtain about 40 mL of distilled water in a 150-mL beaker. Allow the water to sit on the desk while you weigh and record the weight of an empty, dry 50-mL Erlenmeyer flask (tare) to the nearest 0.1 mg. Measure and record the temperature of the water. Using your pipet, pipet exactly 10 mL of water into this flask and weigh the flask with the water in it (gross) to the

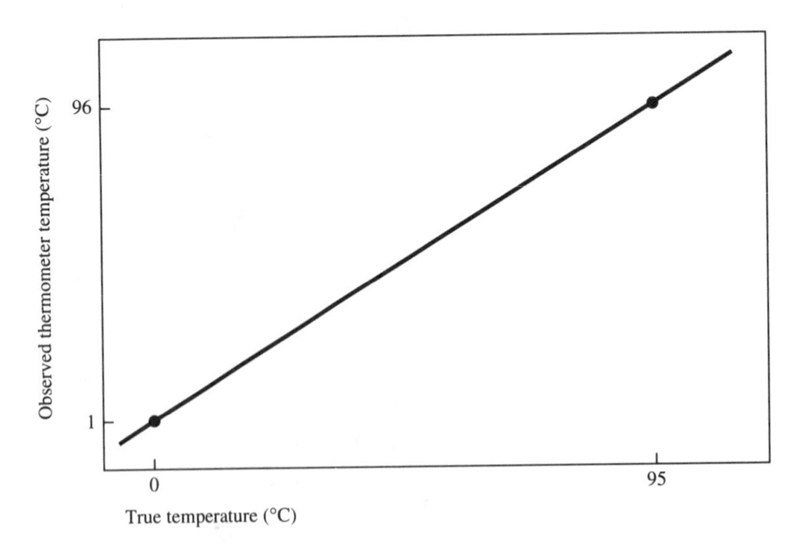

Figure 1.10 Typical thermometer-calibration curve.

nearest 0.1 mg. Obtain the weight of the water by subtraction (gross − tare = net). Using the equation below and the data given in Table 1.6, obtain the volume of water delivered and therefore the volume of your pipet.

$$\text{Density} = \frac{\text{mass}}{\text{volume}} \qquad d = \frac{m}{V}$$

Normally, density is given in units of grams per milliliter (g/mL) for liquids, grams per cubic centimeter (g/cm^3) for solids, and grams per liter (g/L) for gases. Repeat this procedure in triplicate—that is, deliver and weigh exactly 10 mL of water three separate times.

Table 1.6 Density of Pure Water at Various Temperatures

T (°C)	*d* (g/mL)	*T* (°C)	*d* (g/mL)
15	0.999099	22	0.997770
16	0.998943	23	0.997538
17	0.998774	24	0.997296
18	0.998595	25	0.997044
19	0.998405	26	0.996783
20	0.998203	27	0.996512
21	0.997992	28	0.996232

EXAMPLE 1.3

Using the procedure given above, a weight of 10.0025 g was obtained as the weight of the water delivered by one 10-mL pipet at 22°C. What is the volume delivered by the pipet?

Solution: From the density equation given above, we know that

$$V = \frac{m}{d}$$

For mass we substitute our value of 10.0025 g. For the density, consult Table 1.6. At 22°C, the density is 0.997770 g/mL. The calculation is

$$V = \frac{10.0025 \text{ g}}{0.997770 \text{ g/mL}} = 10.0249 \text{ mL}$$

which must be rounded off to 10.02, because the pipet's precision can be determined only to within ±0.02 mL.

The precision of a measurement is a statement about the internal agreement among repeated results; it is a measure of the reproducibility of a given set of results. The arithmetic mean (average) of the results is usually taken as the "best" value. The simplest measure of precision is the *average deviation from the mean*. The average deviation is calculated by first determining the mean of the measurements, then calculating the deviation of each individual measurement from the mean and, finally, averaging the deviations (treating each as a positive quantity). Study Example 1.4 and then, using your own experimental results, calculate the mean volume delivered by your 10-mL pipet. Also calculate for your three trials the individual deviations from the mean and then state your pipet's volume with its average deviation.

EXAMPLE 1.4

The following values were obtained for the calibration of a 10-mL pipet: 10.10, 9.98, and 10.00 mL. Calculate the mean value and the average deviation from the mean.

Solution:

$$\text{Mean} = \frac{10.10 + 9.98 + 10.00}{3} = 10.03$$

Deviations from the mean: $|\text{value} - \text{mean}|$

$$|10.10 - 10.03| = 0.07$$
$$|9.98 - 10.03| = 0.05$$
$$|10.00 - 10.03| = 0.03$$

Average deviation from the mean

$$= \frac{0.07 + 0.05 + 0.03}{3} = 0.05$$

The reported value is therefore 10.03 ± 0.05 mL.

REVIEW QUESTIONS

You should be able to answer the following questions before beginning this experiment:

1. What are the basic units of length, mass, volume, and temperature in the SI system?
2. A liquid has a volume of 1.25 liters. What is its volume in mL? in cm^3?
3. If an object weighs 1.52 g, what is its weight in mg?
4. Why should you never weigh a hot object?
5. Why is it necessary to calibrate a thermometer and volumetric glassware?
6. What is precision?
7. What is the definition of density? Can it be determined from a single measurement?
8. What is the density of an object with a mass of 9.67 g and a volume of 0.2236 mL?
9. Weighing an object three times gave the following results: 10.4 g, 10.1 g, and 10.2 g. Find the mean weight and the average deviation from the mean.

10. Normal body temperature is 98.6°F. What is the corresponding Celsius temperature?

11. What is the weight in kilograms of 1000 mL of a substance that has a density of 1.349 g/mL?

12. An object weighs exactly six grams on an analytical balance that has an accuracy of ±0.1 mg. To how many significant figures should this weight be recorded?

NOTES AND CALCULATIONS

Identification of Substances by Physical Properties

EXPERIMENT 2

OBJECTIVE

To become acquainted with procedures used in evaluating physical properties and the use of these properties in identifying substances.

APPARATUS AND CHEMICALS

balance
250-mL beaker
burner and hose
dropper
50-mL Erlenmeyer flask
10-mL graduated cylinder
large test tubes (2)
small test tubes (6)
test-tube rack
spatula
10-mL pipet
5-mL pipet
tubing with right-angle bend
ring stand and ring
utility clamp
wire gauze
thermometer
no. 3 two-hole stopper with one of the holes slit to the side or a buret clamp
two-hole stopper
stirring rod
small rubber bands (or small sections of $\frac{1}{4}$-in. rubber tubing)
20 mL cyclohexane
boiling chips
15 mL ethyl alcohol
1 g naphthalene
5 mL toluene
two unknowns (one liquid, one solid)
capillary tubes (5)
50-mL beakers (2)
small watch glass
soap solution

DISCUSSION

Properties are those characteristics of a substance that enable us to identify it and to distinguish it from other substances. Direct identification of some substances can readily be made by simply examining them. For example, we see color, size, shape, and texture, and we can smell odors and discern a variety of tastes. Thus, copper can be distinguished from other metals on the basis of its color.

Physical properties are those properties that can be observed without altering the composition of the substance. Whereas it is difficult to assign definitive values to such properties as taste, color, and odor, other physical properties, such as melting point, boiling point, solubility, density, viscosity, and refractive index, can be expressed quantitively. For example, the melting point of copper is 1087°C, and its density is 8.96 g/cm^3. As you probably realize, a specific combination of properties is unique to a given substance, thus making it possible to identify most substances just by careful determination of several properties. This is so important that large books have been compiled listing characteristic properties of most known substances. Many scientists, most notably several German scientists during the latter part of the nineteenth century and earlier part of the twentieth, spent their entire lives gathering data of this sort. Two of the most complete references of this type that are readily available today are The Chemical Rubber Company's *Handbook of Chemistry and Physics* and N. A. Lange's *Handbook of Chemistry*.

In this experiment you will use the following properties to identify a substance whose identity is unknown to you: solubility, density, melting point, and boiling point. The *solubility* of a substance in a solvent at a specified temperature is the maximum weight of that substance that dissolves in a given volume (usually 100 mL or 1000 mL) of a solvent. It is tabulated in handbooks in terms of grams per 100 mL of solvent; the solvent is usually water.

In the preceding experiment you learned that the density of a substance is defined as the mass per unit volume:

$$d = \frac{m}{V}$$

Melting or freezing points correspond to the temperature at which the liquid and solid states of a substance are in equilibrium. These terms refer to the same temperature but differ slightly in their meaning. The *freezing point* is the equilibrium temperature when approached from the liquid phase, that is, when solid begins to appear in the liquid. The *melting point* is the equilibrium temperature when approached from the solid phase, that is, when liquid begins to appear in the solid.

A liquid is said to boil when bubbles of vapor form within it, rise rapidly to the surface, and burst. Any liquid in contact with the atmosphere will boil when its vapor pressure is equal to atmospheric pressure—that is, the liquid and gaseous states of a substance are in equilibrium. Boiling points of liquids depend upon atmospheric pressure. A liquid will boil at a higher temperature at a higher pressure or at a lower temperature at a lower pressure. The temperature at which a liquid boils at 760 mm Hg is called the *normal* boiling point. To account for these pressure effects on boiling points, people have studied and tabulated data for boiling point versus pressure for a large number of compounds. From these data, nomographs have been constructed. A *nomograph* is a set of scales for connected variables (see Figure 2.5 for an example); these scales are so placed that a straight line connecting the known values on some scales will provide the unknown value at the straight line's intersection with other scales. A nomograph allows you to find the correction necessary to convert the normal boiling point of a substance to its boiling point at any pressure of interest.

PROCEDURE

A. Solubility

CAUTION: *Cyclohexane is highly flammable and must be kept away from open flames*. Qualitatively determine the solubility of naphthalene (mothballs) in three solvents: water, cyclohexane, and ethyl alcohol. Determine the solubility by adding a few crystals of napthalene to 2 to 3 mL (it is not necessary to measure either the solute weight or solvent volume) of each of these three solvents in separate, clean, *dry* test tubes. Make an attempt to keep the amount of napthalene and solvent the same in each case. Place a cork in each test tube and shake briefly. Cloudiness indicates insolubility. Record your conclusions on the report sheet using the abbreviations s (soluble), sp (sparingly soluble), and i (insoluble). Into each of three more clean, *dry* test tubes place 2 or 3 mL of these same solvents and add 4 or 5 drops of toluene in place of naphthalene. Record your observations. The formation of two layers indicates immiscibility (lack of solubility). Now repeat these experiments using each of the three solvents (water, cyclohexane, and ethyl alcohol) with your solid and liquid unknowns and record your observations.

B. Density

Determine the densities of your two unknowns in the following manner.

The Density of a Solid Weigh about 1.5 g of your solid unknown to the nearest 0.001 g and record the weight. Using a pipet or a wash bottle, half fill a *clean, dry* 10-mL graduated cylinder with a solvent in which your unknown is *insoluble*. Be *careful* not to get the liquid on the inside walls, because you do not want your solid to adhere to the cylinder walls when you add it in a subsequent step. Read and record this volume to the nearest 0.1 mL. Add the weighed solid to the liquid in the cylinder, being careful not to lose any of the material in the transfer process and ensuring that all of the solid is beneath the surface of the liquid. Carefully tapping the sides of the cylinder with your fingers will help settle the material to the bottom. Do not be concerned about a few crystals that do not settle, but if a large quantity of the solid resists settling, add one or two drops of a soap solution and continue tapping the cylinder with your fingers. Now read the new volume to the nearest 0.1 mL. The difference in these two volumes is the volume of your solid (see Figure 2.1). Calculate the density of your solid unknown.

You may recall that by measuring the density of metals in this way Archimedes proved to the king that the charlatan alchemists had in fact not transmuted lead into gold. He did this after observing that he weighed less in the bathtub than he did normally by an amount equal to the weight of the fluid displaced. According to legend, upon making his discovery Archimedes

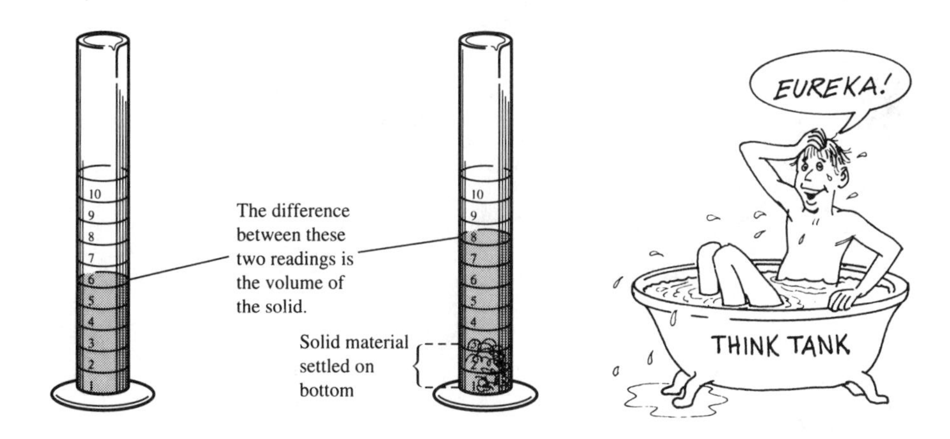

Figure 2.1

emerged from his bath and ran naked through the streets shouting *Eureka!* (I have found it).

The Density of a Liquid Weigh a clean, *dry* 50-mL Erlenmeyer flask to the nearest 0.0001 g. Obtain at least 15 mL of the unknown liquid in a clean, *dry* test tube. Using a 10-mL pipet, pipet exactly 10 mL of the unknown liquid into the 50 mL Erlenmeyer flask and quickly weigh the flask containing the 10 mL of unknown to the nearest 0.0001 g. Using the calibration value for your pipet (from Experiment 1) and the weight of this volume of unknown, calculate its density. Record your results and show how (with units) you performed your calculations. (SAVE THE LIQUID FOR YOUR BOILING-POINT DETERMINATION.)

C. Melting Point (for Solid Unknown)

Obtain a capillary tube and a small rubber band. Seal one end of the capillary tube by carefully heating the end in the edge of the flame of a Bunsen burner until the end *completely* closes. Rotating the tube during heating will help you to avoid burning yourself (see Figure 2.2).

Pulverize a small portion of your solid-unknown sample with the end of a test tube on a clean watch glass; partially fill the capillary with your unknown by gently tapping the pulverized sample with the open end of the capillary to force some of the sample inside. Drop the capillary into a glass tube about 38 to 50 cm in length, with the sealed end down to pack the sample into the bottom of the capillary tube. Repeat this procedure until the sample column is roughly 5 mm in height. Now set up a melting-point apparatus as illustrated in Figure 2.3.

Place the rubber band about 5 cm above the bulb on the thermometer and out of the liquid. Carefully insert the capillary tube under the rubber band with the closed end at the bottom. Place the thermometer with attached capillary into the beaker of water so that the sample is covered by water, the thermometer does not touch the bottom of the beaker, and the open end of the capillary tube is above the surface of the water. Heat the water slowly while gently agitating the water with a stirring rod. Observe the sample in the capillary tube

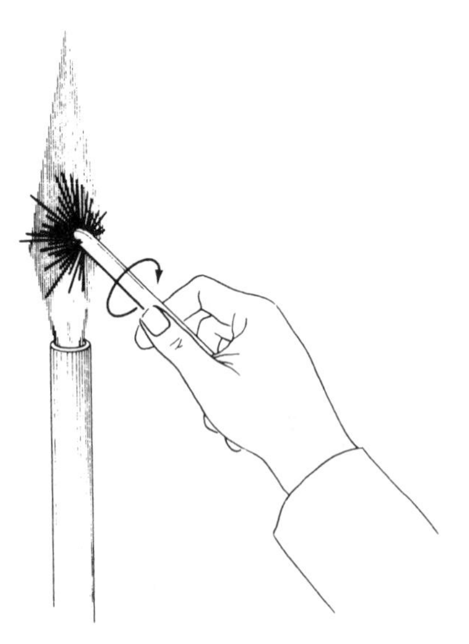

Figure 2.2 Sealing one end of a capillary tube.

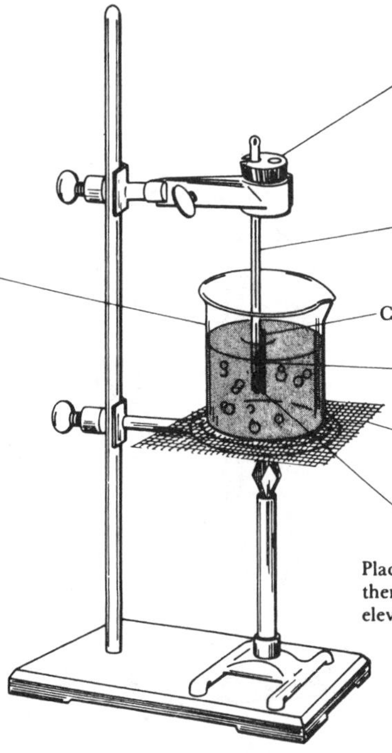

Figure 2.3 Apparatus for melting-point determination.

while you are doing this. At the moment that the solid melts, record the temperature. Also record the melting-point range, which is the temperature range between the temperature at which the sample begins to melt and the temperature at which all of the sample has melted. Using your thermometer-calibration curve (from Experiment 1), correct these temperatures to the true temperatures and record the melting point and melting-point range. These temperatures may differ by only 1°C or less.

D. Boiling Point (for Liquid Unknown)

Determine the boiling point of your liquid unknown (use some of the same material you used to determine the density) by adding about 3 mL to a clean, dry test tube. Fit the test tube with a two-hole rubber stopper that has one slit; insert your thermometer into the hole with the slit and one of your right-angle-bend glass tubes into the other hole, as shown in Figure 2.4. Add one or two small boiling chips to the test tube to ensure even boiling of your sample. Position the thermometer so that it is about 1 cm above the surface of the unknown liquid. Clamp the test tube in the ring stand and connect to the right-angle-bend tubing a length of rubber tubing that reaches to the sink. Assemble your apparatus as shown in Figure 2.4. (CAUTION: *Be certain that there are no constrictions in the rubber tubing. Your sample is flammable. Keep it away from open flames.*)

Heat the water gradually and watch for changes in temperature. The temperature will become constant at the boiling point of the liquid. Record the observed boiling point. Correct the observed boiling point to the true boiling point at room atmospheric pressure using your thermometer-calibration curve. The normal boiling point (b.p. at 1 atm = 760 mm Hg) can now be calculated (see Example 2.1, below) using the nomograph provided in Figure 2.5. Your boiling-point correction should not be more than +5°C.

Figure 2.4 Apparatus for boiling-point determination.

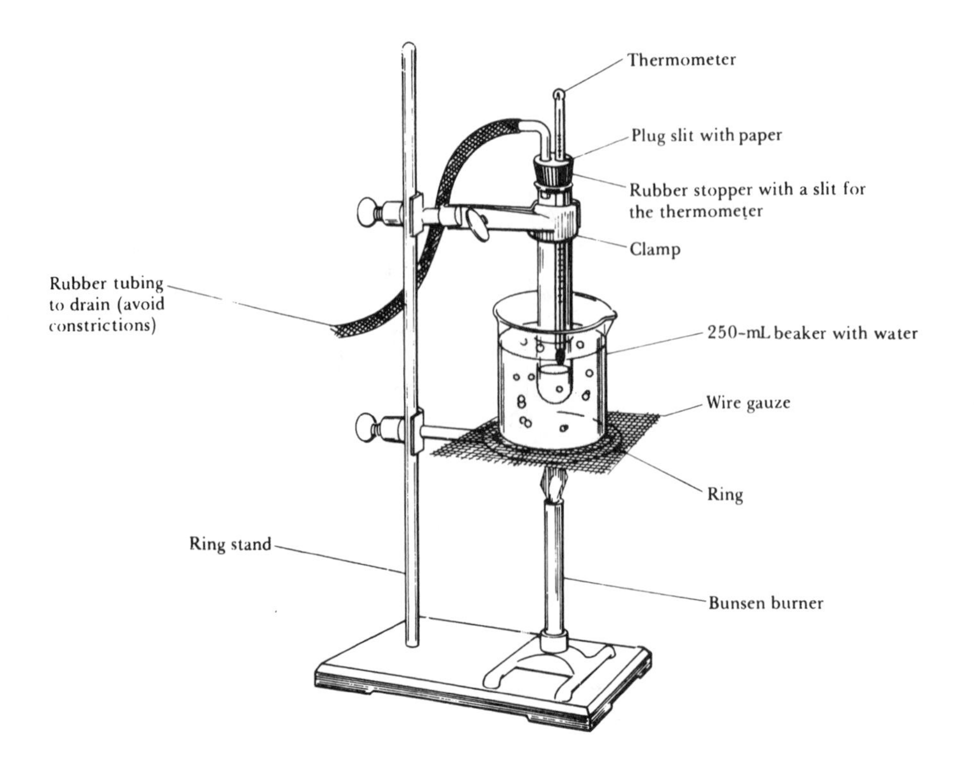

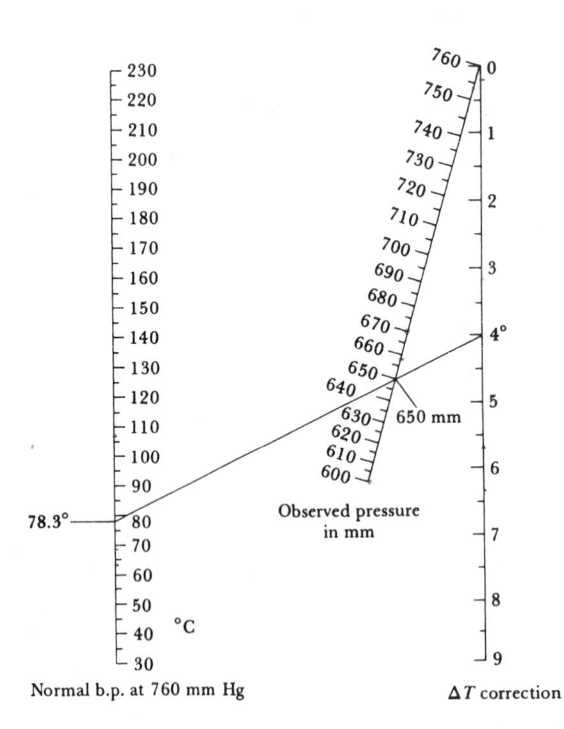

Figure 2.5 Nomograph for boiling point correction to 760 mm Hg.

EXAMPLE 2.1

What will be the boiling point of ethanol at 650 mm Hg when its normal boiling point at 760 mm Hg is known to be 78.3°C.

Solution: The answer is easily found by consulting the nomograph in Figure 2.5. A straight line drawn from 78.3°C on the left scale of normal boiling points through 650 mm Hg on the pressure scale intersects the temperature correction scale at 4°C. Therefore,

$$\text{Normal b.p.} - \text{correction} = \text{observed b.p.}$$

$$78.3°\text{C} - 4.0°\text{C} = 74.3°\text{C}$$

Similar calculations could be done for the compounds in Table 2.1 at any pressure listed on the nomograph in Figure 2.5. In this experiment you will observe a boiling point at a pressure other than at 760 mm Hg, and you wish to know its normal boiling point. In order to estimate its normal boiling point, assume that your observed boiling point is 57.0°C and the observed pressure is 650 mm Hg. Use your observed boiling point of 57.0°C as if it were the normal boiling point and find the correction

Table 2.1 Physical Properties of Pure Substances

Substance	Density (g/mL)	Melting point (°C)	Boiling point (°C)	Solubility[a] in		
				Water	Cyclohexane	Alcohol
Acetanilide	1.22	114	304	sp	sp	s
Acetone	0.79	−95	56	s	s	s
Benzophenone	1.15	48	306	i	s	s
Bromoform	2.89	8	150	i	s	s
2,3-Butanedione	0.98	−2.4	88	s	s	s
t-Butyl alcohol	0.79	25	83	s	s	s
Cadmium nitrate · $4H_2O$	2.46	59	132	s	i	s
Chloroform[b]	1.49	−63.5	61	i	s	s
Cyclohexane	0.78	6.5	81.4	i	s	s
p-Dibromobenzene	1.83	86.9	219	i	s	s
p-Dicholorobenze	1.46	53	174	i	s	s
m-Dinitrobenzene	1.58	90	291	i	s	s
Diphenyl	0.99	70	255	i	s	s
Diphenylamine	1.16	53	302	i	s	s
Diphenylmethane	1.00	27	265	i	s	s
Ether, ethyl propyl	1.37	−79	64	s	s	s
Hexane	0.66	−94	69	i	s	s
Isopropyl alcohol	0.79	−98	83	s	s	s
Lauric acid	0.88	43	225	i	s	s
Magnesium nitrate · $6H_2O$	1.63	89	330[c]	s	i	s
Methyl alcohol	0.79	−98	65	s	sp	s
Methylene chloride[b]	1.34	−97	40.1	i	s	s
Naphthalene	1.15	80	218	i	s	sp
α-Naphthol	1.22	94	288	i	i	s
Phenyl benzoate	1.23	71	314	i	s	s
Propionaldehyde	0.81	−81	48.8	s	i	s
Sodium acetate · $3H_2O$	1.45	58	123	s	i	sp
Stearic acid	0.85	70	291	i	s	sp
Thymol	0.97	52	232	sp	s	s
Toluene	0.87	−95	111	i	s	s
p-Toluidine	0.97	45	200	sp	s	s
Zinc chloride	2.91	283	732	s	i	s

[a] s = soluble; sp = sparingly soluble; i = insoluble.
[b] Toxic. Most organic compounds used in the lab are toxic.
[c] Boils with decomposition.

for a pressure of 650 mm Hg. Using the nomograph, you can see that the correction is 3.8°C. You would then add this correction to your observed boiling point to obtain an approximate normal boiling point:

$$57°C + 3.8°C = 60.8°C, \text{ or } 61°C$$

By consulting Table 2.1, you can find the compound that best fits your data; in this example, the data are for chloroform.

E. Unknown Identification

Your unknowns are substances contained in Table 2.1. Compare the properties that you have determined for your unknowns with those in the table. Identify your unknowns and record your results.

REVIEW QUESTIONS

Before beginning this experiment in the laboratory, you should be able to answer the following questions:

1. List five physical properties.
2. A 8.792-mL sample of an unknown weighed 10.12 g. What is the density of the unknown?
3. Are the substances methylene chloride and diphenyl solids or liquids at room temperature?
4. Could you determine the density of benzophenone using water? Why or why not?
5. What would be the boiling point of hexane at 670 mm Hg?
6. Why do we calibrate thermometers and pipets?
7. Is bromoform miscible with water? with cyclohexane?
8. When water and toluene are mixed, two layers form. Is the bottom layer water or toluene? (See Table 2.1.)
9. What solvent would you use to determine the density of zinc chloride?
10. The density of a solid with a melting point of 52 to 54°C was determined to be 1.45 ± 0.02 g/mL. What is the solid?
11. The density of a liquid whose boiling point is 63–65°C was determined to be 1.36 ± 0.05 g/mL. What is the liquid?
12. Which has the greater volume, 10 g of chloroform or 10 g of acetone? What is the volume of each?

Separation of the Components of a Mixture

EXPERIMENT 3

OBJECTIVE

To become familiar with the methods of separating substances from one another using decantation, extraction, and sublimation techniques.

APPARATUS AND CHEMICALS

- balance
- Bunsen burner and hose
- clay triangles (2)
- evaporating dishes (2)
- glass stirring rod
- 50- or 100-mL graduated cylinder
- iron rings (2)
- ring stands (2)
- tongs
- watch glass
- unknown mixture of sodium chloride, ammonium chloride, and silicon dioxide (2–3g)

DISCUSSION

Materials that are not uniform in composition are said to be impure or heterogeneous and are called *mixtures*. Most of the materials we encounter in everyday life, such as cement, wood, and soil, are mixtures. When two or more substances that do not react chemically are combined, a mixture results. Mixtures are characterized by two fundamental properties: First, each of the substances in the mixture retains its chemical integrity; second, mixtures are separable into these components by physical means. If one of the substances in a mixture is preponderant—that is, if its amount far exceeds the amounts of the other substances in the mixture—then we usually call this mixture an impure substance and speak of the other substances in the mixture as impurities.

The preparation of compounds usually involves their separation or isolation from reactants or other impurities. Thus the separation of mixtures into their components and the purification of impure substances are frequently encountered problems. You are probably aware of everyday problems of this sort. For example, our drinking water usually begins as a mixture of silt, sand, dissolved salts, and water. Since water is by far the largest component in this mixture, we usually call this impure water. How do we purify it? The separation of the components of mixtures is based upon the fact that each component has different physical properties. The components of mixtures are always pure substances, either compounds or elements, and each pure substance possesses a unique set of properties. The properties of every sample of a pure substance are identical under the same conditions of temperature and pressure. This means that once we have determined that a sample of sodium chloride, $NaCl$, is water-soluble and a sample of silicon dioxide (sand), SiO_2, is not, we realize that all samples of sodium chloride are water-soluble and all samples of silicon dioxide are not.

Likewise, every crystal of a pure substance melts at a specific temperature, and at a given pressure, every pure substance boils at a specific temperature.

Although there are numerous physical properties that can be used to identify a particular substance (as you learned in the previous experiment), we will be concerned in this experiment merely with the separation of the components and not with their identification. The methods we will use for the separation depend upon differences in physical properties, and they include the following:

1. *Decantation*. This is the process of separation of a liquid from a solid (sediment) by gently pouring the liquid from the solid so as not to disturb the solid (see Figure 3.1).
2. *Filtration*. This is the process of separating a solid from a liquid by means of a porous substance, a filter, which allows the liquid to pass through but not the solid (see Figure 3.1). Common filter materials are paper, layers of charcoal, and sand. Silt and sand can be removed from our drinking water by this process.
3. *Extraction*. This is the separation of a substance from a mixture by preferentially dissolving that substance in a suitable solvent. By this process a soluble compound is usually separated from an insoluble compound.
4. *Sublimation*. This is the process in which a solid passes directly to the gaseous state and back to the solid state without the appearance of the liquid state. Not all substances possess the ability to be sublimed. Iodine, naphthalene, and ammonium chloride (NH_4Cl) are common substances that easily sublime.

The mixture that you will separate contains three components: $NaCl$, NH_4Cl, and SiO_2. Their separation will be accomplished by heating the mixture to sublime the NH_4Cl, extracting the $NaCl$ with water, and finally drying the remaining SiO_2, as illustrated in the scheme shown in Figure 3.2.

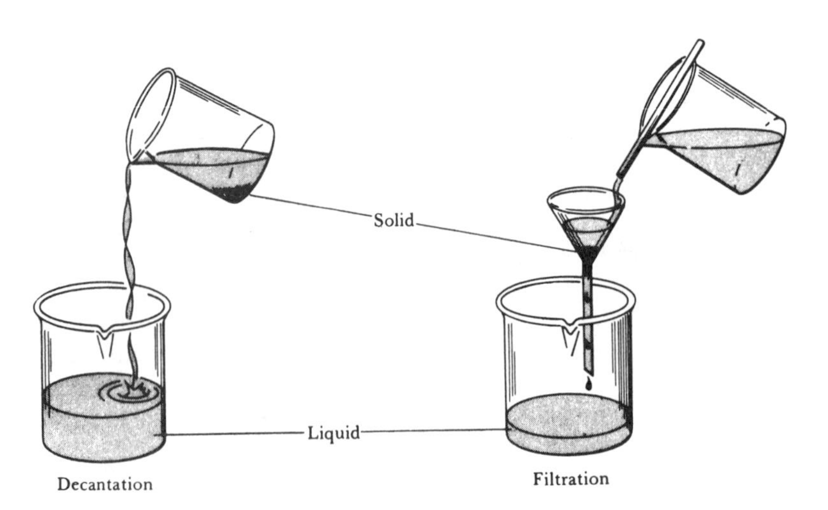

Figure 3.1

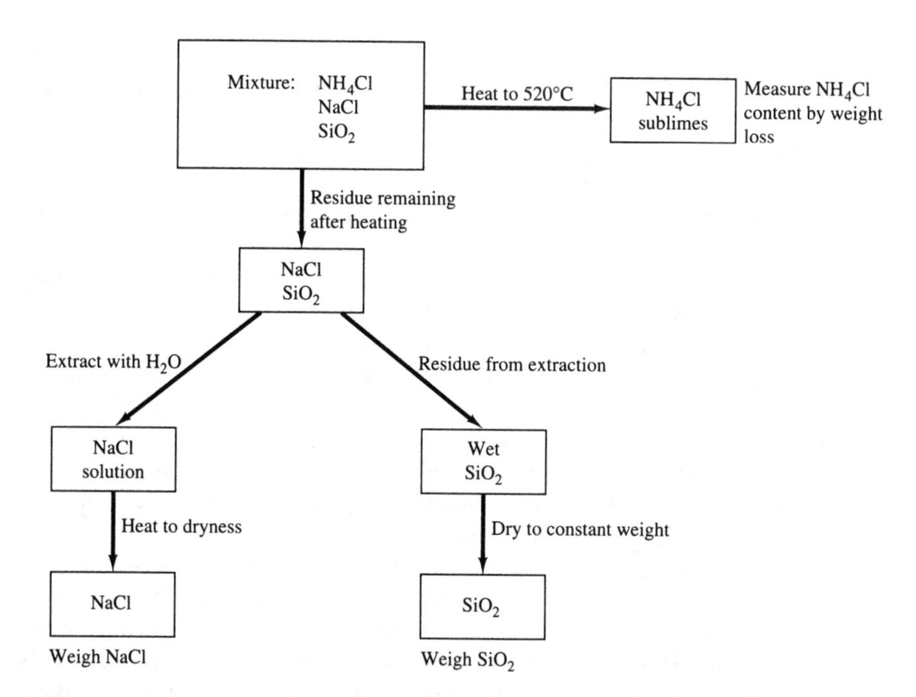

Figure 3.2 Flow diagram for the separation of the components of a mixture.

PROCEDURE

Carefully weigh a clean, dry evaporating dish to the nearest 0.01 g. Then obtain from your instructor a 2- to 3-g sample of the unknown mixture in the evaporating dish. Weigh the evaporating dish containing the sample and calculate the sample weight.

Place the evaporating dish containing the mixture on a clay triangle, ring, and ring-stand assembly IN THE HOOD as shown in Figure 3.3. Heat the evaporating dish with a burner until white fumes are no longer formed (a total

NOTE: To sublime NH_4Cl do not use the watch glass; but to dry $NaCl$ and SiO_2, use the watch glass on top of the evaporating dish. Heat slowly–do not flame the edges of the watch glass that extend beyond the edge of the evaporating dish.

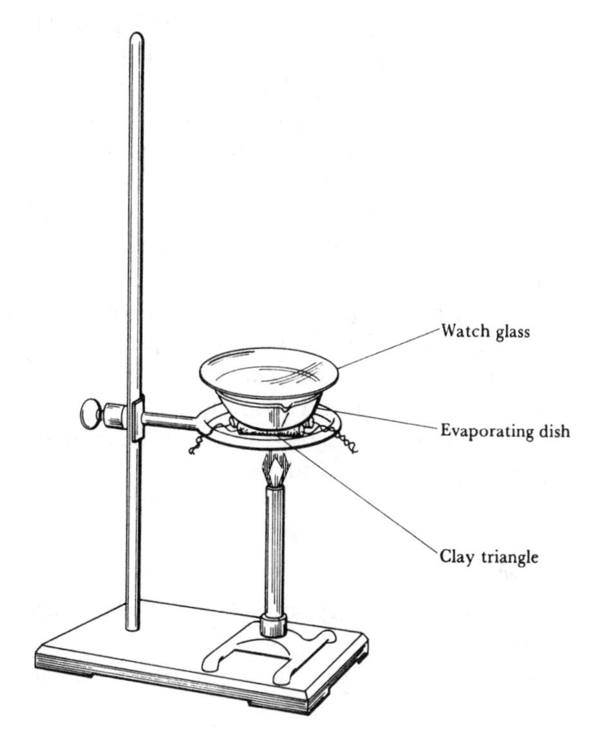

Figure 3.3

of about 15 min). Heat carefully to avoid spattering, especially when liquid is present. After the first 10 min remove the flame and gently stir the mixture with your glass stirring rod; then apply the heat again.

Allow the evaporating dish to cool until it reaches room temperature and then weigh the evaporating dish with the contained solid. NEVER WEIGH HOT OR WARM OBJECTS! The loss in weight represents the amount of NH_4Cl in your mixture; calculate this.

Add 25 mL of water to the solid in this evaporating dish and stir gently for 5 min. Next, weigh another clean, dry evaporating dish and watch glass. Decant the liquid carefully into the second evaporating dish, *which you have weighed,* being careful not to transfer any of the solid into the second evaporating dish. Add 10 mL more of water to the solid in the first evaporating dish, stir, and decant this liquid into the second evaporating dish as before. Repeat with still another 10 mL of water. This process extracts the soluble NaCl from the sand. You now have two evaporating dishes—one containing wet sand and the second, a solution of sodium chloride.

Place the evaporating dish containing the sodium chloride solution carefully on the clay triangle on the ring stand. Begin gently heating the solution to evaporate the water. Take care to avoid boiling or spattering, especially when liquid is present. Near the end, cover the evaporating dish with the watch glass that was weighed with this evaporating dish, and reduce the heat to prevent spattering. While the water is evaporating you may proceed to dry the SiO_2 in the other evaporating dish as explained in the next paragraph, if you have another Bunsen burner available. When you have dried the sodium chloride completely, no more water will condense on the watch glass, and it, too, will be dry. Let the evaporating dish and watch glass cool to room temperature and weigh them. The difference between this weight and the weight of the empty evaporating dish and watch glass is the weight of the NaCl. Calculate this weight.

Place the evaporating dish containing the wet sand on the clay triangle on the ring stand and cover the evaporating dish with a clean, dry watch glass. Heat slowly at first until the lumps break up and the sand appears dry. Then heat the evaporating dish to dull redness and maintain this heat for 10 min. Take care not to overheat, or the evaporating dish will crack. When the sand is dry, remove the heat and let the dish cool to room temperature. Weigh the dish after it has cooled to room temperature. The difference between this weight and the weight of the empty dish is the weight of the sand. Calculate this weight.

Calculate the percentage of each substance in the mixture using an approach similar to that shown in Example 3.1.

The accuracy of this experiment is such that the combined total of your three components should be in the neighborhood of 99 percent. If it is less than this, you have been sloppy. If it is more than 100 percent, you have not sufficiently dried the sand and salt.

EXAMPLE 3.1

What is the percentage of SiO_2 in a 7.69-g sample mixture if 3.76 g of SiO_2 has been recovered?

Solution: The percentage of each substance in such a mixture can be calculated as follows:

$$\% \text{ Component} = \frac{\text{wt component in grams} \times 100}{\text{wt sample in grams}}$$

Therefore, the percentage of SiO_2 in this particular sample mixture is

$$\% \; SiO_2 = \frac{3.76 \text{ g} \times 100}{7.69\text{g}} = 48.9\%$$

REVIEW QUESTIONS

Before beginning this experiment in the laboratory, you should be able to answer the following questions:

1. What distinguishes a mixture from an impure substance?
2. Define the process of sublimation.
3. How do decantation and filtration differ? Which should be faster?
4. Why does one never weigh a hot object?
5. How does this experiment illustrate the principle of conservation of matter?
6. A mixture was found to contain 2.10 g of SiO_2, 0.38 g of cellulose, and 7.52 g of calcium carbonate. What is the percentage of SiO_2 in this mixture?
7. How could you separate a mixture of zinc chloride and cyclohexane? Consult Table 2.1 for physical properties.
8. How could you separate zinc chloride from SiO_2?
9. A student found that her mixture was 15 percent NH_4Cl, 20 percent NaCl, and 75 percent SiO_2. Assuming her calculations are correct, what did she most likely do incorrectly in her experiment?
10. Why is the SiO_2 extracted with water three times as opposed to only once?

NOTES AND CALCULATIONS

EXPERIMENT

4 Chemical Reactions

OBJECTIVE

To observe some typical chemical reactions, identify some of the products, and summarize the chemical changes in terms of balanced chemical equations.

APPARATUS AND CHEMICALS

Bunsen Burner
crucible and cover
ring stand, ring, wire triangle
powdered sulfur
0.1 *M* sodium oxalate, $Na_2C_2O_4$
0.1 *M* $KMnO_4$
10 *M* NaOH
0.1 *M* $Pb(NO_3)_2$
0.1 *M* $BaCl_2$
1 *M* K_2CrO_4
0.1 *M* $NaHSO_3$ (freshly prepared)
thistle tube or long-stem funnel
6 *M* NH_3
2-in. length of copper wire, 14, 16 or 18 gauge
6 *M* HCl
copper oxide
mossy zinc
0.01 *M* $CuSO_4$
6 *M* H_2SO_4
conc. HNO_3
3 *M* $(NH_4)_2CO_3$
$KMnO_4$ (solid)
Na_2CO_3 (solid)
Na_2SO_3 (solid)
ZnS (solid)
6-in. test tube
glass tubing

DISCUSSION

Chemical equations represent what transpires in a chemical reaction. For example, the equation

$$2KClO_3(s) \xrightarrow{\Delta} 2KCl(s) + 3O_2(g)$$

means that potassium chlorate, $KClO_3$, decomposes on heating (Δ is the symbol used for heat) to yield potassium chloride, KCl, and oxygen, O_2. *Before* an equation can be written for a reaction, someone must establish what the products are. How does one decide what these products are? Products are identified by their chemical and physical properties as well as by analyses. That oxygen and not chlorine gas is produced in the above reaction can be established by the fact that oxygen is a colorless, odorless gas. Chlorine, on the other hand, is a pale, yellow-green gas with an irritating odor.

In this experiment you will observe that in some cases gases are produced, precipitates are formed, or color changes occur during the reactions. These are all indications that a chemical reaction has occured. To identify some of the products of the reactions, you can consult Table 4.1, which lists

Table 4.1 Properties of Reaction Products

Water-soluble solids	Water-insoluble solids	Manganese oxyanions	Gases
KCl, white (colorless solution) NH_4Cl, white (colorless solution) $KMnO_4$, purple $MnCl_2$, pink $Cu(NO_3)_2$, blue	CuS, very dark blue or black Cu_2S, black $BaCrO_4$, yellow $BaCO_3$, white $PbCl_2$, white MnO_2, black or brown	MnO_4^-, purple MnO_4^{2-}, dark green MnO_4^{3-}, dark blue	H_2, colorless; odorless NO_2, brown; pungent odor (TOXIC) NO, colorless; slight, pleasant odor CO_2, colorless; odorless Cl_2, pale yellow-green; pungent odor (TOXIC) SO_2, colorless; choking odor (as from matches) (TOXIC) H_2S, colorless; rotten-egg odor (TOXIC)

some of the properties of the substances that could be formed in these reactions.

PROCEDURE

A. A Reaction Between the Elements Copper and Sulfur

PERFORM THIS EXPERIMENT IN THE HOOD WITH A PARTNER. Obtain about a 5-cm (2-in.) length of copper wire and note its properties. Observe that its surface is shiny, that it can be easily bent, and that it has a characteristic color. Make a small coil of the wire by wrapping it around your pencil and place the wire coil in a crucible. Add sufficient powdered sulfur to the crucible. Cover and place it on a clay triangle on an iron ring for heating. THIS APPARATUS MUST BE SET UP IN THE HOOD, because some sulfur will burn to form noxious sulfur dioxide. Heat the crucible initially with a low heat on all sides and then use the hottest flame to heat the bottom of the crucible to red heat. CONTINUE heating until no more smoking occurs, indicating that all the sulfur is burned off. Using the crucible tongs, remove the crucible from the clay triangle without removing the cover and place it on a heat-resistant pad or wire gauze, *not on the desk top,* to cool. After the crucible has cooled, remove the cover and inspect the substance. Note its properties. Be sure to record your answers on the Report Sheet.

1. Does the substance resemble copper?

2. Is is possible to bend the substance without breaking it?

3. What color is it?

4. Has a reaction occurred?

Copper(II) sulfide, CuS, is insoluble in aqueous ammonia, NH_3, (that is, it does not react with NH_3), while copper(I) sulfide, Cu_2S, dissolves (that is, reacts) to give a blue solution with NH_3. Place a small portion of your product in a test tube and add 2 mL of 6 *M* NH_3 in the hood. Heat gently with a Bunsen burner.

5. Does your product react with NH_3?

6. Suggest a possible formula for the product.
7. Write a reaction showing the formation of your proposed product:

$$Cu(s) + S_8(s) \longrightarrow ?$$

B. Oxidation-Reduction Reactions

Many metals react with acids to liberate hydrogen and form the metal salt of the acid. The noble metals do not react with acids to produce hydrogen. Some of the unreactive metals do react with nitric acid, HNO_3; however, in these cases gases that are oxides of nitrogen are formed rather than hydrogen.

Add a small piece of zinc to a test tube containing 2 mL of 6 *M* HCl and note what happens.

8. Record your observations.
9. Suggest possible products for the observed reaction: $Zn(s) + HCl(aq) \longrightarrow ?$

Place a 1-in. piece of copper wire in a clean test tube; add 2 mL of 6 *M* HCl and note if a reaction occurs.

10. Record your observations.
11. Is Cu an active or an inactive metal?

WHILE HOLDING A CLEAN TEST TUBE IN THE HOOD, place a 1-in. piece of copper wire in it and add 1 mL of concentrated nitric acid, HNO_3.

12. Record your observations.
13. Is the gas colored?
14. Suggest a formula for the gas.
15. After the reaction has proceeded for 5 min, carefully add 5 mL of water. Based upon the color of the solution what substance is present in solution?

Potassium permanganate, $KMnO_4$, is an excellent oxidizing agent in acidic media. The permanganate ion is purple and is reduced to the manganous ion, Mn^{2+}, which has a very faint, pink color. To 1 mL of 0.1 *M* sodium oxalate, $Na_2C_2O_4$, in a clean test tube add 10 drops of 6 *M* sulfuric acid. Mix thoroughly. To the resulting solution add 1 to 2 drops of 0.1 M $KMnO_4$ and stir. If there is no obvious indication that a reaction has occurred, warm the test tube gently in a hot water bath.

16. Record your observations. Was the $KMnO_4$ reduced to Mn^{2+}?

Place 3 mL of 0.1 *M* sodium hydrogen sulfite, $NaHSO_3$, solution in a test tube. Add 1 mL of 10 *M* sodium hydroxide, NaOH, solution and stir. To the mixture in the test tube add 1 drop of 0.1 *M* $KMnO_4$ solution.

17. Record your observations. Was the $KMnO_4$ reduced? Identify the manganese compound formed.

Add additional 0.1 *M* $KMnO_4$ solution, one drop at a time, observing the effect of each drop until 10 drops have been added.

18. Record your observations.
19. Suggest why the effect of additional potassium permanganate changes as more is added.

WHILE HOLDING A TEST TUBE IN THE FUME HOOD, add one or two crystals of potassium permanganate, $KMnO_4$, to 1 mL of 6 *M* HCl.

20. Record your observations.
21. Note the color of the gas evolved.
22. Based on the color of the gas, what is the gas?

C. Metathesis Reactions

Additional observations are needed before equations can be written for the reaction above, but we see that we can identify some of the products. The remaining reactions are simple, and you will be able, from available information, not only to identify products but also to write equations. A number of reactions may be represented by equations of the following type:

$$AB + CD \longrightarrow AD + CB$$

These are called double-decomposition, or *metathesis,* reactions. This type of reaction involves the exchange of atoms or groups of atoms between interacting substances. The following is a specific example:

$$NaCl(aq) + AgNO_3(aq) \longrightarrow AgCl(s) + NaNO_3(aq)$$

Place a small sample of sodium carbonate, Na_2CO_3, in a test tube and add several drops of 6 *M* HCl.

23. Record your observations.
24. Note the odor and color of the gas that forms.
25. What is the evolved gas?
26. Write an equation for the reaction $HCl(aq) + Na_2CO_3(s) \longrightarrow ?$ (NOTE: In this reaction the products must have H, Cl, Na, and O atoms in some new combinations, but no other elements can be present.)

Note that H_2CO_3 and H_2SO_3 readily decompose as follows:

$$H_2CO_3(aq) \longrightarrow H_2O(l) + CO_2(g)$$

$$H_2SO_3(aq) \longrightarrow H_2O(l) + SO_2(g)$$

IN THE HOOD, repeat the same test with sodium sulfite, Na_2SO_3.

27. Record your observations.

28. What is the gas?

29. Write an equation for the following reaction (note the similarity to the equation above): $HCl(aq) + Na_2SO_3(s) \longrightarrow ?$

IN THE HOOD, repeat this test with zinc sulfide, ZnS.

30. Record your observations.

31. What is the gas?

32. Write an equation for the reaction: $HCl(aq) + ZnS(s) \longrightarrow ?$

To 1 mL of 0.1 *M* lead nitrate, $Pb(NO_3)_2$, solution in a clean test tube add a few drops of 6 *M* HCl.

33. Record your observations.

34. What is the precipitate?

35. Write an equation for the reaction: $Pb(NO_3)_2(aq) + HCl(aq) \longrightarrow ?$

To 1 mL of 0.1 *M* barium chloride, $BaCl_2$, solution add 2 drops of 1 *M* potassium chromate, K_2CrO_4, solution.

36. Record your observations.

37. What is the precipitate?

38. Write an equation for the reaction: $BaCl_2(aq) + K_2CrO_4(aq) \longrightarrow ?$

To 1 mL of 0.1 *M* barium chloride, $BaCl_2$, solution add several drops of 3 *M* ammonium carbonate, $(NH_4)_2CO_3$, solution in a test tube.

39. What is the precipitate?

40. Write an equation for the reaction: $(NH_4)_2CO_3(aq) + BaCl_2(aq) \longrightarrow ?$

After the precipitate has settled somewhat, carefully decant (that is, pour off) the excess liquid. Add 1 mL of water to the test tube, shake it, allow the precipitate to settle, and again carefully pour off the liquid. To the remaining solid, add several drops of 6 *M* HCl.

41. Record your observations.

42. Note the odor.

43. What is the evolved gas? (Recall the reaction in step 26 of this experiment.)

REVIEW QUESTIONS

Before beginning this experiment in the laboratory, you should be able to answer the following questions:

1. Before a chemical equation can be written, what must you know?
2. What observations might you make that suggest that a chemical reaction has occurred?
3. How could you distinguish between NO_2 and NO?
4. Define metathesis reactions. Give an example.
5. What is a precipitate?
6. Balance these equations:

$$KBrO_3(s) \xrightarrow{\Delta} KBr(s) + O_2(g)$$

$$ZnCl_2(aq) + AgNO_3(aq) \longrightarrow Zn(NO_3)_2(aq) + AgCl(s)$$

7. How could you distinguish between the gases H_2 and H_2S?
8. Using water, how could you distinguish between the white solids KCl and $PbCl_2$?
9. Write equations for the decomposition of $H_2CO_3(aq)$ and $H_2SO_3(aq)$.

EXPERIMENT 5

Chemical Formulas

OBJECTIVE

To become familiar with chemical formulas and how they are obtained.

APPARATUS AND CHEMICALS

balance	crucible and cover
250-mL beaker	clay triangle
Bunsen burner	6 *M* HCl
evaporating dish	granular zinc, about 1 g
50-mL graduated cylinder	copper wire, about 2 g
ring stand and ring	powdered sulfur, about 3 g
stirring rod	carborundum boiling chips
wire gauze	

DISCUSSION

Just as a secretary uses shorthand in order to take dictation at a greater speed, chemists use a shorthand notation of their own to indicate the exact chemical composition of compounds (chemical formulas). We then use these chemical formulas to indicate how new compounds are formed by chemical combinations of other compounds (chemical reactions). However, before we can learn how chemical formulas are written, we must first acquaint ourselves with the symbols used to denote the elements from which these compounds are formed.

Symbols

We generally use two letters to symbolize a chemical element. (There are only 14 exceptions.) These symbols are derived, as a rule, from the first two letters or first syllable of the elements name. For elements that were known to the early alchemists, these symbols come from the Latin names for the elements. For elements discovered at a later date, the symbols come from the Latin, Greek, English, German, or French names as they appear on current periodic tables. Some examples are given in Table 5.1. Note that the first letter of the symbol is always capitalized and that the second is lowercase.

Formulas

We can combine the symbols to denote the formulas for compounds in the following fashion. The formula CO (read "carbon monoxide") means that one atom of carbon, C, combines with one of oxygen, O, to form the compound carbon monoxide. Similarly, in the compound $CHCl_3$, chloroform, one atom of carbon, C, is combined with one atom of hydrogen, H, and three atoms of

Table 5.1 Symbols of Some Elements

Symbol	Name	Symbol	Name
H	Hydrogen	Pb	Lead, from *plumbum*
Li	Lithium	Ag	Silver, from *argentum*
Ca	Calcium	Au	Gold, from *aurum*
Ni	Nickel	Cu	Copper, from *cuprum*
B	Boron	Sn	Tin, from *stannum*
C	Carbon	Sb	Antimony, from *stibium*
P	Phosphorus	W	Tungsten, from *wolfram*
Fe	Iron, from *ferrum*		

chlorine. The subscripts, 3 on the chlorine and 1 (understood) on both the carbon and hydrogen, imply exactly this. Thus, our chemical shorthand indicates the precise combining ratios of the chemical elements that form the molecule. Each and every molecule of carbon monoxide contains one atom of carbon and one of oxygen, just as each and every molecule of chloroform contains exactly one atom each of carbon and hydrogen and three of chlorine. Every chemical compound has this property; that is, *every compound is composed of definite numbers of whole atoms in fixed proportions.*

Atomic Weights

It is important to know something about masses of atoms and molecules. It has been determined by experiment that an atom of hydrogen (the lightest of all elements) weighs 1.67×10^{-24} g (0.00000000000000000000000167 g). It is very inconvenient to express atomic weights in grams, and a smaller unit, the *atomic mass unit* (amu) is used: 1 amu = 1.66053×10^{-24} g. On this scale the atomic weight of hydrogen is 1.0079 amu. The atomic weights given in a periodic table are in amu. For example, the atomic weight of carbon is 12.011 amu.

Moles and Avogadro's Number

It is convenient to work with gram quantities in the laboratory. That quantity of an element corresponding to the atomic weight in grams is called a *mole*, abbreviated mol. This quantity is also called the *gram-atomic weight*. When we use a chemical symbol we can take it to mean an atom, its corresponding atomic weight in atomic mass units, or a mole of that element. For example, C represents one atom or 12.011 amu of carbon as well as one gram mole, 12.011 g, of carbon atoms.

It has been determined experimentally that one mole of atoms of any element contains 6.022×10^{23} atoms. This number, known as *Avogadro's number*, is very important to chemists. Knowing this number and the atomic weight, we can easily calculate the mass of an individual atom. For example, consider magnesium, whose atomic weight is 24.312 amu. The weight of one magnesium atom is:

$$\left(\frac{24.312 \text{ g Mg}}{1 \text{ mol Mg}}\right)\left(\frac{1 \text{ mol Mg}}{6.022 \times 10^{23} \text{ atoms}}\right) = 4.037 \times 10^{-23} \text{ g Mg/atom}$$

It should be clear to you by now that atoms are indeed very small particles. In fact, more atoms than there are people in the world could be placed on the head of an ordinary pin.

The sum of the atomic weights of the atoms shown by a chemical formula is the formula weight. Thus for NaCl

$$\begin{array}{ll} \text{1 atomic weight Na} & = 23.0\text{ amu} \\ \underline{\text{1 atomic weight Cl}} & \underline{= 35.5\text{ amu}} \\ \text{1 formula weight NaCl} & = 58.5\text{ amu} \end{array}$$

Similarly, for $LiNO_3$

$$\begin{array}{llll} \text{1 atomic weight Li} & = 1 \times 6.9 & = & 6.9\text{ amu} \\ \text{1 atomic weight N} & = 1 \times 14.0 & = & 14.0\text{ amu} \\ \text{3 atomic weights O} & = 3 \times 16.0 & = & 48.0\text{ amu} \\ \hline \text{1 formula weight LiNO}_3 & & = & 68.9\text{ amu} \end{array}$$

We can also speak of a mole of these substances. For example, a mole of NaCl weighs 58.5 g and a mole of $LiNO_3$ weighs 68.9 g. A mole of a compound is also called a *gram-formula weight* and, sometimes, gram-molecular weight. Many compounds consist of ions rather than discrete molecules. For ionic compounds, *gram-molecular weight* and *molecular weight* are not meaningful terms. We will say more about this later. We will use the term "mole" throughout this book to mean gram-formula weight. Just be apprised of the distinction between formula weight and molecular weight.

It is a simple matter to calculate the number of moles of any substance whose weight we can obtain. For example, suppose we have one quart of rubbing alcohol (generally isopropyl alcohol) and know its density to be 0.785 g/mL and we want to know how many moles this is. First we need to convert the volume to our system of units. Since 1 qt is 0.946 L and 1 L contains 1000 mL, our quart of isopropanol is

$$(1\ \cancel{\text{qt}})\left(\frac{0.946\ \cancel{\text{L}}}{\cancel{\text{qt}}}\right)\left(\frac{1000\text{ mL}}{\cancel{\text{L}}}\right) = 946\text{ mL}$$

we can now calculate the weight:

$$946\text{ mL} \times 0.785\text{ g/mL} = 743\text{ g}$$

We next need the chemical formula for isopropyl alcohol. This is C_3H_7OH. The molecular weight is, therefore,

$$\begin{array}{llll} \text{Weight carbon} & 3 \times 12.0 & = & 36.0\text{ amu} \\ \text{Weight hydrogen} & 8 \times 1.0 & = & 8.0\text{ amu} \\ \text{Weight oxygen} & 1 \times 16.0 & = & 16.0\text{ amu} \\ \hline \text{Molecular weight C}_3\text{H}_7\text{OH} & & = & 60.0\text{ amu} \end{array}$$

Hence,

$$\text{Moles of } C_3H_7OH = (743 \text{ g } C_3H_7OH)\left(\frac{1 \text{ mol } C_3H_7OH}{60.0 \text{ g } C_3H_7OH}\right) = 12.4 \text{ mol}$$

Thus our quart of rubbing alcohol contains 12.4 mol of isopropyl alcohol. It should now be apparent to you how much information is contained in a chemical formula. After you have some feeling for the nature of different chemical substances, you should be able to read the labels of many household items and know what the chemical contents are and why they are there. We have often heard the phrase "What's in a name?" The answer to this question when the name is that of a chemical compound takes on considerable meaning as we gain familiarity with that compound. For example the simple name carbon dioxide tells us precisely

1. The elements present, carbon and oxygen;
2. The combining ratios of these elements, one atom of carbon for every two atoms of oxygen in each molecule of carbon dioxide;
3. The molecular weight;
4. And, after we learn some chemistry, that we are speaking of a colorless gas, heavier than air, which will not support combustion and which is lethal simply because it will not allow us to take in oxygen when there is a lot of it around. We will also learn that it can be solidified by compression to produce the familiar dry ice.

All of this dialogue is simply to illustrate that chemical formulas are tremendously important, and it behooves us in the very beginning to have a firm understanding of just exactly what they are. It should also be evident that we must have some way of determining them precisely if they do contain so much information. Before we delve into this area, let's look at some additional information contained in chemical formulas.

Percentage Composition

Since we now have access to molecular and atomic weights, it is a simple matter to calculate the weight-percentage composition of a chemical compound. This is useful in determining chemical formulas, and thus we will discuss it first. Suppose you wished to know the percentage composition of water and hydrogen peroxide (a household bleach and mild antiseptic). The chemical formulas for these compounds are H_2O and H_2O_2, respectively. The weight percentages of the elements in H_2O are calculated as follows:

$$\text{Wt \% H} = \frac{2 \text{ g H/mol } H_2O}{18 \text{ g } H_2O\text{/mol}} \times 100 = 11\%$$

$$\text{Wt \% O} = 100 - 11 = 89\%$$

The weight percentages of the elements in H_2O_2 are

$$\text{Wt \% H} = \frac{2 \text{ g H/mol } H_2O_2}{34 \text{ g } H_2O_2\text{/mol}} \times 100 = 5.9\%$$

$$\text{Wt \% O} = 100 - 5.9 = 94\%$$

All pure samples of these two compounds, no matter what their source, contain exactly these percentage compositions. This is one of the fundamental concepts of chemistry formulated early in the development of the field and is known as the *law of definite proportions or composition*. It can be simply stated as follows: *Different samples of a pure compound always contain the same elements in the same proportions by weight*. This law, among others, convinced John Dalton of the atomic nature of matter and led him to outline his atomic theory. Now let's investigate how information such as this enables us to determine chemical formulas.

Derivation of Formulas

Suppose you were working in a hospital laboratory today, and suppose further that this morning the emergency ward admitted a patient complaining of severe stomach cramps and labored respiration and that this patient died within minutes of being admitted. Relatives of the patient later told you that he may have ingested some rat poison. You therefore had his stomach pumped to verify this or simply to determine the cause of death. One of the more logical things to do would be to attempt to isolate the agent that caused death and perform chemical analyses on it. Let's suppose that this was done, and the analyses showed that the isolated chemical compound contained, by weight, 60.0 percent potassium, 18.5 percent carbon, and 21.5 percent nitrogen. What is the chemical formula for this compound? One simple and direct way of making the necessary calculations is as follows.

Assume you had 100 g of the compound. This 100 g therefore would contain

$$(100 \text{ g})(0.600) = 60.0 \text{ g potassium}$$

$$(100 \text{ g})(0.185) = 18.5 \text{ g carbon}$$

$$(100 \text{ g})(0.215) = 21.5 \text{ g nitrogen}$$

Chemical formulas tell what elements are present and the ratio of the number of atoms of the constituent elements. Hence, the next step is to determine the number of moles of each element present:

$$\text{Moles potassium} = \frac{60.0 \text{ g K}}{39.0 \text{ g K/mol K}} = 1.54 \text{ mol K}$$

$$\text{Moles carbon} = \frac{18.5 \text{ g C}}{12.0 \text{ g C/mol C}} = 1.54 \text{ mol C}$$

$$\text{Moles nitrogen} = \frac{21.5 \text{ g N}}{14.0 \text{ g N/mol N}} = 1.54 \text{ mol N}$$

Hence the chemical formula is $K_{1.54}C_{1.54}N_{1.54}$. But molecules are not formed from partial atoms; therefore, the above numbers must be changed to whole numbers. This is accomplished by division of all subscripts by the smallest subscript. In this case, they are all the same:

$$K_{1.54/1.54}C_{1.54/1.54}N_{1.54/1.54} = KCN$$

The smallest whole-number mole ratio is 1 : 1 : 1. Since KCN is a common rat poison, we may justifiably conclude that the relatives' suggestion of rat-poison ingestion as the probable cause of death is correct.

The above calculation has given us what is known as the *empirical* formula. There is another type of chemical formula the *molecular* formula. The distinction between these two is simply that the empirical formula represents the *smallest* whole-number ratio of the combining atoms in a chemical compound, whereas the molecular formula gives the *actual* number of atoms in a molecule. Recall, however, as we stated earlier, that not all compounds exist as discrete molecules. This is true for most ionic compounds, whereas most covalent compounds do exist as discrete molecules. The distinction between empirical and molecular formulas may be clarified by the following example.

A chemical compound was found by elemental analyses to contain 92.3 percent carbon and 7.7 percent hydrogen by weight and to have a molecular weight of 78. The empirical formula may be obtained just as in the previous example—that is, in 100 g of the compound there are 92.3 g C and 7.7 g H. Hence,

$$\text{Moles C} = \frac{92.3 \text{ g C}}{12 \text{ g C/mol C}} = 7.7 \text{ mol C}$$

$$\text{Moles H} = \frac{7.7 \text{ g H}}{1.0 \text{ g H/mol H}} = 7.7 \text{ mol H}$$

The empirical formula is then $C_{7.7}H_{7.7}$, or CH, whose formula weight is 12 + 1 = 13. But the molecular weight of the compound is 78. Therefore, there are 78/13 = 6 empirical-formula weights in the molecular weight. The molecular formula is then C_6H_6.

In this experiment you will determine the empirical formulas of two chemical compounds. One is copper sulfide, which you will prepare according to the following chemical reaction:

$$x\text{Cu}(s) + y\text{S}_8(s) \longrightarrow \text{Cu}_x\text{S}_y(s)$$

The other is zinc chloride, which you will prepare according to the chemical reaction

$$x\text{Zn}(s) + y\text{HCl}(aq) \longrightarrow \text{Zn}_x\text{Cl}_y(s) + \frac{y}{2}\text{H}_2(g)$$

The objective is to determine the combining ratios of the elements (that is, to determine x and y) and to balance the chemical equations given above.

PROCEDURE

A. Zinc Chloride

CAUTION: *Zinc chloride is caustic and must be handled carefully in order to avoid any contact with your skin. Should you come in contact with it, immediately wash the area with copious amounts of water*. Clean and dry your evaporating dish and place it on the wire gauze resting on the iron ring. Heat the dish with your Bunsen burner, gently at first, and then more strongly, until all of the condensed moisture has been driven off. This should require heating for about 5 min. Allow the dish to cool to room temperature on a heat-resistant pad (do not place the hot dish on the counter top) and weigh it. Record the weight of the empty evaporating dish.

Obtain a sample of granular zinc from your laboratory instructor and add about 0.5 g of it to the weighed evaporating dish. Weigh the evaporating dish containing the zinc and record the total weight. Calculate the weight of the zinc.

Slowly, and with constant stirring, add 15 mL of 6 *M* HCl to the evaporating dish containing the zinc. A vigorous reaction will ensue, and hydrogen gas will be produced. NO FLAMES ARE PERMITTED IN THE LABORATORY WHILE THIS REACTION IS TAKING PLACE, SINCE WET HYDROGEN GAS IS VERY EXPLOSIVE. If any undissolved zinc remains after the reaction ceases, add an additional 5 mL of acid. Continue to add 5-mL portions of acid as needed until all the zinc has dissolved.

Set up a steam bath as illustrated in Figure 5.1 using a 250-mL beaker, and place the evaporating dish on the steam bath. Heat the evaporating dish very carefully on the steam bath until most of the liquid has disappeared. Then remove the steam bath and heat the dish on the wire gauze. During this last stage of heating, the flame must be carefully controlled or there will be spattering, and some loss of product will occur. DO NOT HEAT TO THE POINT THAT THE COMPOUND MELTS, OR SOME WILL BE LOST DUE TO SUBLIMATION. Leave the compound looking somewhat pasty while hot.

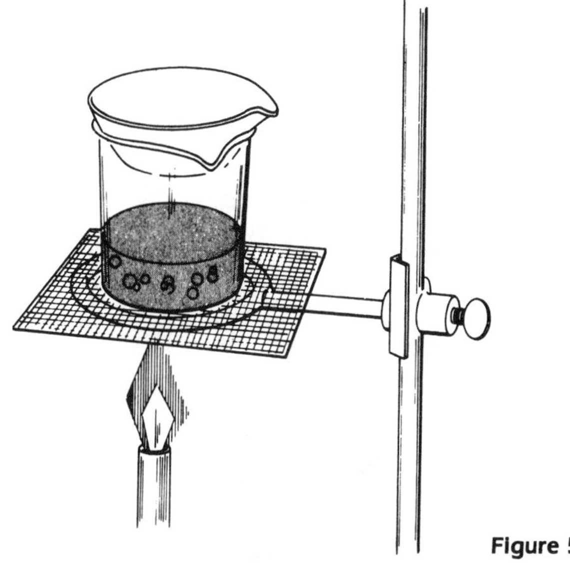

Figure 5.1 Steam bath.

Allow the dish to cool to room temperature and weigh it. Record the weight. After this first weighing, heat the dish again very gently; cool it and reweigh it. If these weighings do not agree within 0.02 g, repeat the heating and weighing until two successive weighings agree. This is known as *drying to constant weight* and is the only way to be certain that all the moisture is driven off. Zinc chloride is very deliquescent and so should be weighed quickly.

Calculate the weight of zinc chloride. The difference in weight between the zinc and zinc chloride is the weight of chlorine. Calculate the weight of chlorine in zinc chloride. From this information you can readily calculate the empirical formula for zinc chloride and balance the chemical equation for its formation. Perform these operations on the report sheet.

B. Copper Sulfide

Support a clean, dry porcelain crucible and cover on a clay triangle and dry by heating to a dull red in a Bunsen flame, as illustrated in Figure 5.2. Allow the crucible and cover to cool to room temperature and weigh them. Record the weight.

Place 1.5–2.0 g of tightly wound copper wire or copper turnings in the crucible and weigh the copper, crucible, and lid. Calculate the weight of copper. Record your results.

In the hood, add sufficient sulfur to cover the copper, place the crucible with cover in place on the triangle, and heat the crucible gently until sulfur ceases to burn (blue flame) at the end of the cover. Do not remove the cover while the crucible is hot. Finally, heat the crucible to dull redness for about 5 min.

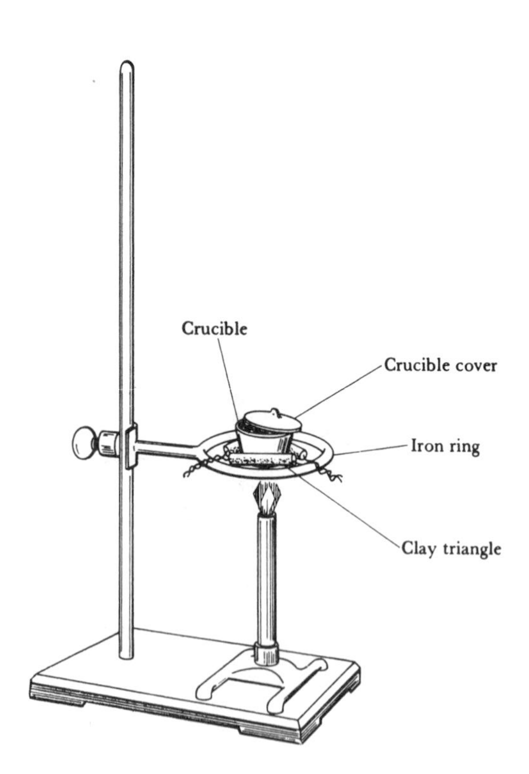

Figure 5.2 Setup for copper sulfide determination.

Allow the crucible to cool to room temperature. This will take about 10 min. Then weigh with the cover in place. Record the weight. Again cover the contents of the crucible with sulfur and repeat the heating procedure. Allow the crucible to cool and reweigh it. Record the weight. If the last two weighings do not agree to within 0.02 g, the chemical reaction between the copper and sulfur is incomplete. If this is found to be the case, add more sulfur and repeat the heating and weighing until a constant weight is obtained.

Calculate the weight of copper sulfide obtained. The difference in weight between the copper sulfide and copper is the weight of sulfur in copper sulfide. Calculate this weight. From this information the empirical formula for copper sulfide can be obtained, and the chemical equation for its production can be balanced. Perform these operations on your report sheet.

REVIEW QUESTIONS

Before beginning this experiment in the laboratory, you should be able to answer the following questions:

1. Define the term "compound".
2. Why are atomic weights relative weights?
3. How do formula weights and molecular weights differ?
4. What is the percentage composition of $BaCO_3$?
5. A substance was found by analysis to contain 65.95 percent barium and 34.05 percent chlorine. What is the empirical formula for the substance?
6. What is the law of definite proportions?
7. How do empirical and molecular formulas differ?
8. What is the weight of one sodium atom?
9. Soda-lime glass is prepared by fusing sodium carbonate, Na_2CO_3; limestone, $CaCO_3$; and sand, SiO_2. The composition of the glass varies, but the commonly accepted reaction for its formation is

   $$Na_2CO_3(s) + CaCO_3(s) + 6SiO_2(s) \longrightarrow Na_2CaSi_6O_{14}(s) + 2CO_2(g)$$

   Using this equation, how many kilograms of sand would be required to produce enough glass to make five thousand 400-g wine bottles?
10. Caffeine, a stimulant found in coffee and tea contains 49.5 percent C, 5.15 percent H, 28.9 percent N, and 16.5 percent O by mass. What is the empirical formula of caffeine? If its molar mass is about 195 g, what is its molecular formula?
11. An analysis of an oxide of nitrogen with a molecular weight of 92.02 amu gave 69.57 percent oxygen and 30.43 percent nitrogen. What are the empirical and molecular formulas for this nitrogen oxide? Complete and balance the equation for its formation from the elements nitrogen and oxygen.
12. How many sodium atoms are present in 0.01456 g of sodium?

NOTES AND CALCULATIONS

Chemical Reactions of Copper and Percent Yield

EXPERIMENT 6

OBJECTIVE

To gain some familiarity with basic laboratory procedures, some chemistry of a typical transition element, and the concept of percent yield.

APPARATUS AND CHEMICALS

- 0.5-g piece of no. 16 or no. 18 copper wire
- 250-mL beakers (2)
- conc. HNO_3
- graduated cylinder
- 3.0 *M* NaOH
- carborundum boiling chips
- stirring rod
- iron ring and ring stand
- wire gauze
- Bunsen burner
- evaporating dish
- weighing paper
- 6.0 *M* H_2SO_4
- granular zinc
- methanol
- acetone
- towel
- balance
- aluminum foil cut in 1-in. squares
- conc. HCl

DISCUSSION

Most chemical syntheses involve separation and purification of the desired product from unwanted side products. Some methods of separation, such as filtration, sedimentation, decantation, extraction, and sublimation, were described in Experiment 3. This experiment is designed as a quantitative evaluation of your individual laboratory skills in carrying out some of these operations. At the same time you will become more acquainted with two fundamental types of chemical reactions like those described in Experiment 4—redox reactions and metathesis reactions. By means of these reactions, you will carry out several chemical transformations involving copper, and you will finally recover the copper sample with maximum efficiency. The chemical reactions involved are the following:

$$Cu(s) + 4HNO_3(aq) \longrightarrow Cu(NO_3)_2(aq) + 2NO_2(g) + 2H_2O(l) \quad \text{Redox} \quad [1]$$

$$Cu(NO_3)_2(aq) + 2NaOH(aq) \longrightarrow Cu(OH)_2(s) + 2NaNO_3(aq) \quad \text{Metathesis} \quad [2]$$

$$Cu(OH)_2(s) \xrightarrow{\Delta} CuO(s) + H_2O(g) \quad \text{Dehydration} \quad [3]$$

$$CuO(s) + H_2SO_4(aq) \longrightarrow CuSO_4(aq) + H_2O(l) \quad \text{Metathesis} \quad [4]$$

$$CuSO_4(aq) + Zn(s) \longrightarrow ZnSO_4(aq) + Cu(s) \quad \text{Redox} \quad [5]$$

Each of these reactions proceeds to completion. Metathesis reactions proceed to completion whenever one of the components is removed from the solution, such as in the formation of a gas or an insoluble precipitate. This is the case for reactions [1], [2], and [3], where in reactions [1] and [3] a gas and in reaction [2] an insoluble precipitate are formed. Reaction [5] proceeds to completion because zinc has a lower ionization energy or oxidation potential than copper. More discussion of this type of reaction can be found in Experiment 14.

The object in this experiment is to recover all of the copper you begin with in analytically pure form. This is the test of your laboratory skills.

The percent yield of the copper can be expressed as the ratio of the recovered weight to initial weight, multipled by 100:

$$\% \text{ yield} = \frac{\text{recovered wt of Cu}}{\text{initial wt of Cu}} \times 100$$

PROCEDURE

Weigh approximately 0.500 g of no. 16 or no. 18 copper wire (1) to the nearest 0.0001 g and place it in a 250-mL beaker. Add 4–5 mL of concentrated HNO_3 to the beaker IN THE HOOD. After the reaction is complete, add 100 mL distilled H_2O. Describe the reaction (6) as to color change, evolution of a gas, and change in temperature (exothermic or endothermic) on the report sheet.

Add 30 mL of 3.0 *M* NaOH to the solution in your beaker and describe the reaction (7). Add two or three boiling chips and carefully heat the solution—while stirring with a stirring rod—just to the boiling point. Describe the reaction on your report sheet (8).

Allow the black CuO to settle; then decant the supernatant liquid. Add about 200 mL of very hot distilled water, stir, and then allow the CuO to settle. Decant once more. What are you removing by the washing and decantation (9)?

Add 15 mL of 6.0 *M* H_2SO_4. What copper compound is present in the beaker now (10)?

Your instructor will tell you whether you should use zinc or aluminum for the reduction of Cu(II) in the following step.

A. Zinc

In the hood, add 2.0 g of 30-mesh zinc metal all at once and stir until the supernatant liquid is colorless. Describe the reaction on your report sheet (11). What is present in solution (12)? When gas evolution has become *very* slow, heat the solution gently (but do not boil) and allow it to cool. What gas is formed in this reaction (13)? How do you know (14)?

B. Aluminum

In the hood, add several 1-in. squares of aluminum foil and a few drops of concentrated HCl. Continue to add pieces of aluminum until the supernatant liquid is colorless. Describe the reaction on your report sheet (11). What is present in solution (12)? What gas is formed in this reaction (13)? How do you know (14)?

When gas evolution has ceased, decant the solution and transfer the precipitate to a preweighed porcelain evaporating dish (3). Wash the precipitated

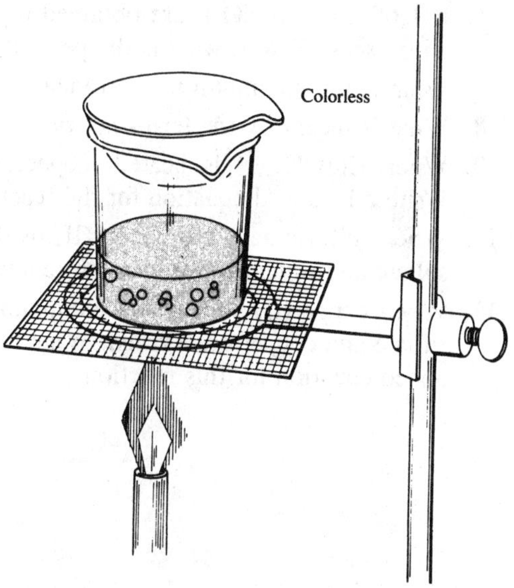

Figure 6.1 Steam bath.

copper with about 5 mL of distilled water, allow it to settle, decant the solution, and repeat the process. What are you removing by washing (15)? Wash the precipitate with about 5 mL of methanol (KEEP THE METHANOL AWAY FROM FLAMES—IT IS FLAMMABLE!!) Allow the precipitate to settle, and decant the methanol. (METHANOL IS ALSO EXTREMELY TOXIC: AVOID BREATHING THE VAPORS AS MUCH AS POSSIBLE.) Finally, wash the precipitate with about 5 mL of acetone (KEEP THE ACETONE AWAY FROM FLAMES—IT IS EXTREMELY FLAMMABLE!!), allow the precipitate to settle, and decant the acetone from the precipitate. Prepare a steam bath as illustrated in Figure 6.1 and dry the product on your steam bath for at least 5 min. Wipe the bottom of the evaporating dish with a towel, remove the boiling chips and weigh the evaporating dish plus copper (2). Calculate the final weight of copper (4). Compare the weight with your initial weight and calculate the percent yield (5). What color is your copper sample (16)? Is it uniform in appearance (17)? Suggest possible sources of error in this experiment (18).

REVIEW QUESTIONS

Before beginning this experiment in the laboratory, you should be able to answer the following questions:

1. Give an example, other than the ones listed in this experiment, of redox and methathesis reactions.
2. When will reactions proceed to completion?
3. Define percent yield in general terms.
4. Name six methods of separating materials.
5. Give criteria in terms of temperature changes for exothermic and endothermic reactions.

6. If 1.65 g of $Cu(NO_3)_2$ are obtained from allowing 0.93 g of Cu to react with excess HNO_3, what is the percent yield of the reaction?
7. What is the maximum percent yield in any reaction?
8. What is meant by the terms *decantation* and *filtration*?
9. When $Cu(OH)_2(s)$ is heated, copper(II) oxide and water are formed. Write a balanced equation for the reaction.
10. When sulfuric acid and copper(II) oxide are allowed to react, copper(II) sulfate and water are formed. Write a balanced equation for this reaction.
11. When copper(II) sulfate and aluminum are allowed to react, aluminum sulfate and copper are formed. What kind of reaction is this? Write a balanced equation for this reaction.

Name ______________________________ Desk ________

Date ________________ Laboratory Instructor ______________________________

REPORT SHEET FOR EXPERIMENT 6

CHEMICAL REACTIONS OF COPPER AND PERCENT YIELD

1. Weight copper initial ____________

2. Weight of copper and evaporating dish ____________

3. Weight of evaporating dish ____________

4. Weight copper final ____________

5. % yield (show calculations) ____________

6. Describe the reaction $Cu(s) + HNO_3(aq) \rightarrow$

7. Describe the reaction $Cu(NO_3)_2(aq) + NaOH(aq) \rightarrow$

8. Describe the reaction $Cu(OH)_2(s) \xrightarrow{\Delta}$

9. What are you removing by this washing? ______________________________

10. What copper compound is present in the beaker? ______________________________

11. Describe the reaction $CuSO_4(aq) + Zn(s)$, or $CuSO_4(aq) + Al(s)$

12. What is present in solution? ______________________________

13. What is the gas? ______________________________

14. How do you know? ______________________________

15. What are you removing by washing? ______________________________

16. What color is your copper sample? ______________________________

17. Is it uniform in appearance? ______________________________

18. Suggest possible sources of error in this experiment.

QUESTIONS

1. If your percent yield of copper was greater than 100%, what are two plausible errors you may have made?

2. Consider the combustion of methane, CH_4:

$$CH_4(g) + 2O_2(g) \longrightarrow CO_2(g) + 2H_2O(g)$$

Suppose 2 mol of methane is allowed to react with 3 mol of oxygen.
(a) What is the limiting reagent?

(b) How many moles of CO_2 can be made from this mixture? How many grams of CO_2?

3. Suppose 8.00 g of CH_4 is allowed to burn in the presence of 6.00 g of oxygen. How much (in grams) CH_4, O_2, CO_2, and H_2O remain after the reaction is complete?

4. How many milliliters of 6.0 *M* H_2SO_4 are required to react with 0.80 g of CuO according to Equation [4]?

5. If 2.00 g of Zn is allowed to react with 1.75 g of $CuSO_4$ according to Equation [5], how many grams of Zn will remain after the reaction is complete?

6. What is meant by the term limiting reagent?

NOTES AND CALCULATIONS

Chemicals in Everyday Life: What Are They and How Do We Know?

EXPERIMENT 7

OBJECTIVE

To observe some reactions of common substances found around the home and to learn how to identify them.

APPARATUS AND CHEMICALS

household ammonia	150-mL beaker
chemical fertilizer	1 *M* NH_4Cl
bleach	8 *M* NaOH
table salt	$(NH_4)_2CO_3$(solid)
baking soda	18 *M* H_2SO_4
Epsom salts	2 *M* HCl
vinegar	$Ba(OH)_2$ (sat. soln.)
chalk	3 *M* HNO_3
NaI	0.1 *M* $AgNO_3$
mineral oil	0.2 *M* $BaCl_2$
3-in. test tubes (6)	medicine droppers (3)
solid unknown containing CO_3^{2-}, Cl^-, SO_4^{2-}, or I^-	red and blue litmus paper

DISCUSSION

One important aspect of chemistry is the identification of substances. The identification of minerals—for example, "fool's gold" as opposed to genuine gold—was and still is of great importance to prospectors. The rapid identification of a toxic substance ingested by an infant may expedite the child's rapid recovery, or in fact it may be the determining factor in saving a life. Substances are identified either by the use of instruments or by reactions characteristic of the substance, or both. Reactions that are characteristic of a substance are frequently referred to as a *test*. For example, one may test for oxygen with a glowing splint; if the splint bursts into flame, oxygen is probably present. One tests for chloride ions by adding silver nitrate to an acidified solution. The formation of a white precipitate suggests the presence of chloride ions. Since other substances may yield a white precipitate under these conditions, one "confirms" the presence of chloride ions by observing that this precipitate dissolves in ammonium hydroxide. The area of chemistry concerned with identification of substances is termed *qualitative analysis*.

In this experiment you will perform tests on or with substances that you are apt to encounter in everyday life, substances such as table salt, bleach, smelling salts, and baking soda. You probably don't think of these as "chemicals," and yet they are, even though we don't refer to them around the home by their chemical names (which are sodium chloride, chlorine, ammonium carbonate, and sodium bicarbonate, respectively), but rather by a trade name. After observing some reactions of these household chemicals, you will partially identify an unknown. Your task will be to determine whether the substance contains the carbonate (CO_3^{2-}), chloride (Cl^-), sulfate (SO_4^{2-}), or iodide, (I^-) ion.

(CAUTION: *Even though household chemicals may appear innocuous,* NEVER *mix them unless you are absolutely certain you know what you are doing. Innocuous chemicals, when combined, can sometimes produce severe explosions or other hazardous reactions.*)

PROCEDURE

A. Household Ammonia

Obtain 5 mL of household ammonia in a 150-mL beaker and hold a dry piece of red litmus paper over the beaker, being careful not to touch the sides of the beaker or the solution with the paper. Record your observations (1). Repeat the operation using a piece of red litmus paper that has been moistened with tap water. Do you note any difference in the time required for the litmus to change colors or the intensity of the color change? Record your observations (2).

Ammonium salts are converted to ammonia, NH_3, by the action of strong bases; hence, one can test for the ammonium ion, NH_4^+, by adding sodium hydroxide, NaOH, and noting the familiar odor of NH_3 or by the use of red litmus. The *net* reaction is as follows:

$$NH_4^+(aq) + OH^-(aq) \rightleftharpoons NH_3(aq) + H_2O(l)$$

Place about 1 mL of 1 *M* NH_4Cl, ammonium chloride, in a test tube and hold a moist piece of red litmus in the mouth of the tube. Record your observations (3). Now add about 1 mL of 8 *M* NaOH and repeat the test. (Do not allow the litmus to touch the sides of the tube, because it may come in contact with NaOH, which will turn it blue.) If the litmus does not change color, gently warm the test tube, but do not boil the solution. Record your observations (4).

You may suspect that ordinary garden fertilizer contains ammonium compounds. Confirm your suspicions by placing some solid fertilizer, an amount about the size of a pea, in a test tube; add 1 mL of 8 *M* NaOH and test as above with moist litmus paper. Does the fertilizer contain ammonium salts (5)?

What is the active ingredient in "smelling salts"? Hold a moist piece of red litmus over the mouth of an open jar of ammonium carbonate, $(NH_4)_2CO_3$; carefully, by fanning your hand over the jar, see if you can detect a familiar odor. Record you observations (6). Most ammonium salts are stable; for example, the ammonium chloride solution that you tested above should not have had any effect on litmus *before* you added the sodium hydroxide. However, $(NH_4)_2CO_3$ is quite unstable and decomposes to ammonia and carbon dioxide:

$$(NH_4)_2CO_3(s) \xrightarrow{\Delta} 2NH_3(g) + CO_2(g) + H_2O(g)$$

Smelling salts contain ammonium carbonate that has been moistened with ammonium hydroxide.

B. Baking Soda, $NaHCO_3$

Substances that contain the carbonate ion, CO_3^{2-}, react with acids to liberate carbon dioxide, CO_2, which is a colorless and odorless gas. Carbon dioxide, when released from baking soda by acids (for example, those present in lemon juice or sour milk) helps to "raise" the cake:

$$NaHCO_3(s) + H^+(aq) \longrightarrow CO_2(g) + H_2O(l) + Na^+(aq)$$

Place in a small, dry test tube an amount of solid baking soda about the size of a small pea. (CAUTION: *Concentrated H_2SO_4 causes severe burns. Do not get it on your skin. If you come in contact with it, immediately wash the area with copious amounts of water.*) Then add 1 or 2 drops of 18 M H_2SO_4 and notice what happens. Record your observations (7). Repeate this procedure, but use vinegar in place of the sulfuric acid. Record your observations (8).

A confirmatory test for CO_2 is to allow it to react with $Ba(OH)_2$, barium hydroxide, solution. A white precipitate of $BaCO_3$, barium carbonate, is produced:

$$CO_2(g) + Ba(OH)_2(aq) \longrightarrow BaCO_3(s) + H_2O(l)$$

Many substances, such as eggshells, oyster shells, and limestone, contain the carbonate ion. Decide whether common blackboard chalk contains the carbonate ion as follows: Place a small piece of chalk in a dry test tube; add a few drops of 2 M HCl. Test the escaping gas for CO_2 by carefully holding a drop of $Ba(OH)_2$, suspended from the tip of a medicine dropper or a wire loop, a short distance down into the mouth of the test tube. Clouding of the drop is due to the formation of $BaCO_3$ and proves the presence of carbonate. (NOTE: Breathing on the drop will cause it to cloud, since your breath contains CO_2!) Record your observations (9).

C. Table Salt, NaCl

Chloride salts react with sulfuric acid to liberate hydrogen chloride, which is a pungent and colorless gas that turns moist blue litmus red:

$$2Cl^-(s) + H_2SO_4(aq) \longrightarrow 2HCl(g) + SO_4^{2-}(aq)$$

This reaction will occur independently of whether the substance is $BaCl_2$, KCl, or $ZnCl_2$; the only requirement is that the salt be a chloride. For KCl, the complete equation is

$$2KCl(s) + H_2SO_4(aq) \longrightarrow 2HCl(g) + K_2SO_4(aq)$$

Another reaction characteristic of the chloride ion is the reaction with silver nitrate to form the white, insoluble substance silver chloride, AgCl:

$$Cl^-(aq) + AgNO_3(aq) \longrightarrow AgCl(s) + NO_3^-(aq)$$

Place in a small, dry test tube an amount of sodium chloride about the size of a small pea and add 1 or 2 drops of 18 M H_2SO_4. Very carefully note the color and odor of the escaping gas by fanning the gas with your hand toward your nose. DO NOT PLACE YOUR NOSE DIRECTLY OVER THE MOUTH OF THE TEST TUBE. Record your observations (10). Complete the equation $NaCl + H_2SO_4 \longrightarrow$? (11).

Place a small amount (about the size of a pea) of NaCl in a small test tube and add 15 drops of distilled water and one drop of 3 M HNO_3. Then add 3–4 drops of 0.1 M $AgNO_3$ and mix the contents. Record your observations (12). Why should you use distilled water for this test (13)? Confirm your answer by testing tap water for chloride ions: Add 1 drop of 3 M HNO_3 to about 2 mL of tap water and then add 3 drops of 0.1 M $AgNO_3$. Does this test indicate the presence of chloride ions in tap water (14)?

Sodium ions impart a yellow color to a flame. When potatoes boil over on the gas stove or the campfire, a burst of yellow flames appears because of the presence of sodium ions. Simply handling a utensil will contaminate it sufficiently with sodium ions from the skin so that when the utensil is placed in a hot flame a yellow color will appear. Obtain a few crystals of table salt on the tip of a clean spatula and place the tip in the flame of your burner for a brief moment. Record your observations (15).

D. Epsom Salts, $MgSO_4 \cdot 7H_2O$

Epsom salts are used as a purgative, and solutions of this salt are used to soak "tired, aching feet." The following tests are characteristic of the sulfate ion, SO_4^{2-}. Place a small quantity of Epsom salts in a small, dry test tube and add 1 or 2 drops of 18 M H_2SO_4. Record your observations (16). Note the difference in the behavior of this substance toward sulfuric acid compared with the behavior of baking soda toward sulfuric acid.

Place some Epsom salts (an amount the size of a small pea) in a small test tube and dissolve in 1 mL of distilled water. Add 1 drop of 3 M HNO_3 and then 1 or 2 drops of 0.2 M $BaCl_2$. Record your observations (17). Barium sulfate is a white, insoluble substance that forms when barium chloride is added to a solution of any soluble sulfate salt, such as Epsom salts, as follows:

$$SO_4^{2-}(aq) + BaCl_2(aq) \longrightarrow BaSO_4(s) + 2Cl^-(aq)$$

E. Bleach, Cl_2 Water

Commercial bleach is usually a 5 percent solution of sodium hypochlorite, NaOCl. This solution behaves as though only chlorine, Cl_2, were dissolved in it. Since this solution is fairly concentrated, direct contact with the skin or eyes must be avoided! The element chlorine, Cl_2, behaves very differently from the chloride ion. Chlorine is a pale, yellow-green gas, has an irritating odor, is slightly soluble in water, and is toxic. It is capable of liberating the element iodine, I_2, from iodide salts:

$$Cl_2(aq) + 2I^-(aq) \longrightarrow I_2(aq) + 2Cl^-(aq)$$

Iodine gives a reddish-brown color to water; it is more soluble in mineral oil than in water and imparts a violet color to the mineral oil. Thus chlorine can be used to identify iodide salts.

Dissolve a small amount (about the size of a pea) of sodium iodide, NaI, in 1 mL of distilled water in a small test tube; add 5 drops of bleach. Note the color, then add several drops of mineral oil, shake, and allow to separate, which takes about 20 sec. Note, the mineral oil is the top layer. Record your observations (18).

Another reaction characteristic of iodides is that they form a pale yellow precipitate when treated with silver nitrate solution:

$$I^-(aq) + AgNO_3(aq) \longrightarrow AgI(s) + NO_3^-(aq)$$

Dissolve a small amount of sodium iodide in 1 mL of distilled water and add a drop of 3 *M* HNO_3, then add 3–4 drops of 0.1 *M* $AgNO_3$ solution. Record your observations (19).

Solid iodide salts react with concentrated sulfuric acid by instantly turning dark brown, with the slight evolution of a gas that fumes in moist air and with the appearance of violet fumes of iodine. Place a small amount (about the size of a pea) of sodium iodide in a small, dry test tube and add, in the hood, 1 or 2 drops of 18 *M* H_2SO_4. Record your observations (20).

F. Unknown

Your solid unknown will contain only one of the following ions: carbonate, chloride, sulfate, or iodide. Table 7.1 summarizes the behavior of these ions toward H_2SO_4.

Place a small amount of your unknown (save some for further tests) in a small dry test tube and add a drop of 18 *M* H_2SO_4. Record your observations and the formula for the unknown ion (21). What additional test might you perform to help identify the ion (22)? Consult with your instructor *before* doing the test.

Table 7.1 Reaction of Solid Salts with H_2SO_4

Ion	Reaction
CO_3^{2-}	Colorless, odorless gas, CO_2, evolved
Cl^-	Colorless, pungent gas, HCl, evolved, which turns blue litmus red
SO_4^{2-}	No observable reaction
I^-	Violet vapors of I_2 formed

REVIEW QUESTIONS

Before beginning this experiment in the laboratory, you should be able to answer the following questions:

1. Why is it unwise to haphazardly mix household or other chemicals?
2. How could you detect the presence of the NH_4^+ ion?
3. How could you detect the presence of the CO_3^{2-} ion?
4. How could you detect the presence of the Cl^- ion?
5. How could you detect the presence of the SO_4^{2-} ion?

6. How could you detect the presence of the I^- ion?

7. How could you detect the presence of Ag^+ ion?

8. Complete and balance the following equations:

$$LiCl(s) + H_2SO_4(aq) \longrightarrow$$

$$NH_4^+(aq) + OH^-(aq) \rightleftarrows$$

$$AgNO_3(aq) + I^-(aq) \longrightarrow$$

$$NaHCO_3(s) + H^+(aq) \longrightarrow$$

9. Why should distilled water be used when making chemical tests?

10. Assume you had a mixture of solid Na_2CO_3 and NaCl. Could you use only H_2SO_4 to determine whether or not Na_2CO_3 was present? Explain.

11. Assume you had a mixture of solid Na_2CO_3 and NaCl. How could you show the presence of both carbonate and chloride in this mixture?

12. How could you show the presence of both iodide and sulfate in a mixture? Consult Appendix E for help.

EXPERIMENT

8

Gravimetric Analysis of a Chloride Salt

OBJECTIVE

To illustrate typical techniques used in gravimetric analysis by quantitatively determining the amount of chloride in an unknown.

APPARATUS AND CHEMICALS

250-mL beakers (3)
filter paper
funnels (3)
stirring rods (3)
watch glasses (3)
rubber policeman
weighing paper
Bunsen burner
ring stand, ring, and wire gauze
balance

plastic wash bottle
3 beakers—any combination
  100 mL or larger
funnel support
1.5 g unknown chloride sample
0.5 *M* $AgNO_3$
6 *M* HNO_3
acetone, 60 mL
distilled water

DISCUSSION

Quantitative analysis is that aspect of analytical chemistry which is concerned with determining *how much* of one or more constituents is present in a particular sample of material. We have already seen in Experiment 5 how information such as percentage composition is essential to establishing formulas for compounds. Two common methods used in analytical chemistry are gravimetric and volumetric analysis. *Gravimetric analysis* derives its name from the fact that the constituent being determined can be isolated in some weighable form. *Volumetric analysis,* on the other hand, derives its name from the fact that the method used to determine the amount of a constituent involves measuring the volume of a reagent. Usually, gravimetric analyses involve the following steps:

1. Drying and then accurately weighing representative samples of the material to be analyzed.
2. Dissolving the samples.
3. Precipitating the constituent in the form of a substance of known composition by adding a suitable reagent.
4. Isolating the precipitate by filtration.
5. Washing the precipitate to free it of contaminants.
6. Drying the precipitate to a constant weight (to obtain an analytically weighable form of known composition).

7. Calculating the percentage of the desired constituent from the weights of the sample and precipitate.

Although the techniques of gravimetric analysis are applicable to a large variety of substances, we have chosen to illustrate them with an analysis that incorporates a number of other techniques as well. Chloride ion may be quantitatively precipitated from solution by the addition of silver ion according to the following ionic equation:

$$Ag^+(aq) + Cl^-(aq) \longrightarrow AgCl(s) \qquad [1]$$

Silver chloride is quite insoluble (only about 0.0001 g of AgCl dissolves in 100 mL of H_2O at 20°C); hence, the addition of silver nitrate solution to an aqueous solution containing chloride ion precipitates AgCl quantitatively. The precipitate can be collected on a filter paper, dried, and weighed. From the weight of the AgCl obtained, the amount of chloride in the original sample can then be calculated.

This experiment also illustrates the concept of stoichiometry. *Stoichiometry* is the determination of the proportions in which chemical elements combine and the weight relations in any chemical reaction. In this experiment stoichiometry means specifically the mole ratio of the substances entering into and resulting from the combination of Ag^+ and Cl^-. In the reaction of Ag^+ and Cl^- in Equation [1], it can be seen that 1 mol of chloride ions reacts with 1 mol of silver ions to produce 1 mol of silver chloride. Thus

$$\text{Moles } Cl^- = \text{moles AgCl} = \frac{\text{grams AgCl}}{\text{gram-formula weight AgCl}}$$

$$\begin{aligned}
\text{Grams Cl in sample} &= (\text{moles } Cl^-)(\text{gram-atomic weight Cl}) \\
&= \frac{(\text{gram-atomic weight Cl})(\text{grams AgCl})}{\text{gram-formula weight AgCl}} \\
&= \frac{(35.45 \text{ g Cl})(\text{grams AgCl})}{143.3 \text{ g AgCl}} \\
&= (0.2474 \text{ g Cl/g AgCl})(\text{grams AgCl})
\end{aligned}$$

The number 0.2474 is called a gravimetric factor. It converts grams of AgCl into grams of Cl. Gravimetric factors are used repeatedly in analytical chemistry and are tabulated in handbooks. The percentage of Cl in the sample can be calculated according to the following formula:

$$\% \text{ Cl in sample} = \frac{(\text{grams Cl in sample})(100)}{\text{gram-weight sample}}$$

EXAMPLE 8.1

In a gravimetric chloride analysis it was found that 0.2516 g AgCl was obtained from an unknown which weighed 0.1567 g. What is the percent chloride in this sample?

Solution: The weight of Cl in the sample is g Cl = (0.2474 g Cl/g AgCl) × (0.2516 g AgCl) = 0.06220 g Cl

$$\% \text{ Cl} = \frac{(0.06220 \text{ g Cl})(100)}{(0.1567 \text{ g sample})} = 39.72\%$$

PROCEDURE

Weigh to the nearest 0.0001 g about 0.2 to 0.4 g of your unknown sample. Transfer the sample quantitatively to a clean 250-mL beaker (do not weigh the beaker) and label the beaker #1 with a pencil. Record the sample weight. Add 150 mL of distilled water and 1 mL of 6 *M* HNO_3 to the beaker. Repeat with sample numbers 2 and 3 and label the beakers 2 and 3. Stir each of the solutions with three different glass stirring rods until all of the sample has dissolved. Leave the stirring rods in the beaker. Do not place them on the desktop.

While stirring one of the solutions, add to it about 20 mL of 0.5 *M* $AgNO_3$ solution. Place a watch glass over the beaker. Warm the solution gently with your Bunsen burner and keep it warm for 5 to 10 min. Do not boil the solution.

Obtain a filter paper (three of these will be needed) and weigh it accurately. Fold the paper as illustrated in Figure 8.1 and fit it into a glass funnel.

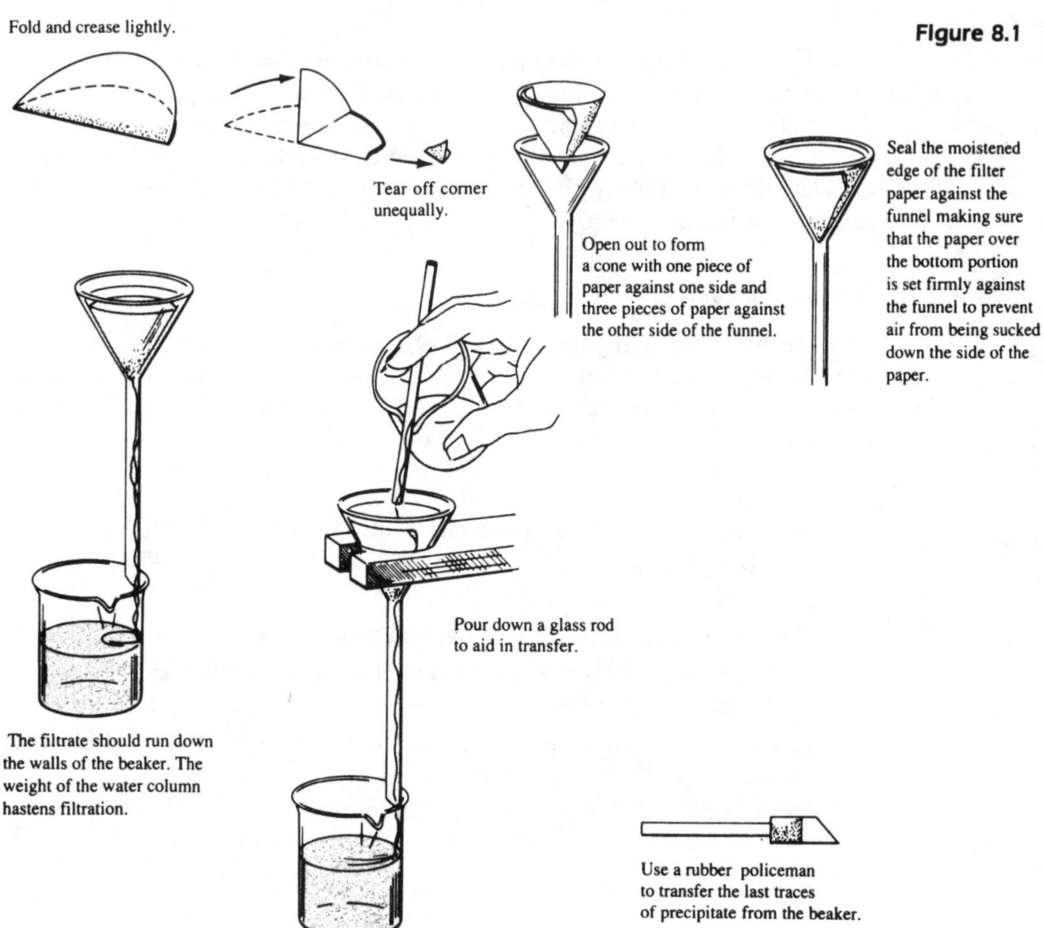

Figure 8.1 Filter paper use.

Be certain that you open the filter paper in the funnel so that one side has three pieces and one side has one piece of paper against the funnel—not two pieces on each side. *Why?* The teaching assistant will also demonstrate this for you. (Be certain that you weigh the paper after it has been folded and torn, not before.) Wet the paper with distilled water to hold it in place in the funnel. Completely and quantitatively transfer the precipitate and all the warm solution from the beaker onto the filter, using a rubber policeman (your laboratory instructor will show you how to use a rubber policeman) and a wash bottle to wash out the last traces of precipitate. The level of solution in the filter funnel should always be *below* the top edge of the filter paper. Wash the precipitate on the filter paper with two or three 5-mL portions of water from the wash bottle. Finally, pour three 5-mL portions of acetone through the filter. KEEP THE ACETONE AWAY FROM OPEN FLAMES, BECAUSE IT IS HIGHLY FLAMMABLE. Remove the filter paper, place it on a numbered watch glass, and store it in your locker until the next period.

Repeat the above processes with your other two samples, being sure that you have numbered your watch glasses so that you can identify the samples. The precipitated AgCl must be kept out of bright light, because it is photosensitive and slowly decomposes in the presence of light as follows:

$$2AgCl(s) \xrightarrow{hv} 2Ag(s) + Cl_2(g)$$

In this equation *hv* is a symbol for electromagnetic radiation; here it represents radiation in the visible and ultraviolet regions of the spectrum. This is the reaction used by Corning to make photosensitive sunglasses. *In the next period* when the AgCl is thoroughly dry, weigh the filter papers plus AgCl and calculate the weight of AgCl. From these data calculate the percentage of chloride in your original sample.

Standard Deviation

As a means of estimating the precision of your results, it is desirable to calculate the standard deviation. Before we illustrate how to do this, however, we will define some of the terms above as well as some additional ones that are necessary.

Accuracy: correctness of a measurement, closeness to the true result.

Precision: internal consistency among one's own results, that is, reproducibility.

Error: difference between the true result and the determined result.

Determinate errors: errors in method of performance that can be discovered and eliminated.

Indeterminate errors: random errors, which are incapable of discovery but which can be treated by statistics.

Mean: arithmetic mean or average (μ), where

$$\mu = \frac{\text{sum of results}}{\text{number of results}}$$

For example, if an experiment's results are 1, 3, and 5, then

$$\mu = \frac{1 + 3 + 5}{3} = 3$$

Median: the midpoint of the results for an odd number of results and the average of the two middle results for an even number of results (m). For example, if an experiment's results are 1, 3, and 5, then $m = 3$. If results are 1.0, 3.0, 4.0, and 5.0, then

$$m = \frac{3.0 + 4.0}{2} = 3.5$$

The scatter about the mean or median—that is, the deviations from the mean or median—are measures of precision. Thus the less the deviation, the more precise the measurements.

EXAMPLE 8.2

If an experiment's results are 1.0, 2.0, 3.0, and 4.0, calculate the mean, the deviations from the mean, the average deviation from the mean, and the relative deviation from the mean.

Solution: The mean is calculated as follows:

$$\mu = \frac{1.0 + 2.0 + 3.0 + 4.0}{4} = \frac{10.0}{4} = 2.5$$

The deviations from the mean are

$$|2.5 - 1.0| = 1.5$$
$$|2.5 - 2.0| = 0.5$$
$$|2.5 - 3.0| = 0.5$$
$$|2.5 - 4.0| = 1.5$$

The symbol $|\ |$ means absolute value, so all differences are positive. The average deviation from the mean is therefore

$$\frac{1.5 + 0.5 + 0.5 + 1.5}{4} = 1.0$$

The relative average deviation from the mean is calculated by dividing the average deviation from the mean by the mean. Thus

$$\text{Relative deviation} = \frac{1.0}{2.5} = 0.40$$

This can be expressed as 40 percent, 400 parts/thousand (ppt), or 40,000 parts/million (ppm).

EXAMPLE 8.3

If an experiment's results are 1.0, 1.5, 2.0, and 2.5, calculate the mean, the deviations from the mean, the average deviation from the mean, and the relative deviation from the mean.

Solution: The mean is calculated as follows:

$$\mu = \frac{1.0 + 1.5 + 2.0 + 2.5}{4}$$

$$= \frac{7.0}{4} = 1.75, \text{ or } 1.8 \text{ to two significant figures}$$

The deviations from the mean are

$$|1.8 - 1.0| = 0.8$$
$$|1.8 - 1.5| = 0.3$$
$$|1.8 - 2.0| = 0.2$$
$$|1.8 - 2.5| = 0.7$$

The average deviation from the mean is therefore

$$\frac{0.8 + 0.3 + 0.2 + 0.7}{4} = 0.5$$

The relative average deviation from the mean is

$$\frac{0.5}{1.8} = 0.28, \text{ or } 0.3$$

$$= 30\%, \text{ or } 300 \text{ ppt, or } 30{,}000 \text{ ppm}$$

Obviously, the data in Example 8.3 are internally more consistent than the data in Example 8.2 and hence are more precise, since the deviations are smaller. Thus the average deviation and relative deviation measure precision.

Standard deviation (s) is a better measure of precision and is calculated using the formula

$$s = \sqrt{\frac{\text{sum of the squares of the deviations from the mean}}{\text{number of observations} - 1}}$$

$$= \sqrt{\frac{\Sigma_i |\chi_i - \mu|^2}{N - 1}}$$

where s = standard deviation from the mean, χ_i = members of the set, μ = mean, and N = number of members in the set of data. The symbol Σ_i means to sum over the members.

EXAMPLE 8.4

An experiment's results are 1, 3, and 5. Calculate the mean, the deviations from the mean, the standard deviation, and the relative standard deviation for the data.

Solution: The mean is as follows:

$$\mu = \frac{1 + 3 + 5}{3} = 3$$

The deviations from the mean are

$$|x_i - \mu| = \text{deviation}$$
$$|1 - 3| = 2$$
$$|3 - 3| = 0$$
$$|5 - 3| = 2$$

$$s = \sqrt{\frac{2^2 + 0^2 + 2^2}{3 - 1}}$$
$$= \sqrt{\frac{4 + 0 + 4}{2}}$$
$$= \sqrt{\frac{8}{2}}$$
$$= \sqrt{4} = 2$$

The results of this experiment would be reported as 3 ± 2. The relative standard deviation is

$$\frac{2}{3} = 0.66$$
$$= 0.7, \text{ or } 70\%$$

EXAMPLE 8.5

The results of an experiment are 2.100, 2.110, and 2.105. Calculate the mean, the deviations from the mean, the standard deviation, and the relative standard deviation.

Solution: The mean is as follows.

$$\mu = \frac{2.100 + 2.110 + 2.105}{3} = 2.105$$

The deviations from the mean are:

$$|2.105 - 2.100| = 0.005$$
$$|2.105 - 2.110| = 0.005$$
$$|2.105 - 2.105| = 0.000$$

The standard deviation is therefore:

$$s = \sqrt{\frac{(0.005)^2 + (0.005)^2 + (0.000)^2}{2}}$$
$$= \sqrt{\frac{5 \times 10^{-5}}{2}}$$
$$= 0.005$$

The results would be reported as 2.105 ± 0.005. The relative standard deviation is

$$\frac{0.005}{2.105} = 0.002 \text{ or } 0.2\%$$

Obviously, the data in Example 8.5 are more precise, though not necessarily more accurate, than the data in Example 8.4, because both the deviations and the standard deviation are smaller in Example 8.5.

Calculate the standard deviation of your data and report the results on your report sheet.

The standard deviation may be used to determine whether a result should be retained or discarded. As a rule of thumb, you should discard any result that is more than two standard deviations from the mean. For example, if you had a result of 49.65 percent and you had determined that your percentage of chloride was 49.25 ± 0.09 percent, this result (49.65 percent) should be discarded. This is because $s = 0.09$ and $|49.25 - 49.65| = 0.40$, which is greater than 2×0.09. This result is more than two standard deviations from the mean.

REVIEW QUESTIONS

Before beginning this experiment in the laboratory, you should be able to answer the following questions:

1. What is the fundamental difference between gravimetric and volumetric analysis?
2. What does stoichiometry mean?
3. Why should silver chloride be protected from light? Will your result be high or low if you don't protect your silver chloride from light?
4. Can you eliminate indeterminate errors from your experiment?
5. Does standard deviation give a measure of accuracy or precision?
6. Why don't you open your folded filter paper so that two pieces touch each side of the funnel?
7. If your silver chloride undergoes extensive photodecomposition before you weigh it, will your results be high or low?
8. If an experiment's results are 12.1, 12.4, and 12.6, find the mean, the average deviation from the mean, the standard deviation from the mean, and the relative deviation from the mean.
9. What is meant by the term *gravimetric factor*?

Gravimetric Determination of Copper, Cobalt, Nickel, or Zinc

EXPERIMENT 9

OBJECTIVE

To illustrate the techniques used in gravimetric analysis and to gain some familiarity with coordination chemistry.

APPARATUS AND CHEMICALS

pyridine
NH_4SCN
1 g unknown soluble copper, cobalt, zinc, or nickel sample
250-mL beakers (3)
ashless filter paper
funnels (3)
glass stirring rods (3)
weighing paper
3 beakers—any combination, 200 mL or larger
funnel support
ethanol
anhydrous ether

Four wash solutions:

1. 100 mL of H_2O containing 0.3 g NH_4SCN and 0.3 mL pyridine.
2. 80 mL of 95 percent ethanol, 19.2 mL of H_2O, 0.8 mL of pyridine, and 0.05 g of NH_4SCN.
3. 10 mL absolute ethanol + 2 drops pyridine.
4. 20 mL anhydrous ether + 2 drops pyridine. (CAUTION: *Ether is extremely flammable. Keep it away from open flames.*)

READ EXPERIMENT 8 BEFORE BEGINNING THIS EXPERIMENT.

DISCUSSION

As the science of chemistry has evolved, the boundaries separating such traditional disciplines as organic and inorganic chemistry have become increasingly more difficult to define. Nowhere is this more evident than in the area of coordination chemistry. Compounds involving coordinate covalent bonds between a metal ion and an organic ligand are known to be of critical importance. They occur naturally in flora (for example, in chlorophyll, which is a coordination compound of magnesium) and in fauna (for example, in hemoglobin, which is a coordination compound of iron). In addition, the use of coordination complexes as catalysts frequently permits industrial chemists to make useful and practical alterations in molecular architecture that would have been inconceivable only a few years ago. Coordination compounds have found wide and varied usage in analytical chemistry, where specific reagents have been developed for the identification and chelation of specific metal ions. Industrially, this chelation process is used to soften water by removing calcium and magnesium by the use of chelating ligands such as ethylenediaminetetraacetic acid in ion-exchange columns. Four particular coordination compounds that are useful for gravimetric determinations because they are very insoluble are:

1. $[Cu(C_5H_5N)_2(NCS)_2]$
2. $[Zn(C_5H_5N)_2(NCS)_2]$
3. $[Co(C_5H_5N)_4(NCS)_2]$
4. $[Ni(C_5H_5N)_4(NCS)_2]$

These are prepared in neutral or weakly acidic solution by the following reactions:

$$Zn^{2+}(aq) + 2SCN^-(aq) + 2C_5H_5N(l) \rightleftharpoons [Zn(C_5H_5N)_2(NCS)_2](s)$$
$$Cu^{2+}(aq) + 2SCN^-(aq) + 2C_5H_5N(l) \rightleftharpoons [Cu(C_5H_5N)_2(NCS)_2](s)$$
$$Co^{2+}(aq) + 2SCN^-(aq) + 4C_5H_5N(l) \rightleftharpoons [Co(C_5H_5N)_4(NCS)_2](s)$$
$$Ni^{2+}(aq) + 2SCN^-(aq) + 4C_5H_5N(l) \rightleftharpoons [Ni(C_5H_5N)_4(NCS)_2](s)$$

These compounds involve dative, or coordinate covalent, bonds between the Lewis acid (for example, copper) and the Lewis base pyridine as illustrated below:

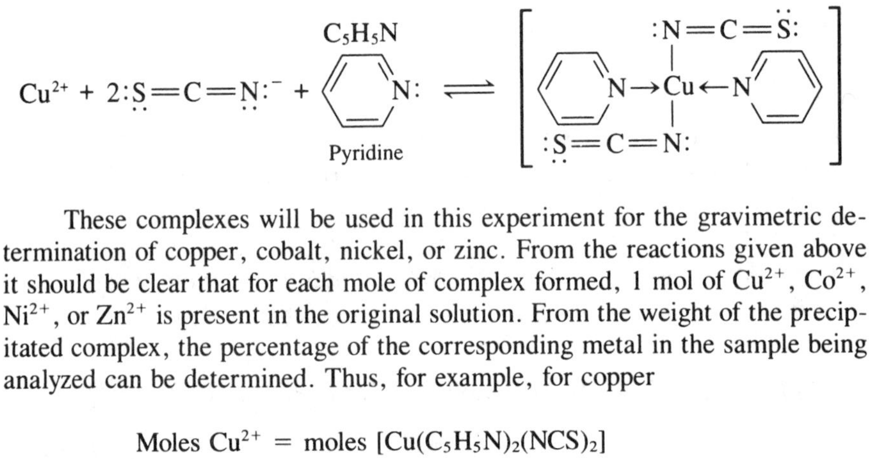

$$Cu^{2+} + 2:\ddot{S}=C=\ddot{N}:^- + \underset{\text{Pyridine}}{\overset{C_5H_5N}{C_5H_5N:}} \rightleftharpoons \left[\begin{array}{c} :N=C=\ddot{S}: \\ | \\ C_5H_5N\rightarrow Cu \leftarrow NC_5H_5 \\ | \\ :\ddot{S}=C=N: \end{array}\right]$$

These complexes will be used in this experiment for the gravimetric determination of copper, cobalt, nickel, or zinc. From the reactions given above it should be clear that for each mole of complex formed, 1 mol of Cu^{2+}, Co^{2+}, Ni^{2+}, or Zn^{2+} is present in the original solution. From the weight of the precipitated complex, the percentage of the corresponding metal in the sample being analyzed can be determined. Thus, for example, for copper

$$\text{Moles } Cu^{2+} = \text{moles } [Cu(C_5H_5N)_2(NCS)_2]$$

$$\begin{aligned} \text{Grams } Cu^{2+} \text{ in sample} &= (\text{grams } [Cu(C_5H_5N)_2(NCS)_2]) \\ &\times \left(\frac{1 \text{ mol } [Cu(C_5H_5N)_2(NCS)_2]}{337.54 \text{ g } [Cu(C_5H_5N)_2(NCS)_2}\right) \\ &\times \left(\frac{1 \text{ mol } Cu^{2+}}{1 \text{ mol } [Cu(C_5H_5N)_2(NCS)_2]}\right)\left(\frac{63.54 \text{ g } Cu^{2+}}{1 \text{ mol } Cu^{2+}}\right) \\ &= \left(0.1882 \frac{\text{g Cu}}{\text{g}[CuC_5H_5N)_2(SCN)_2]}\right) \\ &\times (\text{grams } [Cu(C_5H_5N)_2(NCS)_2]) \end{aligned}$$

$$\% \ Cu^{2+} = \frac{(\text{grams } Cu^{2+} \text{ in sample})(100)}{\text{gram-weight sample}}$$

EXAMPLE 9.1

If a 0.1259-g sample yields 0.2600 g $[Ni(C_5H_5N)_4(NCS)_2]$, what is the percent nickel in the sample?

Solution: The weight of nickel in the sample is

$$\frac{58.93\text{ g Ni/mole}) \times 0.2600\text{ g }[Ni(C_5H_5N)_4(NCS)_2],}{491.05\text{ g/mole }[Ni(C_5H_5N)_4(NCS)_2]} = 0.03120\text{ g Ni}$$

$$\%\text{ Ni} = \frac{0.03120\text{ g Ni}}{0.1259\text{ g sample}} \times 100 = 24.78\%$$

Analogous relations hold for cobalt, nickel, and zinc. In this experiment you will determine only one element. Your lab instructor will tell you which one.

PROCEDURE

NO FLAMES ARE ALLOWED IN THE LABORATORY DURING THE PERFORMANCE OF THIS EXPERIMENT! Weigh by difference to the nearest 0.0001 g about 0.1 to 0.2 g of your unknown sample, using weighing paper. Transfer the sample quantitatively to a 250-mL beaker and record the sample weight. Add 100 mL of distilled water and then, IN THE HOOD, add pure pyridine dropwise until the color of the solution is an intense azure blue for Cu^{2+} or Ni^{2+}, deep red for Co^{2+}, and colorless for Zn^{2+} (1–2 mL should suffice); then add about 0.5 g of solid NH_4SCN and stir vigorously with a glass rod. After a few minutes, collect the precipitate on a preweighed piece of filter paper. (Be certain that you have weighed the filter paper after it has been folded and torn as described in Experiment 8.)

Transfer the precipitate to the filter paper with the aid of solution 1. Wash the precipitate six to eight times with 2 mL of solution 2, then twice with 2 mL of solution 3, and finally several times with small volumes (1–2 mL) of solution 4. *It is important to let the filter drain almost completely between washings and to stir the precipitate gently with a glass rod during the washings.* Dry the precipitate in air to constant weight (about 30 min) and record the weight.

Repeat this process with two more samples.

REVIEW QUESTIONS

Before beginning this experiment in the laboratory, you should be able to answer the following questions:

1. Define and illustrate a coordinate covalent (dative) bond. How does it differ from a covalent bond?
2. Set up the equations necessary to calculate the percentage of cobalt in an unknown sample.
3. Define *chelation*.
4. Define the terms *Lewis acid* and *Lewis base*.
5. A sample was found to contain the following percentages of copper: 9.15 percent, 9.06 percent, and 9.22 percent. (a) What is the average percentage of copper? (b) What is the standard deviation?
6. If a copper sample were being analyzed for Cu according to the proce-

dure of this experiment and the sample was contaminated with nickel, how would the presence of nickel affect the results? Would the measured percentage of copper be too high or too low?

7. How does one know that the precipitate has been thoroughly dried?
8. What are three possible sources of error in this experiment?
9. What is meant by indeterminate errors?
10. What is meant by precision?

Name ______________________________ Desk __________

Date __________________ Laboratory Instructor ______________________

Unknown no. __________________

REPORT SHEET FOR EXPERIMENT 9

GRAVIMETRIC DETERMINATION OF COPPER, COBALT, NICKEL, OR ZINC

Metal determined __________

	Trial 1	Trial 2	Trial 3
Weight of sample	________	________	________
First weight of filter paper and complex	________	________	________
Second weight of filter paper and complex	________	________	________
Weight of filter paper	________	________	________
Weight of complex	________	________	________
Weight of metal in original sample (show calculations)	________	________	________

Average percent metal (show calculations) __________

Standard deviation (show calculations) __________

Do any of your results differ from the mean by more than two standard deviations? __________

Reported percent metal __________ ± __________ %

QUESTIONS

1. Given the formulas for the coordination compounds $[Zn(C_5H_5N)_2(NCS)_2]$, $[Cu(C_2H_5N)_2(NCS)_2]$, $[Co(C_5H_5N)_4(NCS)_2]$, and $[Ni(C_5H_5N)_4(NCS)_2]$, what are the coordination numbers of the metals in these coordination compounds?

2. If a 0.1259-g sample containing only nickel, chloride, and water gives 0.2600 g $[Ni(C_5H_5N)_4(NCS)_2]$, what is the formula for the nickel compound in the original sample?

3. Why do you wash your precipitated C_5H_5N complexes with solutions containing C_5H_5N?

4. Is the ion NCS^- a Lewis base? Write Lewis electron dot formulas for the resonance structures of NCS^-. From these resonance structures, do you think that NCS^- could coordinate to a transition metal through the sulfur end of the ion? Why or why not?

Paper Chromatography: Separation of Cations and Dyes

EXPERIMENT 10

OBJECTIVE

To become acquainted with chromatographic techniques as a method of separation (purification) and identification of substances.

APPARATUS AND CHEMICALS

petri dishes or evaporating dishes or casseroles (2)
4-in. watch glasses (2)
12.5-cm Whatman #1 filter paper
scissors
capillary pipets (6)
sample vials (3)
50- or 100-mL graduated cylinder
metric rule
600-mL beaker
paper clips (3)
0.5 *M* $Cu(NO_3)_2$
0.5 *M* $Fe(NO_3)_3$
0.5 *M* $Ni(NO_3)_2$
25 mL solvent (90 percent acetone, 10 percent 6 *M* HCl) freshly prepared
10 mL 15 *M* ammonium hydroxide
5 mL 1 percent dimethylglyoxime in ethanol
5 mL 1 percent isopropyl alcohol
6-in. glass rod
paper chromatography strips
unknown solutions

DISCUSSION

Chromatography (from the Greek *chrōma,* for color, and *graphein,* to write) is a technique often used by chemists to separate components of a mixture. In 1906, the Russian botanist Mikhail Tsvett separated color pigments present in leaves by allowing a solution of these pigments to flow down a column packed with an insoluble material such as starch, alumina (Al_2O_3), or silica (SiO_2). Because different color bands appeared along the column, he called the procedure chromatography. Color is not a requisite property to achieve separation of compounds by this procedure. Colorless compounds can be made visible by being allowed to react with other reagents, or they can be detected by physical means. Consequently, because of its simplicity and efficiency, this technique has wide applicability for separating and identifying compounds such as drugs and natural products.

The basis of chromatography is the *partitioning* (that is, separation arising from differences in solubility) of compounds between a stationary phase and a moving phase. Stationary phases such as alumina (Al_2O_3), silica (SiO_2), and paper (cellulose) have enormous surface areas. The molecules or ions of the substances to be separated are continuously being adsorbed and then released (desorbed) into the solvent flowing over the surface of the stationary phase. This brings about a separation of the components—that is, they will travel with different speeds in the moving solvent because there are generally different attractions between these components and the stationary phase.

In paper chromatography, a small spot of the mixture to be separated is placed at one end of a strip of paper, and solvent is allowed to move up the paper, through the spot, by capillary action. The solvent and various components of the mixture each travel at different speeds along the paper. In this experiment, Fe^{3+}, Cu^{2+}, and Ni^{2+} will be separated with a solvent that consists of a mixture of acetone, water, and hydrochloric acid. A diagram of a portion of a typical chromatogram strip is shown in Figure 10.1.

The identity of components in a mixture can be deduced by comparing a chromatogram of the unknown with chromatograms of mixtures of components suspected to be present in the unknown. An additional aid in identification of a compound is its R_f value, which is defined as the ratio of the distance traveled by a compound to the distance traveled by the solvent. For example, for Cu^{2+} (see Figure 10.1),

$$R_f(\mathrm{Cu}) = \frac{d_{\mathrm{Cu}}}{d_s} \qquad [1]$$

The R_f value of a compound is a characteristic of the compound, the support, and the solvent used and serves to identify the constitutents of a mixture. An estimate of the relative amount of each of the constituents in a mixture can be made from the relative intensities and sizes of the various bands in a chromatogram.

The three ions studied in this experiment will be discerned in the following manner: Fe^{3+} in water imparts a red-brown or rust color and thus will produce a rust-colored band on the paper. Although Cu^{2+} is blue, the color is faint and not easily detected, especially if copper is present in small amounts. In aqueous solution, however, Cu^{2+} reacts with NH_3 (from ammonium hydroxide) to form a complex ion, $[Cu(NH_3)_4]^{2+}$, which is deep blue and therefore readily observed. Finally, Ni^{2+} reacts with an organic reagent, dimethylglyoxime, to produce a strawberry-red color. The following reactions will occur during the development process on the paper:

$$\underset{\text{Faint yellow}}{Fe^{3+}} + 3OH^- \longrightarrow \underset{\text{Rust colored}}{Fe(OH)_3}$$

$$\underset{\text{Pale blue}}{Cu^{2+}} + 4NH_3 \longrightarrow \underset{\text{Deep blue}}{[Cu(NH_3)_4]^{2+}}$$

$$\underset{\text{Pale green}}{Ni^{2+}} + 2NH_3 + 2\ \mathrm{H_3C{-}C({=}NOH){-}C({=}NOH){-}CH_3} \longrightarrow \underset{\text{Strawberry red}}{\mathrm{Ni[ON{=}C(CH_3){-}C(CH_3){=}NOH]_2}} + 2NH_4^+$$

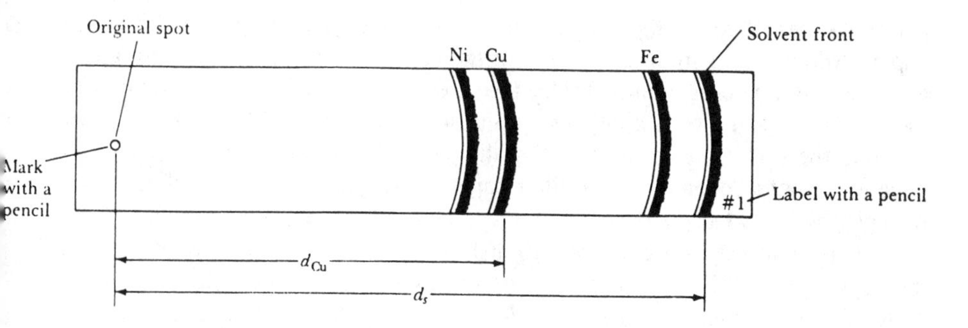

Figure 10.1

PROCEDURE

There are two simple ways of performing this experiment: either by circular horizontal chromatography or by ascending chromatography using paper strips. One half of the class should perform this experiment using one technique, and the other half should use the other technique. The two techniques can then be compared in terms of time required, separation, and ease of identification.

Make six capillary pipets (2 in. long) by drawing out 6-mm glass tubing. Your instructor will demonstrate how this is done, or you an obtain the capillary tubes from your instructor.

Obtain an unknown; you will analyze this at the same time you are conducting the experiment on the known solutions. Your unknown may contain one, two, or three of the ions being studied. Record the colors of the *known solutions* on your report sheet (1).

A. Horizontal Circular Technique

Cut a wick about 1 cm wide on a piece of 12.5-cm Whatman # 1 filter paper, as shown in Figure 10.2. Mark the center of the paper with a pencil dot. Place the cut filter paper on top of another piece of filter paper, which will serve as an absorbing pad. Using the following technique, "spot" the filter paper at the center directly on the pencil dot sequentially with 1 drop of each of the known solutions of Cu^{2+}, Ni^{2+}, and Fe^{3+} (let the spot dry completely between applications): Pour a small quantity (only 1–2 mL) of Cu^{2+} solution into a clean sample vial and fill your capillary pipet by dipping the end into the solution; allow the solution to rise by capillary action. Withdraw the pipet; touch the inside of the vial with the tip of the pipet to remove the hanging drop. Spot the filter pa-

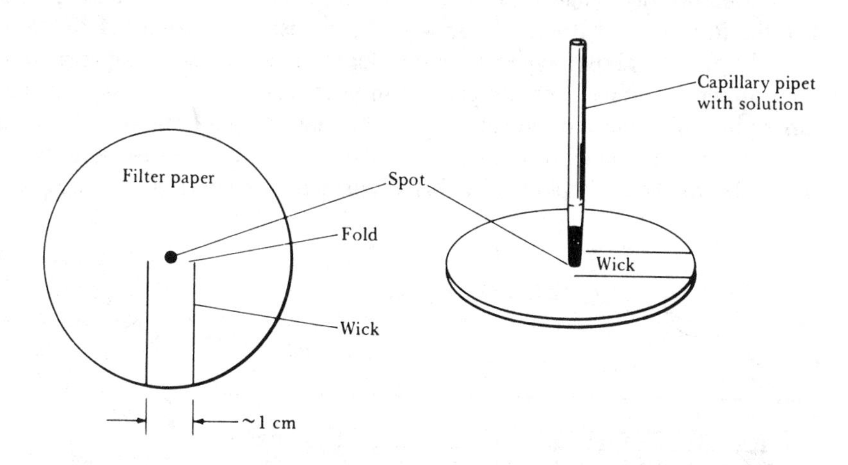

Figure 10.2

per on the pencil dot by touching it with the capillary held perpendicular to the paper. Allow the solution to flow out of the capillary until a spot having about a 5–7 mm diameter is obtained. Dry the filter paper completely by waving it in the air. In the same way, apply a single drop of the Ni^{2+} and then the Fe^{3+} solution to the same spot; be certain that the paper is dry before making each application. Spot a second piece of filter paper which you have prepared for chromatography with 1 drop of your unknown solution in the same manner.

Place two petri (or evaporating) dishes on your desk top away from direct sunlight or heat. Fill the dishes with the solvent mixture to a depth of 5–7 mm. (CAUTION: *Acetone is very flammable and must be kept far from open flames.*) Carefully place the (dry) spotted filter papers on the rims of the petri dishes with the wicks bent down into the solvent (see Figure 10.3). Cautiously place a 4-in. watch glass on top of each filter paper; be careful not to push the papers into the solvent. The purpose of the cover is to prevent uneven evaporation of the solvent from the paper by providing an enclosed atmosphere that is saturated with solvent vapor. These conditions permit the solvent to travel across the paper in a more uniform manner; this is necessary for effective separation of the ions. Do not disturb the systems while the chromatograms are developing. When the solvent front has nearly reached the edge of each dish (15–25 min), carefully remove the watch glass and filter paper. Immediately mark the position of the solvent front with a pencil. Because the solvent evaporates quickly, this marking must be made as soon as possible. Allow each paper to dry by fanning the air with it.

Which of the known ions do you detect without resorting to the use of any other reagents (or development) (2)? Is there any difference between the ring front of this ion and the solvent front (3)? Mark the ring front of this ion.

Developing Reactions IN THE HOOD, pour about 5 mL of 15 *M* ammonium hydroxide into a clean, shallow dish and rest the filter paper containing the knowns on top of the dish. Do not permit the paper to dip into the solution. What color develops (4)? What ion does this indicate (5)? Mark the ring front.

To detect the third ion, dip a new piece of filter paper into a 1 percent solution of dimethylglyoxime; then, using the new piece as a brush, paint the test filter paper. As an alternative procedure, you may spray the paper with a dimethylglyoxime-containing aerosol. What color develops (6)? What ion does this indicate (7)? Mark the ring front.

Measure the distance to each of the ring fronts in millimeters (8). Calculate the R_f values for the three known ions (9) using Equation [1].

Repeat the above sequence of developing reactions with paper containing your unknown. What ions are present in your unknown (10)? Record the distances in millimeters to the ring fronts (11) and calculate the R_f values (12).

Relative amounts of the components of a mixture can be determined by using the methods employed in this experiment. Prepare a mixture of these

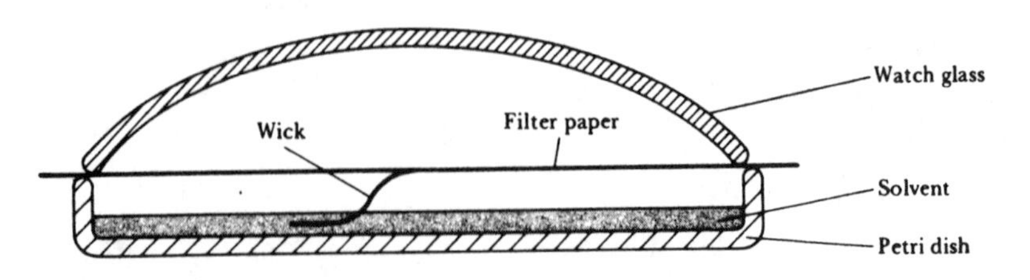

Figure 10.3

ions in the following manner: Mix together 5 mL of the Fe^{3+} and 5 mL of the Ni^{2+} solutions and then add 2 or 3 drops of the Cu^{2+} solution as a trace contaminant. Apply 1 drop of this mixture to a prepared piece of filter paper. Run and develop the chromatogram. Record your observations (13). Compare your results with those of someone who performed the experiment using the other technique (14).

Dyes You may believe that food coloring consists of one substance or that the black ink in felt-tip pens is a single substance. Spot a piece of filter paper heavily with a black felt-tip pen (such as a Flair pen). Make the spot quite dark to be sure that enough ink has been transferred to the paper. Obtain a chromatogram, using as the solvent a solution prepared by diluting 10 mL of isopropyl alcohol with 5 mL of water.

Attach all the dry, developed chromatograms to your report sheet.

B. Ascending Strip Technique

Obtain a piece of chromatography paper strip about 50 cm long. Cut this into 15-cm strips and mark each strip with a pencil dot. Label as diagramed in Figure 10.4. You will be running three chromatograms simultaneously by this technique: the known, the unknown, and the trace Cu^{2+}. Place the developing solvent to a depth of 10–12 mm in the bottom of a 600-mL beaker. Spot the strips as described above for the circles: one strip with the three knowns, one strip with the unknown solution, and one strip with the solution containing a trace of Cu^{2+}. Attach the labeled ends of the strips to a 6-in. glass rod by folding the ends over the rod and clipping the paper together using paper clips (see Figure 10.5). Place the rod with the three strips attached across the top of the beaker. Be certain that the spots on the strips are completely dry before you place the strips in the beaker. Make sure that the strips touch neither one an-

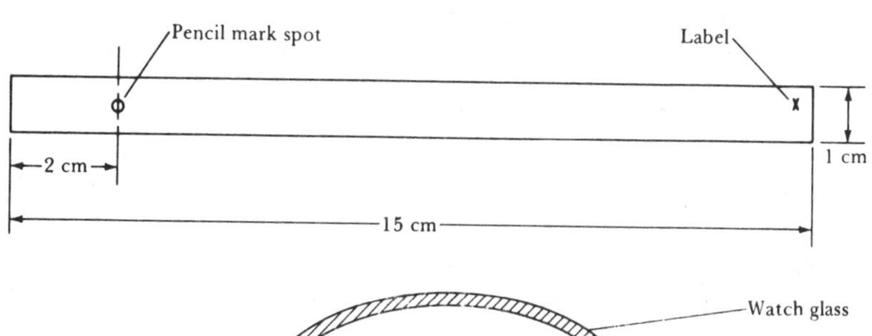

Figure 10.4

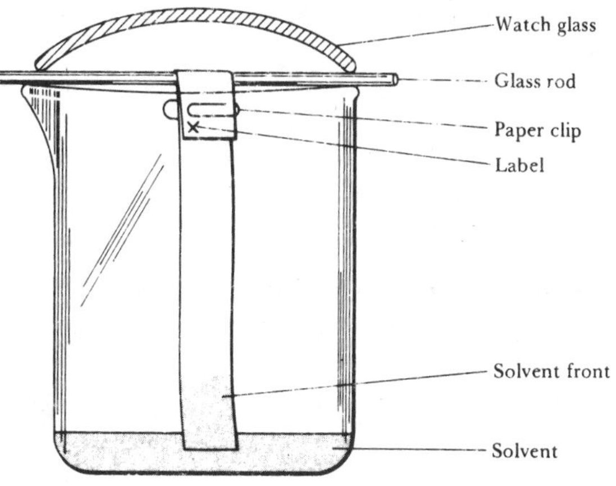

Figure 10.5

other nor the walls of the beaker and that the bottom of each strip is resting in the solution. Cover the beaker carefully with a watch glass. Do not disturb the beaker while the chromatograms are developing. When the solvent front has nearly reached the union of the folded part of the paper, carefully remove the watch glass and the glass rod with the strips. Immediately mark the solvent fronts with a pencil. Allow the strips to dry by fanning the air with them and proceed as directed below with the developing solutions.

Which of the known ions do you detect without resorting to the use of any other reagents (or development) (2)? Is there any difference between the ring front of this ion and the solvent front (3)? Mark the ring front of this ion.

Developing Reactions IN THE HOOD, pour about 5 mL of 15 M ammonium hydroxide into a clean, shallow dish and rest the filter paper containing the knowns on top of the dish. Do not permit the paper to dip into the solution. What color develops (4)? What ion does this indicate (5)? Mark the ring front.

To detect the third ion, dip a new piece of filter paper into a 1 percent solution of dimethylglyoxime; then, using the new piece as a brush, paint the test filter paper. As an alternate procedure, you may spray the paper with a dimethylglyoxime-containing aerosol. What color develops (6)? What ion does this indicate (7)? Mark the ring front.

Measure the distance to each of the ring fronts in millimeters (8). Calculate the R_f values for the three known ions (9) using Equation [1].

Repeat the above sequence of developing reactions with paper containing your unknown. What ions are present in your unknown (10)? Record the distances in millimeters to the ring fronts (11) and calculate the R_f values (12).

Relative amounts of the components of a mixture can be determined by using the methods employed in this experiment. Prepare a mixture of these ions in the following manner: Mix together 5 mL of the Fe^{3+} solution and 5 mL of the Ni^{2+} solution and then add 2 or 3 drops of the Cu^{2+} solution as a trace contaminant. Apply 1 drop of this mixture to a prepared piece of filter paper. Run and develop the chromatogram. Record you observations (13). Compare your results with those of someone who performed the experiment using the other technique (14).

Dyes You may believe that food coloring consists of one substance or that the black ink in felt-tip pens is a single substance. Spot a piece of filter paper heavily with a black felt-tip pen. Make the spot quite dark to be sure that enough ink has been transferred to the paper. Obtain a chromatogram, using as the solvent a solution prepared by diluting 10 mL of isopropyl alcohol with 5 mL water.

Attach all the dry, developed chromatograms to your report sheet.

REVIEW QUESTIONS

Before beginning this experiment in the laboratory, you should be able to answer the following questions:

1. What does the technique of chromatography allow us to do?

2. What is the meaning and utility of an R_f value? What would you expect to influence an R_f value?
3. What are the developing reactions that allow the identification of Ni^{2+} and Cu^{2+}?
4. What forces cause the eluting solution to move along the chromatographic support material?
5. Why should the solvent front be marked immediately?
6. If the solvent front is 65 mm and the ring front of an unknown is 32 mm, what is the *Rf* value?
7. Would you expect changing the solvent to change the *Rf* value? Why?
8. Suggest simple reasons why the Ni^{2+}, and Fe^{3+} ions have different R_f values.
9. Is it necessary that compounds be colored to be separated by chromatography?
10. Why should the petri dish (or beaker) be covered during the development of the chromatogram?

NOTES AND CALCULATIONS

Atomic Spectra and Atomic Structure

EXPERIMENT

11

OBJECTIVE

To gain some understanding of the relationship between emission (line) spectra and atomic structure.

APPARATUS AND CHEMICALS

spectroscope with illuminated scale or diffraction grating	6.0 *M* HCl
high-voltage power supply with lamp holder	0.1 *M* NaCl
hydrogen lamp	0.1 *M* $CaCl_2$
Nichrome wire loop	0.1 *M* $SrCl_2$
mercury-vapor lamp	0.1 *M* LiCl
Bunsen burner and hose	0.1 *M* KCl
unknown solution (one cation from above)	0.1 *M* $BaCl_2$
unknown solution (mixture containing two or more cations from above)	

INTRODUCTION

Our present understanding of atomic structure has come from studies of the properties of light or radiant energy and emission and absorption spectra. Radiant energy is characterized by two variables: its wavelength, λ, and its frequency, ν. The wavelength and frequency are related to one another by the relation

$$\nu\lambda = c \qquad [1]$$

where c is the speed of light, 3.00×10^8 m/s; wavelength is usually expressed in nm (10^{-9} m); and frequency is given in cycles per second, or hertz (Hz).

EXAMPLE 11.1

What is the frequency that corresponds to a wavelength of 500 nm?

Solution

$$\nu = \frac{c}{\lambda} = \left(\frac{3.00 \times 10^8 \text{ m/s}}{500 \text{ nm}}\right)\left(\frac{10^9 \text{ nm}}{1 \text{ m}}\right)$$
$$= 6.00 \times 10^{14} \text{s}^{-1} \text{ or } 6.00 \times 10^{14} \text{Hz}$$

Radiation of different wavelengths affects matter differently. Infrared radiation may cause a "heat burn;" visible and near-ultraviolet light, a sunburn or suntan;

Table 11.1 Wavelength Units for Electromagnetic Radiation

Unit	Symbol	Length (m)	Type of Radiation
Ångstrom	Å	10^{-10}	X ray
Nanometer	nm	10^{-9}	ultraviolet, visible light
Micrometer	μm	10^{-6}	infrared
Millimeter	mm	10^{-3}	infrared
Centimeter	cm	10^{-2}	microwaves
Meter	m	1	TV, radio

and X rays, tissue damage or even cancer. Some wavelength units for various types of radiation are given in Table 11.1.

A particular source of radiant energy may emit a single wavelength, as in the light from a laser, or many different wavelengths, as in the radiations from an incandescent light bulb or a star. Radiation composed of a single wavelength is termed *monochromatic*. When the radiation from a source such as the sun, or other stars, is separated into its components, a *spectrum* is produced. This separation of radiations of differing wavelengths can be achieved by passing the radiation through a prism. Each component of the polychromatic radiation is bent to a different extent by the prism, as shown in Figure 11.1. This rainbow of colors, containing light of all wavelengths, is called a *continuous spectrum*. The most familiar example of a continuous spectrum is the rainbow, produced by the dispersion of sunlight by raindrops or mist.

It was found that by placing a narrow slit between the source of radiation and the prism, the quality of the spectrum was improved and the monochromatic components were more sharply resolved. An instrument used for studying line spectra is called a *spectroscope*. A spectroscope (Figure 11.2) contains the following: a slit for admitting a narrow, collimated beam of light; a prism for dispersing the light into its components; an eyepiece for viewing the spectrum; and an illuminated scale against which the spectrum may be viewed.

Most substances will emit light energy if heated to a high enough temperature. For example, a fireplace poker will glow red if left in the fireplace flame for several minutes. Similarly, neon gas will emit bright red light when excited with a sufficiently high electrical voltage. When energy is absorbed by a substance, electrons within the atoms of the substance may be excited to positions of higher potential energy, farther away from the nucleus. When these electrons return to their normal, or ground-state, position, energy will be emitted. Usually, some of the energy emitted occurs in the visible region of the electromagnetic spectrum (4000–7000 Å, or 400–700 nm).

Excited atoms do not emit a continuous spectrum, but rather emit radiation at only certain discrete, well-defined, fixed wavelengths. For example, if

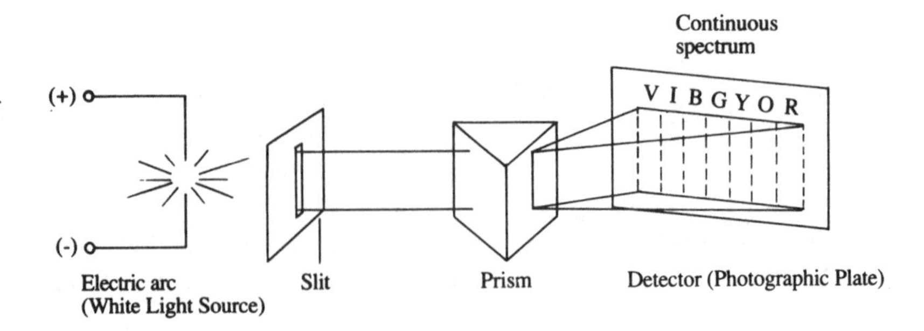

Figure 11.1 The spectrum of "white" light. When light from the sun or from a high-intensity incandescent bulb is passed through a prism, the component wavelengths are spread out into a continuous rainbow spectrum.

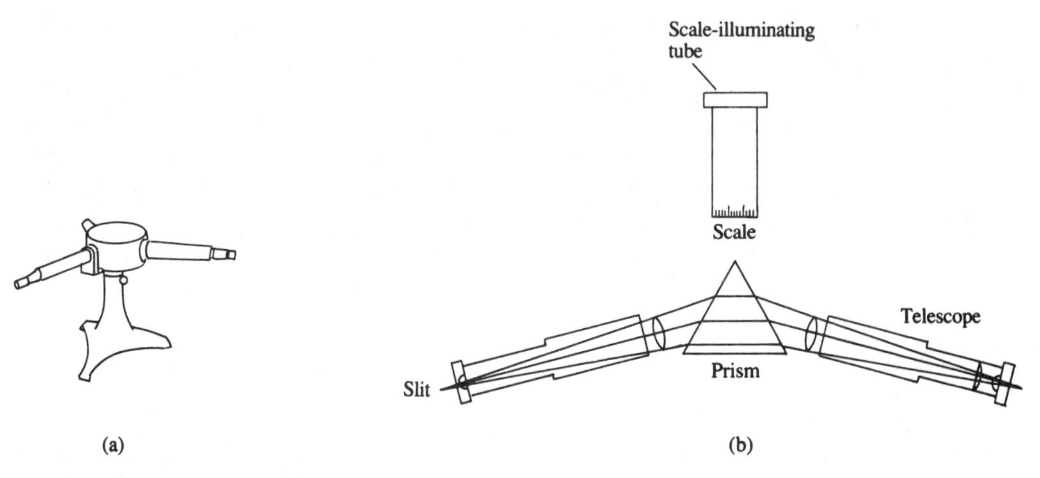

Figure 11.2 (a) A sketch of the spectroscope to be used. (b) A diagram showing its component parts. When viewed through the telescope, the spectrum will appear superimposed on the numerical scale.

you have ever spilled table salt (NaCl) into a flame, you have seen the characteristic yellow emission of excited sodium atoms. If the light emitted by the atoms of a particular element is viewed in a spectroscope, only certain bright-colored lines are seen in the spectrum. See Figure 11.3, which illustrates the emission spectrum of hydrogen.

The observation that a given excited atom emits radiation at only certain fixed wavelengths indicates that the atom can undergo energy changes only of certain fixed, definite amounts. An atom does not emit continuous radiation but rather emits energy corresponding to definite regular changes in the energies of its component electrons. The experimental demonstration of bright-line atomic emission spectra implied a regular, fixed electronic microstructure for the atom and led to the Bohr model for the hydrogen atom.

In this experiment, you will obtain emission spectra for a few elements and explain your results in terms of the Bohr model. You will use a spectroscope like the one illustrated in Figure 11.2 to obtain your data.

Figure 11.3 The emission spectrum of hydrogen.

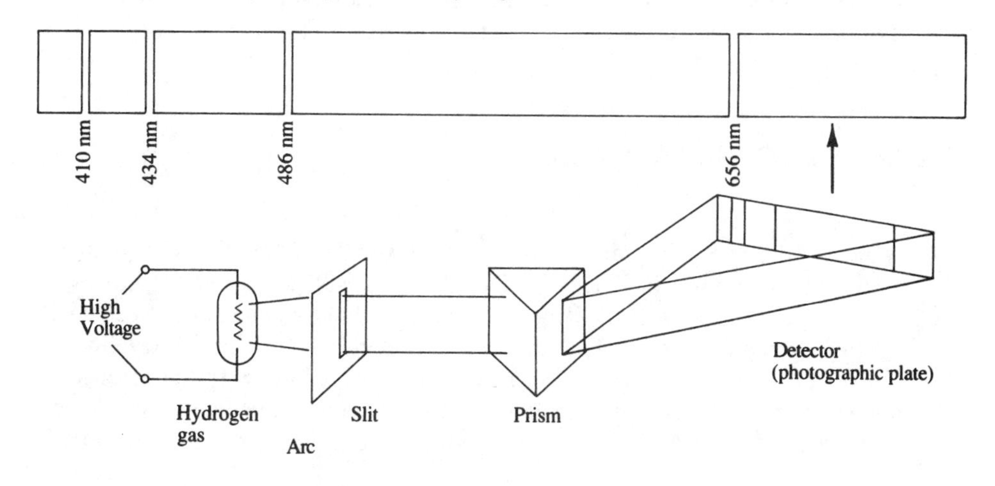

CALIBRATION OF THE SPECTROSCOPE

The scale of the spectroscope has arbitrary divisions and must be calibrated with a known spectrum before the spectrum of an unknown may be obtained. Calibration is accomplished by viewing the emission spectrum of mercury because the emission wavelengths for mercury are very precisely known. You will record the positions on the spectroscope scale of the mercury emission lines and prepare a calibration curve by plotting these positions against the known wavelengths of the lines, which are:

Violet: 404.7 nm
Blue: 435.8 nm
Green: 546.1 nm
Yellow: 579.0 nm

Turn on the illuminated scale of the spectroscope and look through the eyepiece to make sure that the scale is visible but not so brightly lighted that the mercury spectral lines will be obscured when the mercury lamp is turned on. With the power supply unplugged, position the power supply and mercury lamp directly in front of the slit opening of the spectroscope. (CAUTION: *The power supply develops several thousand volts. Do not touch any portion of the power supply, wire leads, or lamps unless the power supply is unplugged from the wall outlet*.) In addition to visible light, the lamps may emit ultraviolet radiation. Ultraviolet radiation is damaging to your eyes. *Wear your safety glasses at all times during this experiment, since they will absorb some of the ultraviolet radiation. Do not look directly at any of the lamps while they are illuminated.*) DO NOT LET THE POWER SUPPLY OR LAMP TOUCH THE SPECTROSCOPE. With your instructor's permission, turn on the power supply and then turn on the power supply switch to illuminate the mercury lamp. Look into the eyepiece, and adjust the slit opening so as to maximize the brightness and sharpness of the emission lines on the scale. If necessary, adjust the position of the illuminated scale so that the numbered divisions are easily read but they do not obscure the mercury spectral lines. Once the slit and scale have been adjusted, do not move them during the course of the rest of the experiment. Record on your report sheet the color and location of the mercury lines on the numbered scale for each line in the visible spectrum of mercury. On the graph paper provided, plot the observed scale reading versus the known wavelength for each line. You will then use this calibration curve to determine the wavelengths of the emission lines for some other atoms.

A. EMISSION SPECTRUM OF ATOMIC HYDROGEN

DISCUSSION

Atoms absorb and emit radiation with characteristic wavelengths. This was one of the observations which led the Danish physicist Niels Bohr to develop a model for the structure of the hydrogen atom. Within this model the electron of the hydrogen atom moves about the central proton in a circular orbit. Only orbits of certain radii and having certain energies are allowed. In the absence of radiant energy, an electron in an atom remains indefinitely in one of the allowed energy states or orbits. When electromagnetic energy impinges upon the atom, the atom may absorb energy, and in the process an electron will be promoted from one energy state to another. The frequency of energy absorbed is related to the energy difference:

$$\Delta E = h\nu = \frac{hc}{\lambda} \qquad [2]$$

In the Bohr model, the radius of the orbit is related to the principal quantum number n:

$$\text{Radius} = n^2(5.3 \times 10^{-11}\ \text{m}) \qquad [3]$$

and the energy of the electron is also related to n:

$$E_n = -R_H\left(\frac{1}{n^2}\right) \qquad [4]$$

Thus, as n increases, the electron moves farther from the nucleus and its energy increases. The constant R_H in Equation [4] is called the *Rydberg constant;* it has the value 2.18×10^{-18} J.

EXAMPLE 11.2

What is the energy of a hydrogen electron when $n = 3$? When $n = 2$?

Solution:

$$E_3 = (-2.18 \times 10^{-18}\ \text{J})\left(\frac{1}{3^2}\right)$$
$$= -2.42 \times 10^{-19}\ \text{J}$$
$$E_2 = (-2.18 \times 10^{-18}\ \text{J})\left(\frac{1}{2^2}\right)$$
$$= -5.45 \times 10^{-19}\ \text{J}$$

The orbital radii and energies are illustrated for $n = 1$, 2, and 3 in Figure 11.4.

According to Bohr's theory, if an electron were to move from an outer orbit to an inner orbit, a photon of light should be emitted having the energy

$$\Delta E = E_{\text{inner}} - E_{\text{outer}} = -R_H\left(\frac{1}{n^2_{\text{inner}}} - \frac{1}{n^2_{\text{outer}}}\right) \qquad [5]$$

Thus, from Example 11.2, an electron moving from $n = 3$ to $n = 2$ would emit light of energy

$$(5.45 - 2.42)(10^{-19}\ \text{J}) = 3.03 \times 10^{-19}\ \text{J}$$

The wavelength of this photon is given by the Planck relation

$$\lambda = \frac{hc}{\Delta E} \qquad [6]$$

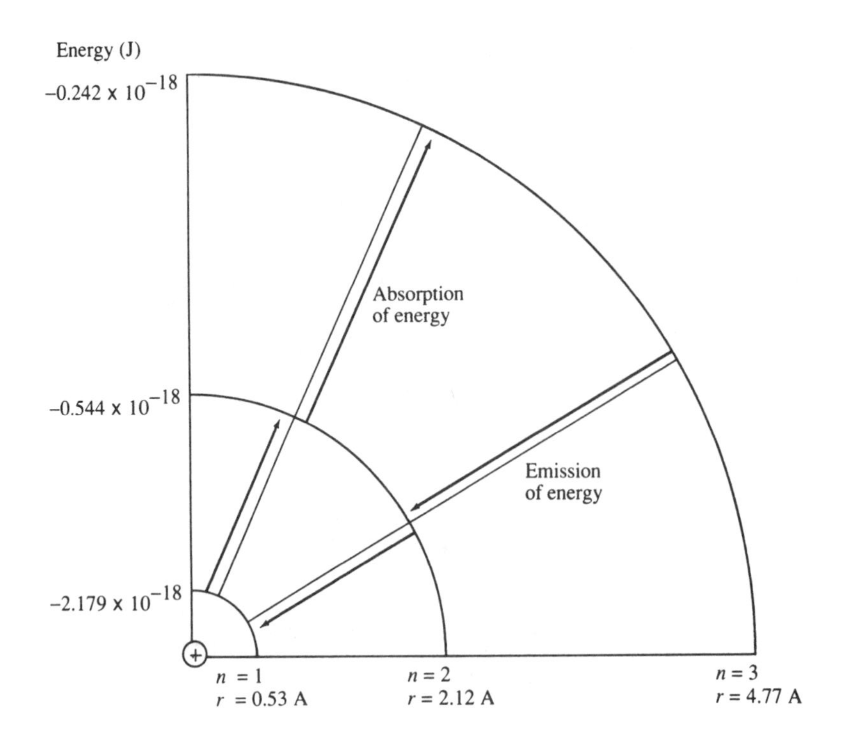

Figure 11.4 Radii and energies of the three lowest-energy orbits in the Bohr model of hydrogen. The arrows refer to transitions of the electron from one allowed energy state to another. When the transition takes the electron from a lower- to a higher-energy state, absorption occurs. When the transition is from a higher- to a lower-energy state, emission occurs.

EXAMPLE 11.3

Calculate the wavelength of light emitted for the $n = 3 \rightarrow n = 2$ transition.

Solution:

$$\lambda = \frac{(6.63 \times 10^{-34} \text{ J-s})(3.00 \times 10^{8} \text{ m/s})}{3.03 \times 10^{-19} \text{ J}}$$

$$= 6.56 \times 10^{-7} \text{ m}$$

$$= 656 \text{ nm}$$

By similar calculations Bohr predicted wavelengths for the hydrogen emission spectrum that agreed exactly with the experimental values. He even predicted emission wavelengths in the infrared and ultraviolet regions of the spectrum that had not yet been measured but were later confirmed.

You will measure the emission wavelengths in the visible region for hydrogen with the spectroscope and assign these wavelengths to their corresponding transitions by calculations similar to those in examples 11.2 and 11.3.

PROCEDURE

Turn on the illuminated scale of the spectroscope and look through the eyepiece to make sure that the scale is visible but not so brightly lighted that the hydrogen spectral lines will be obscured. With the power supply unplugged, position the power supply and hydrogen lamp directly in front of the slit opening of the spectroscope. (CAUTION: *The power supply develops several thousand volts. Do not touch any portion of the power supply, wire leads, or lamps unless the power supply is unplugged from the wall outlet*. In addition to visible light, the lamps may emit ultraviolet radiation. Ultraviolet radiation is damaging to your eyes. *Wear your safety glasses at all times during this experiment, since they will absorb some of the ultraviolet radiation. Do not look directly at any of the lamps while they are illuminated.*) DO NOT LET THE POWER SUPPLY OR LAMP TOUCH THE SPECTROSCOPE. With your instructor's permission, turn on the power supply and then turn on the power supply switch to illuminate the hydrogen lamp. You should have adjusted the slit and the illuminated scale in the calibration step. Should they require further adjustment, adjust the slit so as to maximize the line intensity and sharpness. Record, on your report sheet, the color and location of the hydrogen lines on the numbered scale for each line in the visible spectrum of hydrogen. You should easily observe the red, blue-green, and violet lines. A second, faint violet line may also be visible if the room and scale illumination are not too bright.

Use the calibration curve that you constructed to determine the wavelengths of the lines in the hydrogen emission spectrum. Find the true wavelengths of these lines in your textbook or a handbook and calculate the percent error in your determination of the wavelength of each line:

$$\% \text{ error} = \frac{(\text{true value} - \text{experimental value})}{(\text{true value})} \times 100 \qquad [7]$$

Use equations [5] and [6] to calculate the wavelengths in nanometers for the $n = 3 \rightarrow n = 2$, $n = 4 \rightarrow n = 2$, $n = 5 \rightarrow 2$, and $n = 6 \rightarrow n = 2$ transitions. How do these calculated values compare with your experimental values? assign the transitions.

B. EMISSION SPECTRA OF GROUP 1A AND GROUP 2A ELEMENTS

DISCUSSION

Since the energies of the electrons in the atoms of different elements are different, the emission spectrum of each element is unique. The emission spectrum may be used to detect the presence of an element in both a qualitative and a quantitative way. A number of common metallic elements emit light strongly in the visible region, allowing their detection with a spectroscope. For these elements, the emissions are so intense that the elements may often be recognized by the gross color that they impart to a flame. For example, lithium ions impart a red color to a flame; sodium ions, a yellow color; potassium ions, a violet color; calcium ions, a brick-red color; strontium ions, a bright red color; and barium ions, a green color. If we examine the emission spectra of these ions with a spectroscope, we find that as with mercury and hydrogen, the emission spectra are composed of a series of lines. The series is unique for each metal. Consequently, a flame into which both lithium and strontium, for example, had been placed would be red and we could not tell with our naked eye that both these ions were present. However, with the aid of a spectroscope we could detect the presence of both ions.

PROCEDURE

You will obtain the emission spectra of the ions of each of the elements listed in the preceding discussion. You will use a Bunsen burner as an excitation source and observe the gross color imparted to the flame by these ions. With this information you will determine the contents of two unknown solutions—one containing only one of these metal ions, and the other a mixture containing two or more of these ions.

There are two ways to introduce the metal ions into the flame. First, you can use a wire loop to pick up a drop of the metal-ion solution and then place the drop into the flame for vaporization. Although this method is simple and inexpensive, it produces only a brief burst of color before the sample evaporates completely. Second, you can introduce a fine mist of sample into the flame, by using a spray bottle. This method produces a longer-lived emission that is therefore easier to see. Your instructor will tell you which method to use.

WORK IN PAIRS FOR THIS PART OF THE EXPERIMENT

If you use the wire-loop method, obtain several 6-in. lengths of Nichrome wire and about 10 mL of 6 *M* HCl. Bend the last $\frac{1}{4}$-in. of each wire into a small circular loop for picking up the sample solutions. Dip the loops into the 6 *M* HCl solution to remove any oxides that are present, rinse the loop in distilled water, and then heat the loop in the hottest part of the flame until no color is imparted to the flame by the wires.

If the sprayer is used, check to ensure that the sprayer produces a *fine* mist. If the sprayer nozzle is adjustable, try adjusting it to make a very fine mist. If the nozzle cannot be adjusted, ask your instructor how to clean it to improve the mist that it produces.

Set up a Bunsen burner directly in front of the slit of the spectroscope but at a sufficient distance to avoid damage to the spectroscope. Have your instructor check the burner placement before you ignite the burner. Ignite the burner and adjust the flame so that it is as hot as possible. Adjust the illuminated scale of the spectroscope so that approximate positions of the emission lines may be determined. It is not necessary to make exact measurements of the emission wavelengths.

With either method of introduction (loop or mist), introduce the metal-ion solution into the flame and note the gross color of the flame. Record your observations on your report sheet. Then, while looking through the eyepiece of the spectroscope, have your partner introduce the metal-ion solution into the flame, noting the color, intensity, and approximate scale position of the brightest lines in the emission spectrum of the metal ions. Repeat the above with each of the other metal-ion solutions. If you use the wire-loop method, use a new loop for each solution, or clean the wire loop with HCl and then distilled water and place the wire loop in the flame until it imparts no color to the flame.

After obtaining the emission spectrum of each of the known metal-ion solutions, obtain the emission spectrum of a single metal unknown, and by matching the colors, intensities, and positions of the lines in the emission spectrum to those of a known metal, identify your unknown. Record your results on your report sheet.

Obtain an unknown mixture and its emission spectrum as above. Then compare the color, intensity, and positions of the brightest lines in this spectrum with those of the knowns. In this way *determine which metal ions are present in the unknown mixture*. Record your results on your report sheet.

REVIEW QUESTIONS

You should be able to answer the following questions before beginning this experiment.

1. Name the colors of visible light, beginning with that of lowest energy (longest wavelength).
2. Distinguish between absorption and emission of energy.
3. A system proposed by the U.S. Navy for underwater submarine communication called ELF (for "extremely low frequency") operates with a frequency of 76 Hz. What is the wavelength of this radiation in meters? In miles? (1 mile = 1.61 km).
4. What is the energy in joules of the frequency given in question 3?
5. Red and green light have wavelengths of about 650 nm and 490 nm, respectively. Which light has the higher frequency, red or green? Which light has the higher energy, red or green?
6. Mg emits radiation at 285 nm. Could a spectroscope be used to detect this emission?
7. If boron emits radiation at 518 nm, what color will boron inpart to a flame?
8. From the wavelengths and colors given for the mercury emission spectrum in this experiment, construct a graphical representation of the mercury emission spectrum as it would appear on the scale of a spectroscope.

NOTES AND CALCULATIONS

Behavior of Gases: Molecular Weight of a Vapor

EXPERIMENT 12

OBJECTIVE

To observe how changes in temperature and pressure affect the volume of a fixed amount of a gas; to determine the molecular weight of a gas from a knowledge of its weight, temperature, pressure, and volume.

APPARATUS AND CHEMICALS

- gas-law demonstration apparatus
- balance
- 125-mL Erlenmeyer flask
- 600-mL beaker
- aluminum foil 2-in. square
- boiling chips
- 250-mL graduated cylinder (one per class)
- pins
- Bunsen burner
- ring stand and iron ring
- buret clamp
- wire gauze
- 3-mL sample of a volatile unknown liquid
- rubber band

DISCUSSION

The Effect of Pressure on the Volume of a Gas

The effect of pressure on the volume of a gas can be determined by using a gas buret, as shown in Figure 12.1. The volume of the buret is graduated in terms of cubic centimeters (cm^3), or milliliters (mL). When the stopcock is opened, air can enter the buret, and the level of mercury will be equal in both tubes. If the stopcock is then closed, a fixed volume of air is trapped in the buret at prevailing atmospheric pressure. Raising the leveling bulb increases the pressure on the gas; the new pressure on the gas corresponds to the prevailing atmospheric pressure plus the height that the mercury in the leveling bulb is above the level of mercury in the buret. It is found that when the pressure is doubled, the volume is halved; and when the pressure is halved, the volume is doubled. From such experiments we conclude that, at constant temperature, the volume of a given amount of gas is *inversely* proportional to the pressure. This is *Boyle's law,* which Boyle enunciated in 1662. This may be expressed mathematically as

$$\frac{V_1}{V_2} = \frac{P_2}{P_1}$$

or

$$V = \frac{k}{P}$$

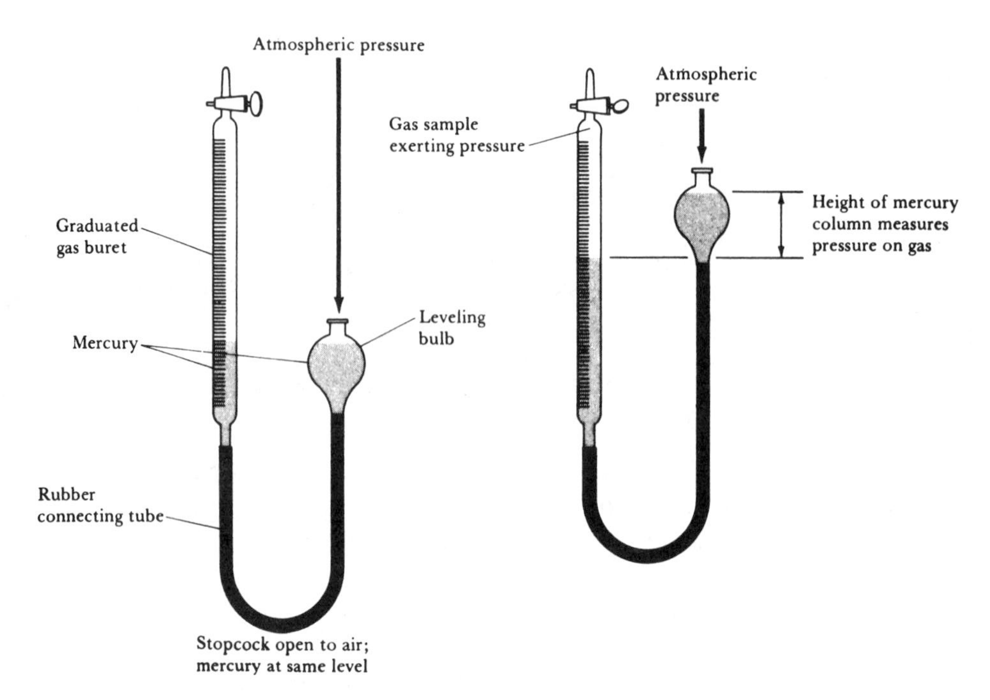

Figure 12.1 A gas buret

where V_1 is the volume at pressure P_1, V_2 is the volume at pressure P_2, and k is a proportionality constant.

By means of this law, we can calculate the volume of a gas at any pressure, provided we know the volume at a given pressure. Example 12.1 is illustrative.

EXAMPLE 12.1

If 100 mL of a gas is enclosed in the buret at 760 mm Hg, what volume would the gas occupy at 1520 mm Hg at the same temperature?

Solution: We could let V_1 = 100 mL, P_1 = 760 mm Hg, and P_2 = 1520 mm Hg, substitute these values into Equation [1], and solve for V_2. A better approach is to reason through the problem. Since the pressure increases from 760 mm Hg to 1520 mm Hg, the new volume must be *less* than 100 mL. The volume decreases in the same ratio that the pressure increases. The new volume is, therefore,

$$V_2 = 100 \text{ mL} \times \frac{760 \text{ mm Hg}}{1520 \text{ mm Hg}} = 50.0 \text{ mL}$$

If the volume had been multiplied by the fraction 1520 mm Hg/760 mm Hg, a fraction larger than 1, the answer would have been larger than 100 mL; and this, we know, must be incorrect.

The Effect of Temperature on the Volume of a Gas

The relation between temperature and volume of a given sample of gas can be studied in a gas buret while the pressure is held constant. We are all familiar with the fact that gases expand when heated. It has been observed that at constant pressure the volume of a given mass of a gas is directly proportional to the absolute (or Kelvin) temperature. This is known as *Charles's law*.

To convert a Celsius temperature to the Kelvin (or absolute) scale, add 273°. Thus 20°C equals (20° + 273°) = 293 K.

Charles's law may be expressed mathematically as follows:

$$\frac{V_1}{V_2} = \frac{T_1}{T_2}$$

or

$$V = kT$$

where V_1 is the volume at temperature T_1, V_2 is the volume at T_2, and k is a proportionality constant. An application of this law is illustrated in Example 12.2.

EXAMPLE 12.2

Suppose a sample of oxygen has a volume of 100 mL at a temperature of 20°C. What will be its volume at 100°C at the same pressure?

Solution: Since the temperature increases, the volume must increase. The volume at 100°C equals the original volume multiplied by a fraction made up of the Kelvin temperatures which is larger than 1:

$$V \text{ at } 100^\circ\text{C} = 100 \text{ mL} \times \frac{373 \text{ K}}{293 \text{ K}} = 127 \text{ mL}$$

Since gases can occupy any volume, it is most informative to compare them under the same conditions. Standard conditions of temperature and pressure, designated as STP, are often used for this purpose. Standard temperature is 0°C, and standard pressure is 760 mm Hg, or one atmosphere (1 atm).

Problems involving changes both in temperature and pressure are worked by combining the effects due to each of these changes. Example 12.3 is illustrative.

EXAMPLE 12.3

Suppose a sample of oxygen at 100°C occupies a volume of 100 mL at 700 mm Hg. What will be its volume at STP?

Solution:

$$V_1 = 100 \text{ mL}$$

$$V_2 = ?$$

$$T_1 = 100^\circ + 273^\circ = 373 \text{ K}$$

$$T_2 = 0^\circ + 273^\circ = 273 \text{ K}$$

$$P_1 = 700 \text{ mm Hg}$$

$$P_2 = 760 \text{ mm Hg}$$

The volume at STP will be equal to the original volume multiplied by a fraction made up of the two temperatures and also another fraction made up of the two pressures. The temperature fraction will be less than 1, because the temperature decreases, and this results in a decrease in volume. The pressure fraction will be less than 1, because the pressure increases, and an increase in pressure also causes the volume to decrease:

$$\text{Volume at STP} = 100 \text{ mL} \times \frac{273 \text{ K}}{373 \text{ K}} \times \frac{700 \text{ mm Hg}}{760 \text{ mm Hg}}$$

$$= 67.4 \text{ mL}$$

Ideal-Gas Law

It is possible to relate the four variables of a gas—pressure, volume, temperature, and number of moles—in one equation, which is referred to as the *ideal-gas law*. Before doing this, we must recognize *Avogadro's law*, which states that *equal volumes of all gases, at the same conditions of temperature and pressure, contain the same number of molecules*. Thus, 6.022×10^{23} molecules (Avogadro's number) of any gaseous substance should occupy the same volume under the same conditions. One mole of an ideal gas at STP occupies 22.4 L, a value known as the *molar volume*.

According to Charles's law, the volume of a fixed number of moles, n, of a gas at a constant pressure is directly proportional to the Kelvin temperature:

$$V \propto T \qquad \text{at constant } P \text{ and } n$$

According to Boyle's law, the volume of a fixed number of moles of a gas at constant temperature is inversely proportional to the pressure:

$$V \propto \frac{1}{P} \qquad \text{at constant } T \text{ and } n$$

And according to Avogadro's law,

$$V \propto n \qquad \text{at constant } P \text{ and } T$$

If a quantity is proportional to two or more variables, it is proportional to the product of those variables:

$$V \propto \text{T} \times \frac{1}{P} \times n \qquad [3]$$

The proportionality symbol, $\propto$, in Equation [3] can be replaced by an equals sign by introducing a proportionality constant, R:

$$V = R \times T \times \frac{1}{P} \times n$$

or

$$PV = nRT \qquad [4]$$

Equation [4] is the ideal-gas law. R, the ideal-gas constant, can be calculated by considering 1 mol of an ideal gas at STP:

$$R = \frac{P \times V}{n \times T} = \frac{760 \text{ mm} \times 22{,}400 \text{ mL}}{1 \text{ mol} \times 273 \text{ K}} = 62{,}400 \text{ mL-mm Hg/mol-K}$$

or, in other units,

$$R = \frac{1 \text{ atm} \times 22.4 \text{ L}}{1 \text{ mol} \times 273 \text{ K}} = 0.0821 \text{ L-atm/mol-K}$$

The number of moles of a substance (n) equals its mass in grams, m, divided by the number of grams per mole (that is, its molecular weight, $\mathcal{M}$): $n = m/\mathcal{M}$. Making this substitution into Equation [4] gives

$$PV = \left(\frac{m}{\mathcal{M}}\right)RT \qquad [5]$$

The units of P, V, T, and m must, of course, be expressed in units consistent with the value of R. Example 12.4 illustrates how this equation may be used to calculate the molecular weight of a gas.

EXAMPLE 12.4

A gaseous sample weighing 0.896 g was found to occupy a volume of 524 mL at 730 mm Hg and 28°C. What is the molecular weight of the gas?

Solution: Solving Equation [5] for molecular weight and substituting the appropriate values, we find

$$\mathcal{M} = \frac{mRT}{PV} \qquad [6]$$

$$= \frac{0.896 \text{ g} \times 62{,}400 \text{ mL-mm Hg/mol-K} \times 30\text{K}}{730 \text{ mm Hg} \times 524 \text{ mL}}$$

$$= 44.0 \text{ g/mol}$$

The first portion of this experiment will be a demonstration of Charles's and Boyle's laws. The laboratory instructor will vary the pressure, temperature, and volume of a sample of air using a gas buret.

In the second portion of this experiment, you will determine the molecular weight of a volatile liquid using Equation [6]. A small quantity of liquid sample is placed in a preweighed flask and vaporized so as to expel all the air from the flask, leaving it filled with the vapor at a known temperature (temperature of boiling water) and atmospheric pressure. The flask plus vapor is then cooled so that the vapor condenses. The flask plus condensed vapor plus air is then weighed. The mass of air, being nearly identical before and after, cancels out and allows one to determine the mass of the vapor. The above data, in conjunction with the volume of the flask, permit the calculation of the molecular weight.

PROCEDURE

A. Demonstration Experiment

Record the data as collected by the laboratory instructor on the report sheet. Calculate the volumes by means of the appropriate gas laws and compare these with the observed volumes. Express the differences as percent error.

Figure 12.2

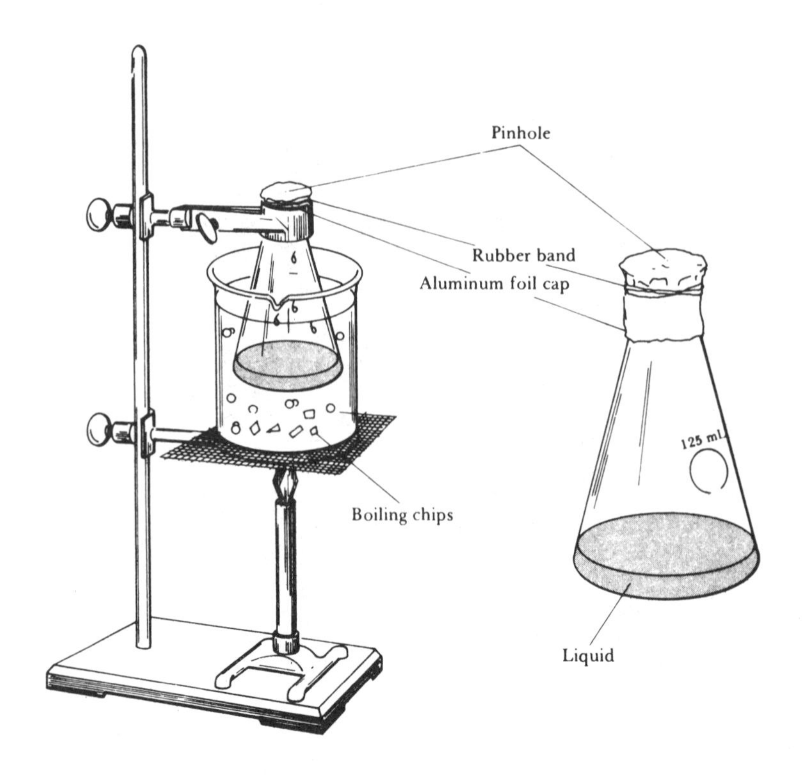

B. Molecular Weight of a Vapor

Record the number of your unknown liquid (1). Place a small square of aluminum foil over the mouth of a clean, dry 125-mL Erlenmeyer flask; fold the foil loosely around the neck and secure with a rubber band (see Figure 12.2). Make a very small hole in the center of the foil with a pin. Weigh the flask, foil cap, and rubber band (2).

Remove the foil and place about 2 mL of your unknown liquid in the flask; then replace the foil and rubber band. Clamp the flask at the top of the neck and immerse it as deeply as possible in a 600-mL beaker nearly full of water, as shown in Figure 12.2. Place some boiling chips in the water and heat to boiling. (Time may be saved if the apparatus is set up at the beginning of the laboratory period and the water brought to a boil). Record the temperature of the boiling water (3) and the barometric pressure (4). Apply your correction to your thermometer reading using the thermometer-calibration curve you previously obtained in Experiment 1.

As the water boils, watch the liquid in the flask. As soon as all of the liquid (including any that has condensed in the neck) has vaporized (about 20–30 min), remove the flask by means of the clamp and set it aside to cool. After the flask has cooled to room temperature, wipe it dry and remove any water that may adhere to the aluminum. Weigh the flask, cap, rubber band, and condensed unknown liquid (5). Calculate the weight of the condensed liquid (6).

Remove the cap and fill the flask completely with water. Measure the volume by pouring the water into a large graduated cylinder (7).

Calculate the molecular weight of the unknown using Equation [6].

REVIEW QUESTIONS

Before beginning this experiment in the laboratory, you should be able to answer the following questions:

1. How does the pressure of an ideal gas at constant volume change as the temperature increases?
2. How does the volume of an ideal gas at constant temperature change as the pressure increases?
3. How does the volume of an ideal gas at constant temperature and pressure change as the number of molecules changes?
4. Write the ideal-gas equation and give the units for each term when R = 0.0821 L-atm/K-mol.
5. Show by mathematical equations how one can determine the molecular weight of a volatile liquid by measurement of the pressure, volume, temperature, and weight of the liquid.
6. If 0.75 g of a gas occupies 300 mL at 27°C and 700 mm Hg of pressure, what is the molecular weight of the gas?
7. A sample of nitrogen occupies a volume of 250 mL at 27°C and 700 mm Hg of pressure. What will be its volume at STP?
8. Consider Figure 12.1. If the height of the mercury column in the leveling bulb is 20 mm greater than that in the gas buret and atmospheric pressure is 650 mm, what is the pressure on the gas trapped in the buret?
9. Consider Figure 12.1. If the level of the mercury in the leveling bulb is lowered, what happens to the volume of the gas in the gas buret?
10. Show that Boyle's law, Charles's law, and Avogadro's law can be derived from the ideal-gas law.
11. Methane burns in oxygen to produce CO_2 and H_2O

$$CH_4(g) + 2O_2(g) \longrightarrow 2H_2O(l) + CO_2(g)$$

    If 5.6 L of gaseous CH_4 is burned at STP, what volume of O_2 is required for complete combustion? What volume of CO_2 is produced?
12. Calculate the density of N_2 at STP, (a) using the ideal-gas law and (b) using the molar volume and molar mass of N_2. How do the densities compare?

NOTES AND CALCULATIONS

Name ______________________________ Desk ________

Date ______________ Laboratory Instructor ______________________

Unknown no. ______________

REPORT SHEET FOR EXPERIMENT 12

BEHAVIOR OF CASES: MOLECULAR WEIGHT OF A VAPOR

A. Demonstration Experiment

1. Boyle's law—effect of pressure at constant temperature

	Trial 1	Trial 2
First pressure, barometer reading, mm Hg	________	________
Difference in mercury levels (+ or −), mm Hg	________	________
Second pressure, mm Hg	________	________
First volume, cm^3	________	________
Second volume, measured, cm^3	________	________
Second volume, calculated, cm^3	________	________
Percent error (show calculations)	________	________

2. Charles's law—effect of temperature at constant pressure

First temperature ________ °C = ________ K

Second temperature ________ °C = ________ K

First volume, cm^3 ________

Second volume, measured, cm^3 ________

Second volume, calculated, cm^3 ________

Percent error ____________
(show calculations)

B. Molecular Weight of a Vapor

1. Unknown liquid number ____________
2. Wt. of flask + cap + rubber band ____________
3. Temperature of boiling water ____________ °C
   Correction ____________ °C
   Correction temperature ____________ °C
4. Barometric pressure ____________ mm Hg
5. Wt. of flask + rubber band + cap + condensed vapor ____________
6. Wt. of condensed liquid ____________
7. Volume of flask ____________ mL
8. Molecular weight of vapor ____________
   (show calculations)

QUESTIONS

1. If an insufficient amount of liquid unknown had been used, how would this have affected the value of the experimental molecular weight?

2. What are the major sources of error in your determination of the molecular weight?

3. If the flask was not thoroughly wiped dry, how would this affect the molecular weight?

4. Isobutyl alcohol has a boiling point of 108°C. How would you modify the procedure used in this experiment to determine its molecular weight?

GAS-LAW PROBLEMS

1. What volume will 200 mL of gas at 20°C and a pressure of 355 mm Hg occupy if the temperature is reduced to −150°C and the pressure increased to 1420 mm Hg?

2. A sample of gas of mass 2.83 g occupies a volume of 428 mL at 27°C and 1.00 atm pressure. What is the molecular weight of the gas?

3. A gas is placed in a storage tank at a pressure of 30.0 atm at 22.3°C. As a safety device, there is a small metal plug in the tank made of a metal alloy that melts at 125°C. If the tank is heated, what is the maximum pressure that will be attained in the tank before the plug will melt and release gas?

4. Sixty liters of a gas were collected over water when the barometer read 650 mm Hg and the temperature was 20°C. What volume would the dry gas occupy at standard conditions? (HINT: Consider Dalton's law of partial pressures.)

5. Six moles of hydrogen gas at 0°C is forced into a steel cylinder with a volume of 200 mL. What is the pressure of the gas in the cylinder?

6. What is the density of He at STP? Why do helium-filled balloons rise in air?

7. What volume in milliliters will 8.8 g of CO_2 occupy at STP?

8. If 16.0 g of O_2 and 1.1 g of CO_2 are placed in a 5.00-L container at 21°C, what is the pressure of this mixture of gases?

9. A mixture of cyclopropane gas and oxygen is used as an anesthetic, Cyclopropane contains 85.7%C, and 14.3%H by mass. At 50.0°C and 0.984 atm pressure 1.56g cyclopropane has a volume of 1L. What is the molecular formula of cyclopropane?

Determination of R: The Gas-Law Constant

EXPERIMENT 13

OBJECTIVE

To gain a feeling for how well real gases obey the ideal-gas law and to determine the ideal-gas-law constant R.

APPARATUS AND CHEMICALS

0.3 g $KClO_3$	250-mL beaker
MnO_2	rubber stoppers (2)
test tube	rubber tubing
ring stand	balance
clamp	125-mL Erlenmeyer flask
pinch clamp	thermometer
8-oz wide-mouth bottle	barometer
rubber bulb	Styrofoam cups

DISCUSSION

Most gases obey the ideal-gas equation, $PV = nRT$, quite well under ordinary conditions, that is, room temperature and atmospheric pressure. Small deviations from this law are observed, however, because real-gas molecules are finite in size and exhibit mutual attractive forces. The van de Waals equation,

$$\left(P + \frac{n^2a}{V^2}\right)(V - nb) = nRT$$

where a and b are constants characteristic of a given gas, takes into account these two causes for deviation and is applicable over a much wider range of temperatures and pressures than is the ideal-gas equation. The term nb in the expression $(V - nb)$ is a correction for the finite volume of the molecules; the correction to the pressure by the term an^2/V^2 takes into account the intermolecular attractions.

In this experiment you will determine the numerical value of the gas-law constant R, in its common units of L-atm/mol-K. This will be done using both the ideal-gas law and the van der Waals equation together with measured values of pressure, P, temperature, T, volume, V, and number of moles, n, of an enclosed sample of oxygen. An error analysis will then be performed on the experimentally determined constant.

The oxygen will be prepared by the decomposition of potassium chlorate using manganese dioxide as a catalyst:

$$2KClO_3(s) \xrightarrow[\Delta]{MnO_2(s)} 2KCl(s) + 3O_2(g)$$

If the $KClO_3$ is accurately weighed before and after the oxygen has been driven off, the weight of the oxygen can be obtained by difference. The oxygen can be collected by displacing water from a bottle, and the volume of gas can be determined from the volume of water displaced. Through use of Dalton's law of partial pressures, the vapor pressure of water, and atmospheric pressure, the pressure of the gas may be obtained. Dalton's law states that the pressure of a mixture of gases in a container is equal to the sum of the pressures that each gas would exert if it were present alone:

$$P_{\text{total}} = \sum_i P_i$$

Since this experiment is conducted at atmospheric pressure, $P_{\text{total}} = P_{\text{atmospheric}}$. Hence,

$$P_{\text{atmospheric}} = P_{O_2} + P_{H_2O\ \text{vapor}}$$

PROCEDURE

Add a small amount of MnO_2 (about 0.02 g) and approximately 0.3 g of $KClO_3$ to a test tube and accurately weigh to the nearest 0.0001 g. Assemble the apparatus illustrated in Figure 13.1 but do not attach the test tube. Be sure that tube *B* does not extend below the water level in the bottle. Fill glass tube *A* and the rubber tubing with water by loosening the pinch clamp and attaching a rubber bulb to and applying pressure through tube *B*. Close the clamp when the tube is filled.

Mix the solids in the test tube by rotating the tube, being certain that none of the mixture is lost from the tube, and attach tube *B* as shown in Figure 13.1. WHEN YOU ATTACH THE TEST TUBE, BE CERTAIN THAT NONE

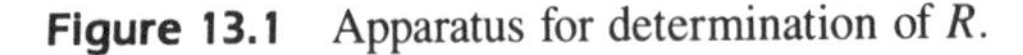

Figure 13.1 Apparatus for determination of *R*.

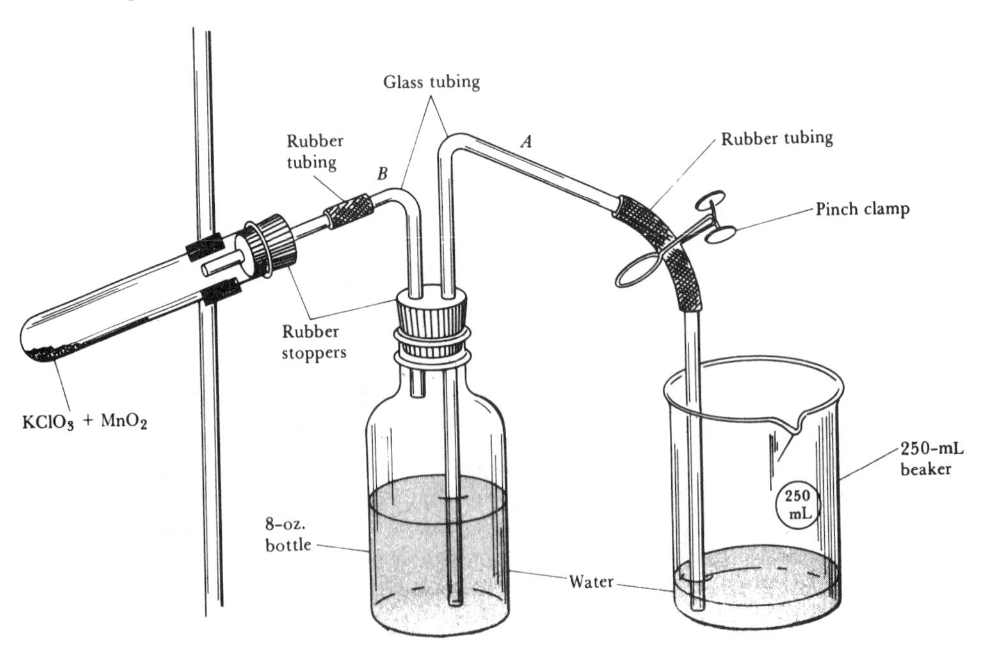

OF THE $KClO_3$ AND MnO_2 COMES INTO CONTACT WITH THE RUBBER STOPPER, OR A SEVERE EXPLOSION MAY RESULT. MAKE CERTAIN THAT THE CLAMP HOLDING THE TEST TUBE IS SECURE SO THAT THE TEST TUBE CANNOT MOVE.

Fill the beaker about half full of water, insert glass tube *A* in it, open the pinch clamp, and lift the beaker until the levels of water in the bottle and beaker are identical; then close the clamp, discard the water in the beaker, and dry the beaker. The purpose of equalizing the levels is to produce atmospheric pressure inside the bottle and test tube.

Set the beaker with tube *A* in it on the desk and open the pinch clamp. A little water will flow into the beaker, but if the system is airtight and has no leaks, the flow will soon stop, and tube *A* will remain filled with water. If this is not the case, check the apparatus for leaks and start over again. Leave the water that has flowed into the beaker in the beaker; at the end of the experiment, the water levels will be adjusted, and this water will flow back into the bottle.

Heat the lower part of the test tube gently (be certain that the pinch clamp is open) so that a slow but steady stream of gas is produced, as evidenced by the flow of water into the beaker. When the rate of gas evolution slows considerably, increase the rate of heating, and heat until no more oxygen is evolved. Allow the apparatus to cool to room temperature, being certain that the end of the glass tube in the beaker is always below the surface of the water. Equalize the water levels in the beaker and the bottle as before and close the clamp. Weigh a 125-mL Erlenmeyer flask* to the nearest 0.01 g and empty the water from the beaker into the flask.* Weigh the flask* with the water in it. Measure the temperature of the water, and using the density of water in Table 1.6 of Experiment 1, calculate the volume of the water displaced. This is equal to the volume of oxygen produced. Remove the test tube from the apparatus and accurately weigh the tube plus the contents. The difference in weight between this and the original weight of the tube plus MnO_2 and $KClO_3$ is the weight of the oxygen produced.

Record the barometric pressure. The vapor pressure of water at various temperatures is given in Table 13.1.

Calculate the gas-law constant, *R*, from your data, using the ideal-gas equation. Calculate *R* using the van der Waals equation $(P + n^2a/V^2)(V - nb) = nRT$ (for O_2, $a = 1.360\ L^2atm/mol^2$), and $b = 31.83\ cm^3/mol$). Be sure to keep your units straight.

Error Analysis

Determine the maximum and the minimum value of *R* consistent with the experimental reliability of your data from the ideal-gas law:

$$R = \frac{PV}{nT} = \frac{(32.00\ \text{g/mol})PV}{mT}$$

Assume that the reliabilities for the various measured quantities in this experi-

* Or styrofoam cup. The volume of water may also be measured directly, but less accurately, with a graduated cylinder.

Table 13.1 Vapor Pressure of Water at Various Temperatures

Temperature (°C)	H_2O vapor pressure (mm Hg)
15	12.8
16	13.6
17	14.5
18	15.5
19	16.5
20	17.5
21	18.6
22	19.8
23	21.1
24	22.4
25	23.8

ment are as follows:

$$P = \pm 0.1 \text{ mm Hg} \qquad T = \pm 1°\text{C}$$

$$V = \pm 0.0001 \text{ L} \qquad m = \pm 0.0001 \text{ g}$$

To determine the maximum value of R, use the maximum values that the pressure and volume may have and the minimum values that the mass and temperature may have. Similarly, calculate the minimum value of R from the minimum values of P and V and the maximum values for m and T. Determine the average value of R and assign an uncertainty range to this average value.

EXAMPLE 13.1

Assume that the measured quantities were as follows: P = 705.5 mm Hg; T = 20°C; V = 242.9 mL; and m = 0.3002 g. What would be the maximum and minimum values of R, the average value of R, and the uncertainty range to be assigned to this average value?

Solution: First, put the measured quantities into proper units as follows:

$$P = \frac{705.5 \text{ mm Hg}}{760 \text{ mm Hg/atm}} = 0.928 \text{ atm}$$

$$V = 242.9 \text{ mL} = 0.2429 \text{ L}$$

$$m = 0.3002 \text{ g}$$

$$T = (20 + 273) \text{ K} = 293 \text{ K}$$

Therefore,

$$\text{Maximum } R = \frac{[705.6 \text{ mm Hg}/(760 \text{ mm Hg/atm})](0.2430 \text{ L})(32.00 \text{ g/mol})}{(0.3001 \text{ g})(292 \text{ K})}$$

$$= 0.0823 \text{ L-atm/mol-K}$$

$$\text{Minimum } R = \frac{[705.4 \text{ mm Hg}/(760 \text{ mm Hg/atm})](0.2428 \text{ L})(32.00 \text{ g/mol})}{(0.3003 \text{ g})(294 \text{ K})}$$

$$= 0.0816 \text{ L-atm/mol-K}$$

The average value for R is, therefore,

$$\text{Average } R = \frac{0.0823 + 0.0816}{2} \text{ L-atm/mol-K}$$

$$= 0.0820 \text{ L-atm/mol-K}$$

Note that the minimum and maximum values of R differ from the average by 0.0004. Consequently, the uncertainty in R can be written as ± 0.0004 L-atm/mol-K and the data would be reported as

$$R = 0.0820 \pm 0.0004 \text{ L-atm/mol-K}$$

REVIEW QUESTIONS

Before beginning this experiment in the laboratory, you should be able to answer the following questions:

1. Under what conditions of temperature and pressure would you expect gases to obey the ideal-gas equation?
2. Calculate the value of R in L-atm/mol-K by assuming that an ideal gas occupies 22.4 L/mol at STP.
3. Why do you equalize the water levels in the bottle and the beaker?
4. Why does the vapor pressure of water contribute to the total pressure in the bottle?
5. What is the value of an error analysis?
6. Suggest reasons, on the molecular level, why real gases might deviate from the ideal-gas law.
7. Newly devised automobile batteries are sealed. When lead storage batteries discharge, they produce hydrogen. Suppose the void volume in the battery is 100 mL at 1 atm of pressure and 25°C. What would be the pressure increase if 0.05 g H_2 were produced by the discharge of the battery? Does this present a problem? Do you know why sealed lead storage batteries have not been used in the past?
8. Why is the corrective term to the volume subtracted and not added to the volume in the van der Waals equation?
9. A sample of pure gas at 18°C and 650 mm Hg occupied a volume of 562 cm^3. How many moles of gas does this represent? (HINT: Use the value of R that you found in question 2.)
10. A certain compound containing only carbon and hydrogen was found to have a vapor density of 2.550 g/L at 100°C and 760 mm Hg. If the empirical formula of this compound is CH, what is the molecular formula of this compound?
11. Which gas would you expect to behave more like an ideal gas He or HCl? Why?

NOTES AND CALCULATIONS

Name ____________________ Desk ________

Date ____________ Laboratory Instructor ____________________

REPORT SHEET FOR EXPERIMENT 13

DETERMINATION OF *R*, THE GAS-LAW CONSTANT

1. Weight of test tube + $KClO_3$ + MnO_2 ____________ g
2. Weight of test tube + contents after reaction ____________ g
3. Weight of oxygen produced ____________ g
4. Weight of 125-mL flask* + water ____________ g
5. Weight of 125-mL flask* ____________ g
6. Weight of water ____________ g
7. Temperature of water ____________
8. Density of water ____________
9. Volume of water ____________ = volume of O_2 gas ____________
10. Barometric pressure ____________ mm Hg
11. Vapor pressure of water ____________
12. Pressure of O_2 gas (show calculations) ____________

13. Gas-law constant, *R*, from ideal-gas law (show calculations) ____________

14. *R* from the van der Waals equation (show calculations) ____________

15. Accepted value of *R* ____________ (Source of *R* value) ____________

*Or Styrofoam cup.

16. Uncertainty in R (show calculations) ________________

QUESTIONS

1. Does your value of R agree with the accepted value within your uncertainty limits?

2. Discuss possible sources of error in the experiment; indicate the ones that you feel are most important.

3. Which gas would you expect to deviate more from ideality, H_2 or HBr? Explain your answer.

4. How does the solubility of oxygen in water affect the value of R you determined? Explain your answer.

5. Calculate the van der Waals correction terms to pressure and volume for Cl_2 at STP. The values of the van der Waals constants a and b are 6.49 L^2-atm/mol^2 and 0.0562 L/mol, respectively, for Cl_2. At STP, which is the major cause of deviation from ideal behavior, the volume of the Cl_2 molecules or the attractive forces between them? Why?

6. How much potassium chlorate is needed to produce 10.0 mL of oxygen gas at 650 mm Hg and 20°C?

7. If oxygen gas were collected over water at 22°C and the total pressure of the wet gas were 650 mm Hg, what would be the partial pressure of the oxygen?

8. An oxide of nitrogen was found by elemental analysis to contain 30.4 percent nitrogen and 69.6 percent oxygen. If 23.0 g of this gas were found to occupy 5.6 L at STP, what are the empirical and molecular formulas for this oxide of nitrogen?

9. The gauge pressure in an automobile tire reads 32 pounds per square inch (psi) in the winter at 32°F. The gauge reads the difference between the tire pressure and the atmospheric pressure (14.7 psi). In other words, the tire pressure is the gauge reading plus 14.7 psi. If the same tire were used in the summer at 110°F and no air had leaked from the tire, what would be the tire gauge reading in the summer? (HINT: Recall that $°C = \frac{5}{9}(°F - 32)$.)

NOTES AND CALCULATIONS

EXPERIMENT 14

Activity Series

OBJECTIVE

To become familiar with the relative activities of metals in chemical reactions.

APPARATUS AND CHEMICALS

small test tubes* (13)
test-tube rack
6 *M* HCl
25 mL of 0.2 *M* $Ca(NO_3)_2$
25 mL of 0.2 *M* $Mg(NO_3)_2$
25 mL of 0.2 *M* $Zn(NO_3)_2$
25 mL of 0.2 *M* $Fe(NO_3)_3$
25 mL of 0.2 *M* $FeSO_4$
25 mL of 0.2 *M* $SnCl_4$
25 mL of 0.5 *M* $Cu(NO_3)_2$
7 small pieces each of calcium, magnesium, zinc, iron wool, tin, copper

DISCUSSION

Chemical elements are usually classified by their properties into three groups: metals, nonmetals, and metalloids. Most of the known elements are metals. Their physical properties include high thermal and electrical conductivity, high luster, malleability (ability to be pounded flat without shattering), and ductility (ability to be drawn out into a fine wire). All common metals are solids at room temperature except mercury, which is a liquid. The periodic table illustrated in Figure 14.1 shows the three classifications of the elements.

All elements to the left of the shaded area are metals; those to the right are nonmetals; and those in the shaded area have intermediate properties and

Figure 14.1 The periodic table.

1A																	8A
H	2A											3A	4A	5A	6A	7A	He
Li	Be											B	C	N	O	F	Ne
Na	Mg					Transition elements						Al	Si	P	S	Cl	Ar
K	Ca	Sc	Ti	V	Cr	Mn	Fe	Co	Ni	Cu	Zn	Ga	Ge	As	Se	Br	Kr
Rb	Sr	Y	Zr	Nb	Mo	Tc	Ru	Rh	Pd	Ag	Cd	In	Sn	Sb	Te	I	Xe
Cs	Ba	La	Hf	Ta	W	Re	Os	Ir	Pt	Au	Hg	Tl	Pb	Bi	Po	At	Rn

*A spot plate may be used in place of test tubes

are called semimetals or metalloids. Families or groups of elements consist of elements in vertical columns in the periodic table. Elements within a group or family have similar chemical properties because they have similar valence electronic structures; that is, the number of valence electrons (electrons in the outermost shell) is the same for all members of a family or group. For historical reasons, most of the groups have names, and some are often referred to by them. These are

1. Group 1, called *alkali metals* because they react with oxygen to form bases;
2. Group 2, called *alkaline earth metals* because their presence makes soils alkaline;
3. Group 3, no common name;
4. Group 4, no common name;
5. Group 5, called *pnictides,* from the Greek word meaning choking suffocation;
6. Group 6, called *chalcogens,* from Greek roots meaning ore former;
7. Group 7, called *halogens,* from Greek roots meaning salt former;
8. Group 8, called *rare, noble,* or *inert gases* because they were thought to be unreactive.

Those most frequently referred to by group name are the alkali metals, the alkaline earth metals, the halogens, and the rare gases.

The three broad categories of the elements also have somewhat similar chemical properties. For example, metals, as compared with the other elements, all have relatively low ionization potentials and enter into chemical combination with nonmetals by *losing* electrons to become cations. This can be symbolized by the following equation:

$$M \longrightarrow M^{n+} + ne^-$$

Nonmetals, as compared with metals, have relatively high electron affinities and enter into chemical combination with metals by *gaining* electrons to become anions. This can be symbolized by the following equation:

$$X + ne^- \longrightarrow X^{n-}$$

Specific examples of these types of reactions can be divided into several useful categories, which will be illustrated by the following examples.

Electron-Transfer Reactions

1 Reactions with Oxygen.

$$2Mg(s) + O_2(g) \longrightarrow 2MgO(s)$$

In this reaction magnesium has been oxidized by oxygen which, in turn, has been reduced by magnesium. This can be better illustrated by breaking down the reaction into fictitious although helpful steps:

$$2(Mg \longrightarrow Mg^{+2} + 2e^-) \quad \text{oxidation}$$

$$O_2 + 4e^- \longrightarrow 2O^{2-} \quad \text{reduction}$$

$$2Mg + O_2 \longrightarrow 2MgO \quad \text{oxidation-reduction (redox) reaction}$$

In *oxidation,* the oxidized element loses electrons and becomes more positive. In *reduction,* the reduced element gains electrons and becomes more negative. Oxidation is always associated with a concomitant reduction.

2 Reactions with Water.

$$2Na(s) + 2H_2O(l) \longrightarrow 2NaOH(aq) + H_2(g)$$

$$Ca(s) + 2H_2O(l) \longrightarrow Ca(OH)_2(aq) + H_2(g)$$

The *ionic* equations for these reactions better illustrate the electron-transfer process:

$$2Na(s) + 2H_2O(l) \longrightarrow 2Na^+(aq) + 2OH^-(aq) + H_2(g)$$

$$Ca(s) + 2H_2O(l) \longrightarrow Ca^{2+} + (aq) + 2OH^-(aq) + H_2(g)$$

3 Reactions with Acids.

$$Zn(s) + 2HCl(aq) \longrightarrow ZnCl_2(aq) + H_2(g)$$

or

$$Zn(s) + 2H^+(aq) + 2Cl^-(aq) \longrightarrow Zn^{2+}(aq) + 2Cl^-(aq) + H_2(g)$$

Since the chloride ion is merely a spectator, that is, it does not participate in the reaction, it may be omitted, yielding the *net ionic equation:*

$$Zn(s) + 2H^+(aq) \longrightarrow Zn^{2+}(aq) + H_2(g)$$

or simply

$$Zn + 2H^+ \longrightarrow Zn^{2+} + H_2$$

4 Electron Transfer Among Metals.

$$Zn(s) + Cu(NO_3)_2(aq) \longrightarrow Zn(NO_3)_2(aq) + Cu(s)$$

or

$$Zn(s) + Cu^{2+}(aq) \longrightarrow Zn^{2+}(aq) + Cu(s)$$

or simply

$$Zn + Cu^{2+} \longrightarrow Zn^{2+} + Cu$$

Note once again that in the ionic equation the spectator ion (NO_3^-) has been omitted because it takes no active part in the reaction and serves only to provide electrical neutrality. Therefore, any other soluble salt of copper (II), such as chloride, sulfate, or acetate, could perform the same function.

In this game of "musical electrons," there are only enough electrons for one atom of the system. In order to achieve the lowest energy level for the system, the more active metal of a pair will lose electrons to the more passive metal or react more vigorously with water, acids, or oxygen. In some cases no reaction will occur at all. Without prior knowledge we have no way of predicting these events.

PROCEDURE

A. Reactions of Metals with Acid

To each of six test tubes containing 5 mL of dilute 6 *M* HCl, add a small piece of the metals Ca, Cu, Fe, Mg, Sn, and Zn. Observe the test tubes and note any changes that occur (such as the evolution of a gas, whether it is vigorous or not, and any color changes). Enter your observations on the report sheet and write both complete and ionic equations for each reaction noted.

B. Reactions of Metals with Solutions of Metal Ions

(WORK IN PAIRS FOR THIS STEP). To each of seven test tubes containing, respectively, about 4 mL of $Ca(NO_3)_2$, $Cu(NO_3)_2$, $FeSO_4$, $Fe(NO_3)_3$, $Mg(NO_3)_2$, $SnCl_4$, and $Zn(NO_3)_2$ solutions add a small piece of calcium metal. Note any reaction that occurs by observing whether a color change occurs on the surface of the metal or in the solution or whether a gas is evolved. Record your observations on the report sheet. Write both complete and ionic equations for any reaction that occurs.

Repeat the foregoing process by adding a small piece of copper to another 4 mL of each of the metal-cation solutions. Do the same for iron, magnesium, tin, and zinc and record all your observations. Write both complete molecular and ionic equations for all reactions.

C. Relative-Activity Series

From the information contained in the table you constructed in Part B, you can rank these six metals according to their relative chemical reactivities. This can be done in the following manner: One of the metals will replace all others in solution. For example, say calcium metal is oxidized by solutions containing cations of each of the other metals; it, therefore, is the most reactive. One of the other metals will replace all but calcium; it, therefore, is the next most reactive. Finally, one of the metals will not replace any of the other metal cations from solution. Therefore, it is the least reactive. List the metals on your report sheet in terms of decreasing reactivity starting with the most reactive (1) and terminating with the least reactive (6).

REVIEW QUESTIONS

Before beginning this experiment in the laboratory, you should be able to answer the following questions:

1. What distinguishes a metal from a nonmetal?

2. What does ionization potential measure?

3. What does electron affinity measure?

4. Why must oxidation be accompanied by a reduction?

5. How does one determine the relative reactivites of metals?

6. Complete and balance the following:

$$Mg + O_2 \longrightarrow$$

$$Zn + HCl \longrightarrow$$

$$Zn + Cu^{2+} \longrightarrow$$

7. Balance the following reactions and identify the species which have been oxidized and the species which have been reduced.

Reaction	Species oxidized	Species reduced
$Cl_2 + I^- \longrightarrow I_2 + Cl^-$		
$WO_2 + H_2 \longrightarrow W + H_2O$		
$Ca + H_2O \longrightarrow H_2 + Ca(OH)_2$		
$Al + O_2 \longrightarrow Al_2O_3$		

8. If the following redox reactions are found to occur spontaneously, identify the more active metal in each reaction.

Reaction	More-active metal
$2Li + Cu^{2+} \longrightarrow 2Li^+ + Cu$	
$Cr + 3V^{3+} \longrightarrow 3V^{2+} + Cr^{3+}$	
$Cd + 2Ti^{3+} \longrightarrow 2Ti^{2+} + Cd^{2+}$	

NOTES AND CALCULATIONS

Electrolysis, The Faraday, and Avogadro's Number

EXPERIMENT

15

OBJECTIVE

To determine the values for the faraday and Avogadro's number by electrolysis.

APPARATUS AND CHEMICALS

3 M H_2SO_4	buret clamps
DC source of electricity	250-mL beaker
insulated copper wires (2)	thermometer
50-mL buret	barometer
ring stand	millimeter ruler or meterstick

DISCUSSION

The passage of an electric current through a solution is accompanied by chemical reactions at the electrodes. Oxidation (loss of electrons) occurs at the anode; reduction (gain of electrons) occurs at the cathode. The amount of reaction that occurs at the electrodes is directly proportional to the number of electrons transferred. A *faraday* is defined as the total charge carried by Avogadro's number of electrons; or, in other words, 1 faraday represents the charge on 1 mol of electrons.

In this experiment you will determine the value of the faraday by measuring the amount of charge required to reduce 1 mol of H^+ ions according to the reaction

$$2H^+(aq) + 2e^- \longrightarrow H_2(g) \quad [1]$$

Electric charge is conveniently measured in coulombs. A *coulomb* (C) is the amount of electrical charge that passes a point in a circuit when a current of 1 ampere (A) flows for 1 second (s):

$$1\ C = 1\ A \times 1\ s$$

Therefore, the number of coulombs passing through a solution in a cell can be obtained by multiplying the current in amperes by the time in seconds for which it flows. The charge on the electron can also be measured in coulombs and is equal to 1.60×10^{-19} C.

EXAMPLE 15.1

A current of 3.00 A was passed through a solution of sulfuric acid for exactly 20.0 min. How many electrons and how many coulombs were passed through the solution?

Solution:

$$\text{Coulombs} = 3.00 \text{ A} \times 20.0 \text{ min} \times \frac{60 \text{ s}}{\text{min}} \times \frac{1 \text{ C}}{\text{A-s}}$$

$$= 3600 \text{ C}$$

$$\text{Electrons} = \frac{3600 \text{ C}}{1.60 \times 10^{-19} \text{ C/electron}}$$

$$= 2.25 \times 10^{22} \text{ electrons}$$

From Equation [1] we note that one hydrogen ion is reduced for every electron passed through the solution and that one molecule of H_2 is produced for every two electrons. Thus in Example 15.1 the 2.25×10^{22} electrons would produce 1.125×10^{22} molecules of H_2.

If we were to measure the volume, pressure, and temperature of the hydrogen gas associated with the electrolysis in Example 15.1, we could calculate the number of moles of H_2 produced using the relation $n = PV/RT$. Example 15.2 illustrates how this information can be used to calculate Avogadro's number.

EXAMPLE 15.2

A current of 3.00 A was passed through a solution of sulfuric acid for 20 min. The hydrogen produced was collected over water at 20°C and 657.5 mm Hg and was found to occupy a volume of 534 mL. How many moles H_2 were produced and what is the value of Avogadro's number, N?

Solution: At 20°C the vapor pressure of H_2O is 17.5 mm Hg. Therefore,

$$P_{\text{total}} = P_{H_2O} + P_{H_2}$$

$$P_{H_2} = 657.5 \text{ mm Hg} - 17.5 \text{ mm Hg}$$

$$= 640.0 \text{ mm Hg}$$

Solving the ideal-gas equation for n gives:

$$n = \frac{PV}{RT} = \frac{[640 \text{ mm Hg}/(760 \text{ mm Hg/atm})](0.534 \text{ L})}{(0.0821 \text{ L-atm/mol-K})(293 \text{ K})}$$

$$= 0.0187 \text{ mol } H_2$$

Since 2 mol of electrons are required for 1 mol of H_2 produced,

$$\text{Moles electrons} = (0.0187 \text{ mol } H_2) \times (2 \text{ mol electrons/mol } H_2)$$

$$= 0.0374 \text{ mol electrons}$$

Avogadro's number is the number of electrons in 1 mol of electrons. From the previous example,

$$2.25 \times 10^{22} \text{ electrons} = 0.0374 \text{ mol electrons}$$

Therefore,

$$N = \frac{2.25 \times 10^{22} \text{ electrons}}{0.0374 \text{ mol}}$$

$$= 60.2 \times 10^{22} \text{ electrons/mol}$$

$$= 6.02 \times 10^{23} \text{ electrons/mol}$$

The faraday, also known as Faraday's constant (abbreviated $\mathcal{F}$), is defined as the number of coulombs that is equivalent to 1 mol of electrons. Thus, in the examples above, the 3600 C corresponds to the charge associated with 0.0374 mol electrons, or

$$\frac{3600\ \text{C}}{0.0374\ \text{mol}} = 96{,}300\ \text{C/mol electrons}$$

This corresponds very closely to the accepted value for the faraday, which is

$$1\ \mathcal{F} = 96{,}500\ \text{C/mol electrons}$$

We should also note from Example 15.2 above that $(2.25 \times 10^{22}/2)$ molecules of H_2 were produced and represent 0.0187 mol H_2:

$$0.0187\ \text{mol H}_2 = (2.25 \times 10^{22}/2)\ \text{molecules H}_2$$

or

$$1\ \text{mol H}_2 = \frac{2.25 \times 10^{22}\ \text{molecules H}_2}{2 \times 0.0187\ \text{mol H}_2} = 6.02 \times 10^{23}\ \text{molecules H}_2/\text{mol H}_2$$

$$= N$$

PROCEDURE

Obtain a DC source with an attached ammeter. Add 100 mL of distilled water and then 50 mL of dilute (3 *M*) sulfuric acid to a 250-mL beaker. Stir with a glass rod to mix well. Fill a 50-mL buret with this solution and invert it in the solution, holding it in place with a ring stand and clamp. Attach the copper-wire cathode to the negative terminal of the DC source and place the other end uninsulated into the inverted mouth of the buret. Be certain that all of the bare part of the wire is wholly inside the buret—otherwise some of the H_2 generated will not be collected in the buret. The anode electrode should be hung over the edge of the beaker and immersed in the acid solution, with the other end attached to the positive electrode of the DC source. The top of the solution in the buret should be within the graduated region of the buret so that the volume may be accurately measured, as illustrated in Figure 15.1.

Record the time, turn on the DC source, and record the ammeter reading. During electrolysis be careful not to move the electrodes, because this may change the current. It is important to maintain a steady current throughout the duration of the electrolysis. If current does fluctuate, it may be necessary to use an average value. Continue the electrolysis until at least 20 mL of hydrogen has been collected. Note the time when the electrolysis is stopped. Measure the volume of H_2 collected.

Measure the height of the water column in the buret above the solution in your beaker. Also measure the temperature of the acid solution and obtain the barometric pressure and the vapor pressure of water at the solution temperature. Note that

$$P_{\text{H}_2} = P_{\text{barometric}} - P_{\text{H}_2\text{O column}} - P_{\text{H}_2\text{O vapor}}$$

$$P_{\text{H}_2\text{O column}} = \frac{\text{height of H}_2\text{O in mm}}{13.6\ \text{mm H}_2\text{O/mm Hg}}$$

If time permits, repeat the experiment.

Figure 15.1

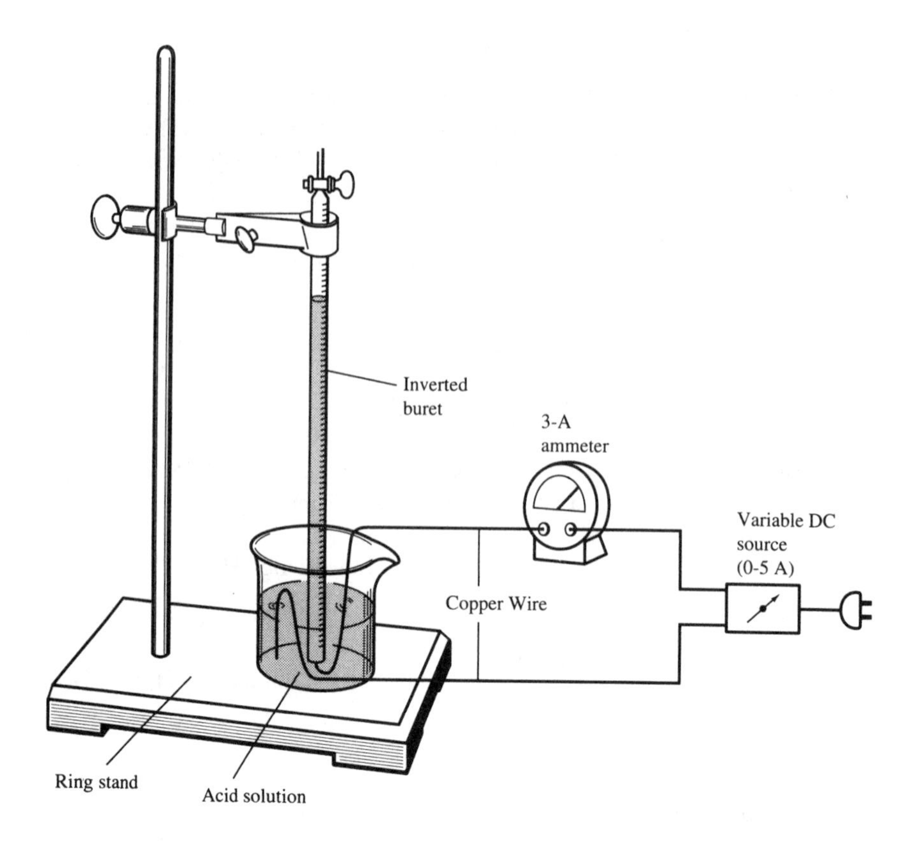

REVIEW QUESTIONS

Before beginning this experiment in the laboratory, you should be able to answer the following questions:

1. What type of chemical reaction is occurring to produce H_2 at the cathode in this experiment?
2. Define the term coulomb.
3. What process is occurring at the anode?
4. Why do you include the height of the water column in the buret in your calculation of the pressure?
5. Why is H_2SO_4 present in the electrolysis solution?
6. If a current of 80.0 μA (80.0×10^{-6} A) is drawn from a solar cell for 3 months, how many faradays are involved?
7. When an aqueous NaCl solution is electrolyzed, how many faradays need to be transferred at the anode to release 0.010 mol of Cl_2 gas?

$$2Cl^-(aq) \longrightarrow Cl_2(g) + 2e^-$$

8. How long must a current of 0.75 A pass through a sulfuric acid solution in order to liberate 0.100 L of H_2 gas at STP?
9. How much silver was there in the solution if all the silver was removed as Ag metal by electrolysis for 0.32 hr with a current of 1.20 mA (1 mA = 10^{-3} A)?
10. Electrolysis of molten NaCl is done in a Downs cell operating at 7.0 volts (V) and 4.0×10^4 A. How much Na(s) and Cl_2 (g) can be produced in four hours in such a cell?

Name ______________________________ Desk __________

Date ______________ Laboratory Instructor ______________________________

REPORT SHEET FOR EXPERIMENT 15

ELECTROLYSIS, THE FARADAY, AND AVOGADRO'S NUMBER

1. Final volume is buret __________
2. Initial volume in buret __________
3. Volume of hydrogen __________
4. Temperature of solution __________
5. Height of water column __________ mm
6. Barometric pressure __________ mm Hg
7. Vapor pressure of H_2O (see Table 13.1 in Experiment 13) __________ mm Hg
8. Pressure of H_2 __________ mm Hg __________ atm (show calculations)

9. Moles of H_2 produced (show calculations) __________

10. Time reaction started __________
11. Time reaction terminated __________
12. Elapsed time __________ min __________ s =
13. Current __________ A
14. Number of coulombs passed __________
15. Value for faraday __________ Accepted value __________ (show calculations)

16. Value for Avogardo's number __________ Accepted value __________ (show calculations)

QUESTIONS

1. Discuss the major sources of error in this experiment.

2. Calculate the percentage error in $\mathcal{F}$ and N.

3. If the amperage in your electrolysis cell were increased by a factor of 2, what effect would this have on the time required to produce the same amount of hydrogen?

4. Would an increase in the concentration of the sulfuric acid increase the rate of hydrogen production? Explain.

5. Electrolysis of an NaCl solution with a current of 2.00 A for a period of 200 s produced 59.6 mL of Cl_2 at 650 mm pressure and 27°C. Calculate the value of the faraday from these data.

6. Why are different products obtained when molten and aqueous NaC1 are electrolyzed? Predict the products in each case.

Electrochemical Cells and Thermodynamics

EXPERIMENT

16

OBJECTIVE

To become familiar with some fundamentals of electrochemistry, including the Nernst equation, by constructing electrochemical (voltaic) cells and measuring their potentials at various temperatures. The quantities ΔG, ΔH, and ΔS are calculated from the temperature variation of the measured emf.

APPARATUS AND CHEMICALS

- lead, tin, and copper strips or wire
- emery cloth
- 1 M $Pb(NO_3)_2$
- 1 M $Cu(NO_3)_2$
- 1 M $SnCl_2$
- DC voltmeter, or potentiometer (to measure mV)
- thermometer
- glass U-tubes (to fit large test tubes) (3)
- glass stirring rods (3)
- cotton
- 0.1 M KNO_3
- agar-agar
- 2 sets alligator clips and lead wires
- 50-mL test tubes (3)
- 600-mL beaker
- Bunsen burner and tubing
- ice
- ring stand, iron ring, and wire gauze
- clamps (2)

DISCUSSION

Background

Electrochemistry is that area of chemistry that deals with the relations between chemical changes and electrical energy. It is primarily concerned with oxidation-reduction phenomena. Chemical reactions can be used to produce electrical energy in cells that are referred to as *voltaic,* or galvanic, cells. Electrical energy, on the other hand, can be used to bring about chemical changes in what are termed *electrolytic* cells. In this experiment you will investigate some of the properties of voltaic cells.

Oxidation-reduction reactions are those that involve the transfer of electrons from one substance to another. The substance that loses electrons is said to be oxidized, while the one gaining electrons is reduced. Thus if a piece of zinc metal were immersed into a solution containing copper(II) ions, zinc would be oxidized by copper(II) ions. Zinc loses electrons and is oxidized, and the copper(II) ions gain electrons and are reduced. We can conveniently express these processes by the following two half-reactions, which add to give the overall reaction:

Oxidation:	$Zn(s)$	$\longrightarrow$	$Zn^{2+}(aq) + 2e^-$	
Reduction:	$Cu^{2+}(aq) + 2e^-$	$\longrightarrow$	$Cu(s)$	[1]
Net reaction:	$Zn(s) + Cu^{2+}(aq)$	$\longrightarrow$	$Cu(s) + Zn^{2+}(aq)$	

In principle, any spontaneous redox reaction can be used to produce electrical energy—that is, the reaction can be used as the basis of a voltaic cell. The trick is to separate the two half-reactions so that electrons will flow through an external circuit. A voltaic cell that is based upon the reaction in Equation [1] and that uses a salt bridge is shown in Figure 16.1.

The cell voltage, or electromotive force (abbreviated emf), is indicated on the voltmeter in units of volts. The cell emf is also called the cell potential. The magnitude of the emf is a quantitative measure of the driving force or thermodynamic tendency for the reaction to occur. In general, the emf of a voltaic cell depends upon the substances that make up the cell as well as on their concentrations. Hence, it is common practice to compare *standard cell potentials,* symbolized by E°_{cell}. These potentials correspond to cell voltages under standard state conditions—gases at 1 atm pressure, solutions at 1 *M* concentration, and temperature at 25°C.

Just as the overall cell reaction may be regarded as the sum of two half-reactions the overall cell emf can be thought of as the sum of two half-cell potentials, that is, the sum of the voltage of the oxidation half-reaction (E_{ox}) and the voltage of the reduction half-reaction (E_{red}):

$$E_{cell} = E_{ox} + E_{red}$$

Because it is impossible to measure directly the potential of an isolated half-cell, the standard hydrogen half-reaction has been selected as a reference and has been assigned a standard reduction potential of exactly 0.000 V:

$$2H^+(1\ M) + 2e^- \longrightarrow H_2(1\text{ atm}) \quad E^\circ_{red} = 0.000\text{V}$$

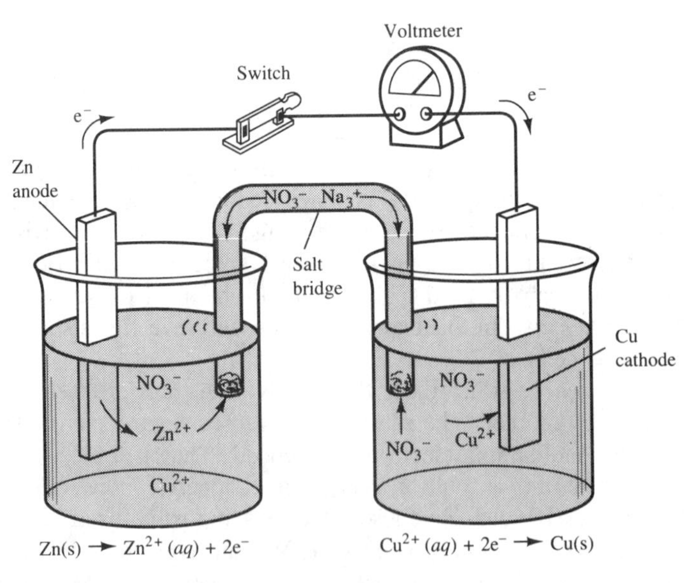

Figure 16.1 Complete and functioning voltaic cell using a salt bridge to complete the electrical circuit.

To demonstrate the consequence of this, let us consider a voltaic cell that utilizes the following reaction:

$$Zn(s) + 2H^+(aq) \longrightarrow Zn^{2+}(aq) + H_2(g)$$

The corresponding half-cell reactions are as follows:

$$Zn(s) \longrightarrow Zn^{2+}(aq) + 2e^- \qquad E^\circ_{ox} = 0.76V$$
$$2H^+(aq) + 2e^- \longrightarrow H_2(g) \qquad E^\circ_{red} = 0.000 \text{ V}$$

The standard cell emf of this cell is 0.76 V (that is, $E^\circ_{red} = 0.76$ V). Because the standard reduction potential of H^+ is 0.000 V, it is possible to calculate the standard oxidation potential, E°_{ox}, of Zn:

$$E^\circ_{cell} = E^\circ_{ox} + E^\circ_{red}$$
$$0.76 \text{ V} = E^\circ_{ox} + 0.000 \qquad [2]$$

Thus the standard oxidation potential of 0.76 V can be assigned to Zn. By measuring other standard-cell emf values, we can establish a series of standard potentials for other half-reactions.

It is important to note that the half-cell potential for any oxidation is equal in magnitude but opposite in sign to that of the reverse reduction. Hence,

$$Zn^{2+}(aq) + 2e^- \longrightarrow Zn(s) \qquad E^\circ_{red} = -0.76 \text{ V}$$

It is customary today to tabulate half-cell potentials as standard reduction potentials and also to refer to them as standard electrode potentials.

EXAMPLE 16.1

The cell in Figure 16.1 may be represented by the following notation:

$$Zn\,|\,Zn^{2+}\,||\,Cu^{2+}\,|\,Cu.$$

The double bar represents the salt bridge. Given that E°_{cell} for this cell is 1.10 V and that E_{ox} is 0.76 V for zinc (see Equation [2]), find the standard electrode potential, E°_{red}, for the reduction of copper ($Cu^{2+} + 2e^- \rightarrow Cu$).

Solution:

$$E^\circ_{cell} = E^\circ_{ox} + E^\circ_{red}$$
$$1.10 \text{ V} = 0.76 \text{ V} + E^\circ_{red}$$
$$E^\circ_{red} = 1.10 \text{ V} - 0.76 \text{ V}$$
$$= 0.34 \text{ V}$$

The free-energy change, ΔG, associated with a chemical reaction is a measure of the driving force or spontaneity of the process. If the free-energy change of a process is negative, the reaction will occur spontaneously in the direction indicated by the chemical equation.

The cell potential of a redox process is related to the free-energy change as follows:

$$\Delta G = -n\mathcal{F}E \qquad [3]$$

In this equation, $\mathcal{F}$ is Faraday's constant, the electrical charge on 1 mol of electrons:

$$1\mathcal{F} = 96{,}500\frac{\text{C}}{\text{mol e}^-} = 96{,}500\frac{\text{J}}{\text{V-mol e}^-}$$

and n represents the number of moles of electrons transferred in the reaction. For the case when both reactants and products are in their standard states, Equation [3] takes the following form:

$$\Delta G^\circ = -n\mathcal{F}E^\circ \qquad [4]$$

EXAMPLE 16.2

Calculate the standard free-energy change associated with the redox reaction $2Ce^{4+} + Tl^{+} \rightleftarrows 2Ce^{3+} + Tl^{3+}$ ($E^\circ = 0.450$ V). Would this reaction occur spontaneously under standard conditions?

Solution:

$$\begin{aligned} \Delta G^\circ &= -n\mathcal{F}E^\circ \\ &= -(2 \text{ mol e}^-)\left(\frac{96{,}500 \text{ J}}{\text{V-mol e}^-}\right)(0.450 \text{ V}) \\ &= -86.9 \times 10^3 \text{ J} \\ &= -86.9 \text{ kJ} \end{aligned}$$

Since $\Delta G^\circ < 0$, this reaction would occur spontaneously.

The standard free-energy change of a chemical reaction is also related to the equilibrium constant for the reaction as follows:

$$\Delta G^\circ = -RT \ln K \qquad [5]$$

where R is the gas-law constant (8.314 J/K-mol) and T is the temperature in kelvin. Consequently, E° is also related to the equilibrium constant. From Equations [4] and [5] it follows that

$$-n\mathcal{F}E^\circ = -RT\ ln\ K$$

$$E^\circ = \frac{RT}{n\mathcal{F}} \ln K \qquad [6]$$

When $T = 298\ K$, $\ln K$ is converted to $\log K$, and the appropriate values of R and $\mathcal{F}$ are substituted, Equation [6] becomes

$$E^\circ = \frac{0.0591}{n} \log K \qquad [7]$$

We can see from this relation that the larger K is, the larger the standard-cell potential will be.

In practice, most voltaic cells are not likely to be operating under standard-state conditions. It is possible, however, to calculate the cell emf, E, under nonstandard-state conditions from a knowledge of E°, temperature, and concentrations of reactants and products:

$$E = E^\circ - \frac{0.0591}{n} \log Q \qquad [8]$$

Q is called the reaction quotient; it has the form of an equilibrium-constant expression, but the concentrations used to calculate Q are not necessarily equilibrium concentrations. The relationship given in Equation [8] is referred to as the Nernst equation (see Example 16.3).

Let us consider the operation of the cell shown in Figure 16.1 in more detail. Earlier we saw that the reaction

$$Cu^{2+} + Zn \rightleftharpoons Zn^{2+} + Cu$$

is spontaneous. Consequently, it has a positive electrochemical potential ($E^\circ = 1.10$ V) and a negative free energy ($\Delta G^\circ = -n\mathcal{F}E^\circ$). As this reaction occurs, Cu^{2+} will be reduced and deposited as copper metal onto the copper electrode. The electrode at which reduction occurs is called the cathode. Simultaneously, zinc metal from the zine electrode will be oxidized and go into solution as Zn^{2+}. The electrode at which oxidation occurs is called the anode. Effectively, then, electrons will flow in the external wire from the zinc electrode through the voltmeter to the copper electrode and be given up to copper ions in solution. These copper ions will be reduced to copper metal and plate out on the copper electrode. Concurrently, zinc metal will give up electrons to become Zn^{2+} ions in solution. These Zn^{2+} ions will diffuse through the salt bridge into the copper solution and replace the Cu^{2+} ions that are being removed.

EXAMPLE 16.3

Calculate the cell potential for the following cell:

$$Zn\,|\,Zn^{2+}(0.6\,M)\,||\,Cu^{2+}(0.2\,M)\,|\,Cu$$

given the following:

$$Cu^{2+} + Zn \rightleftharpoons Cu + Zn^{2+} \qquad E^\circ = 1.10V$$

(HINT: Recall that Q includes expressions for species in solution but not for pure solids.)

Solution:

$$E = E^\circ - \frac{0.059}{n} \log \frac{[Zn^{2+}]}{[Cu^{2+}]}$$

$$= 1.10 \text{ V} - \frac{0.059}{2} \log \frac{[0.6]}{[0.2]}$$

$$= 1.10 - 0.014$$

$$= 1.086$$

$$= 1.09 \text{ V}$$

You can see that small changes in concentrations have small effects on the cell emf.

A list of the properties of electrochemical cells and some definitions of related terms are given in Table 16.1.

Table 16.1 Summary of Properties of Electrochemical Cells and Some Definitions

Voltaic cells: $E > 0$, $\Delta G < 0$; reaction spontaneous, K large (greater than 1)
Electrolytic cells: $E < 0$, $\Delta G > 0$; reaction nonspontaneous, K small (less than 1)
Anode—electrode at which oxidation occurs
Cathode—electrode at which reduction occurs
Oxidizing agent—species accepting electrons to become reduced
Reducing agent—species donating electrons to become oxidized

Chemists have developed a shorthand notation for electrochemical cells, as seen in Example 16.1. The notation for the Cu-Zn cell that explicitly shows concentrations is as follows:

$$\underset{\substack{\text{Anode} \\ \text{(oxidation)}}}{Zn \mid Zn^{2+}(xM)} \parallel \underset{\substack{\text{Cathode} \\ \text{(reduction)}}}{Cu^{2+}(yM) \mid Cu}$$

In this notation, the anode (oxidation half-cell) is written on the left and the cathode (reduction half-cell) is written on the right.

Your objective in this experiment is to construct a set of three electrochemical cells and to measure their cell potentials. With a knowledge of two half-cell potentials and the cell potentials obtained from your measurements, you will calculate the other half-cell potentials and the equilibrium constants for the reactions. By measuring the cell potential as a function of temperature, you may also determine the thermodynamic constants, ΔG, ΔH, and ΔS, for the reactions. This can be done with the aid of Equation [9]:

$$\Delta G = \Delta H - T\Delta S \qquad [9]$$

ΔG may be obtained directly from measurements of the cell potential using the relationship

$$\Delta G = -n\mathcal{F}E$$

A plot of ΔG versus temperature in degrees Kelvin will give $-\Delta S$ as the slope and ΔH as the intercept. A more accurate measure of ΔH can be obtained, however, by substituting ΔG and ΔS back into Equation [9] and calculating ΔH.

EXAMPLE 16.4

For the voltaic cell

$$Pb \mid Pb^{2+}(1\ M) \parallel Cu^{2+}(1\ M) \mid Cu$$

the following data were obtained:

$$E = 0.464 \text{ V} \qquad T = 298 \text{ K}$$
$$E = 0.468 \text{ V} \qquad T = 308 \text{ K}$$
$$E = 0.473 \text{ V} \qquad T = 318 \text{ K}$$

Calculate ΔG, ΔH, and ΔS for this cell.

Solution:

$$\Delta G = -n\mathcal{F}E$$

At 298 K,

$$\Delta G = -(2)(96{,}500)(0.464)$$
$$= -89.6 \text{ kJ}$$

At 308 K,

$$\Delta G = -(2)(96{,}500)(0.468)$$
$$= -90.3 \text{ kJ}$$

At 318 K,

$$\Delta G = -(2)(96{,}500)(0.473)$$
$$= -91.3 \text{ kJ}$$

Since $\Delta G = \Delta H - T\Delta S$, a plot of ΔG versus T in Kelvin gives $\Delta H = -64.2$ kJ/mol and $\Delta S = 85.0$ J/mol-K.

PROCEDURE

Before setting up pairs of half-cells, make a complete cell of the type shown in Figure 16.1 in the following manner: Place 30 mL of 1 *M* $Pb(NO_3)_2$ and 30 mL of 1 *M* $Cu(NO_3)_2$ in separate large test tubes. Obtain a lead strip and a copper strip and clean the surfaces of each with emery cloth or sandpaper. Boil 100 mL of 0.1 *M* KNO_3. Remove this solution from the heat and add to the boiling solution 1 g of agar-agar, stirring constantly until all the agar-agar dissolves. Invert a U-tube and fill the U-tube with this solution before it cools, leaving about a half inch of air space at each end of the U-tube as shown in Figure 16.2. The cotton plugs must protrude from the ends of the U-tube. Construct two additional agar-agar-filled U-tubes in the same manner. Place a U-tube into the test tubes as a salt bridge, as shown in Figure 16.3.

Insert the lead strip into the $Pb(NO_3)_2$ solution and the copper strip into the $Cu(NO_3)_2$ solution. Obtain a voltmeter and attach the positive lead to the copper strip and the negative lead to the lead strip using alligator clips. Read

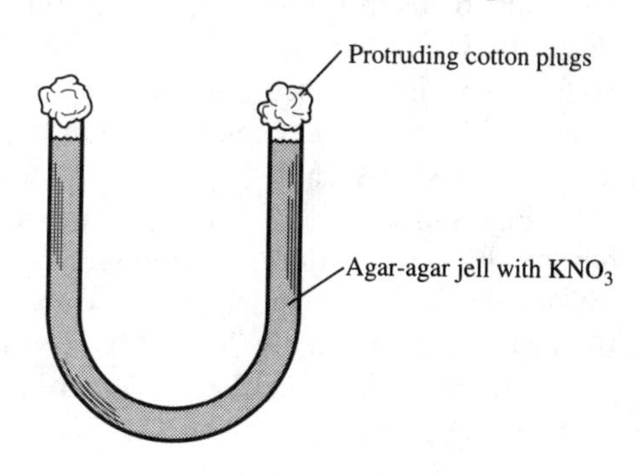

Figure 16.2

Figure 16.3

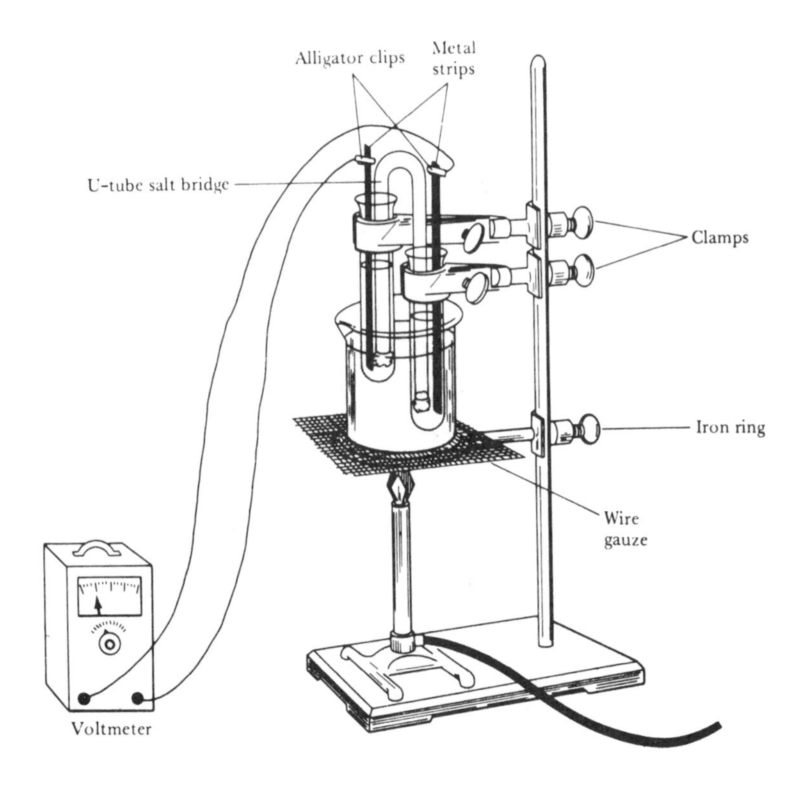

the voltage. Be certain that the alligator clips make good contact with the metal strips. Record this voltage and the temperature of the cells on your report sheet. If your measured potential is negative, reverse the wire connection. Now construct the following cells and measure their voltages in the same manner:

$$Sn \mid Sn^{2+}(1\ M) \parallel Cu^{2+}(1\ M) \mid Cu$$

$$Pb \mid Pb^{2+}(1\ M) \parallel Sn^{2+}(1\ M) \mid Sn$$

Record the voltages and temperature of each cell on your report sheet. From the measured voltages, calculate the half-cell potentials for the lead and tin half-cells and the equilibrium constants for these two reactions. In these calculations use E° = 0.34 V for the Cu^{2+}/Cu couple.

Now choose any one of the three cells and measure the cell potentials as a function of temperature as follows: Insert the metal strips into each of two 50-mL test tubes containing their respective 1 *M* cation solutions and place these test tubes in 600-mL beaker containing about 200 mL of distilled water. Place the U-tube into the two test tubes and connect the metallic strips to the voltmeter as before (see Figure 16.3). Beacuse the voltage changes are of the order of 30 mV for the temperature range you will study, be certain that the voltmeter you use is sensitive enough to detect these small changes.

Begin heating the water in the 600-mL beaker, using your Bunsen burner. Be certain that the test tubes are firmly clamped in place. DO NOT MOVE any part of the cell, beacuse the voltage will fluctuate if you do. Heat the cell to approximately 70°C. Measure the temperature and record it on your report sheet. Determine the cell potential at this temperature and record it on your report sheet. Remove the Bunsen burner and record the temperature and voltage at 15° intervals as the cell cools to room temperature. Finally, replace the beaker of water with a beaker containing an ice-water mixture; be careful not to move the test tubes and their contents. After the cell has been in the ice-water mixture for about 10 min, measure the temperature of the ice-water mixture in the 600-mL beaker and record it and the cell potential. You now have determined the cell potential at various temperatures. Calculate ΔG for the cell at each of these temperatures and plot ΔG versus temperature. The slope of the plot is $-\Delta S$. From the values of ΔG and ΔS, calculate ΔH at 298 K. If time permits, determine the temperature dependence of E for another cell.

REVIEW QUESTIONS

Before beginning this experiment in the laboratory, you should be able to answer the following questions:

1. Define the following: faraday, anode, cathode, voltaic cell, electrolytic cell.
2. Write a chemical equation for the reaction that occurs in the following cell: $Ag|Ag^+||Cu^{2+}|Cu$.
3. Given the following $E°$'s, calculate the standard-cell potential for the cell in question 2.

$$Cu^{2+}(aq) + 2e^- \longrightarrow Cu(s) \qquad E° = +0.34\ V$$
$$Ag^+(aq) + e^- \longrightarrow Ag(s) \qquad E° = +0.80\ V$$

4. Calculate the voltage of the following cell:

$$Zn\,|\,Zn^{2+}(0.10\ M)\,||\,Cu^{2+}(0.40\ M)\,|\,Cu$$

5. Calculate the cell potential, the equilibrium constant, and the free energy for the following cell:

$$Ba(s) + Mn^{2+}(aq)(1\ M) \rightleftharpoons Ba^{2+}(aq)(1\ M) + Mn(s)$$

given the following $E°$ values:

$$Ba^{2+}(aq) + 2e^- \rightleftharpoons Ba(s) \qquad E° = -2.90\ V$$
$$Mn^{2+}(aq) + 2e^- \rightleftharpoons Mn(s) \qquad E° = -1.18\ V$$

6. Would you normally expect $\Delta H°$ to be positive or negative for a voltaic cell? Justify your answer.
7. Predict whether the following reactions are spontaneous or not.

$$Pd^{2+} + H_2 \rightleftharpoons Pd + 2H^+ \qquad Pd^{2+} + 2e^- \rightleftharpoons Pd \qquad E° = 0.987\ V$$
$$Sn^{4+} + H_2 \rightleftharpoons Sn^{2+} + 2H^+ \qquad Sn^{4+} + 2e^- \rightleftharpoons Sn^{2+} \qquad E° = 0.154\ V$$

$$Ni^{2+} + H_2 \rightleftharpoons Ni + 2H^+ \qquad Ni^{2+} + 2e^- \rightleftharpoons Ni \qquad E^\circ = -0.250\ V$$

$$Cd^{2+} + H_2 \rightleftharpoons Cd + 2H^+ \qquad Cd^{2+} + 2e^- \rightleftharpoons Cd \qquad E^\circ = -0.403\ V$$

From your answers decide which of the above metals could be reduced by hydrogen.

8. Identify the oxidizing agents and reducing agents in the reactions in question 7.

Name ______________________________ Desk __________

Date ______________ Laboratory Instructor ______________________________

REPORT SHEET FOR EXPERIMENT 16

ELECTROCHEMICAL CELLS AND THERMODYNAMICS

	Shorthand cell designation	Temperature(°C)	$E°$ cell (measured)	$\Delta G°$ (calculated)	K (calculated)
1.					
2.					
3.					

Show calculations for $\Delta G°$ and K for an exemplary pair.

	Half-cell equation	$E°$ half-cell (calculated)
3.		
4.		
5.		

Effect of Temperature on Cell Potential

Cell designation: $E°$ (measured)	Temperature (°C)	Temperature (K)	$\Delta G°$ (calculated)

$\Delta S°$ determined from the slope of a plot of $\Delta G°$ versus T __________

$\Delta H°$ calculated at 298 K __________ (show calculations)

Is the cell reaction endothermic or exothermic? ______________________________

____

QUESTIONS

1. Write the chemical equations that occur in the following cells:

$Pb|Pb(NO_3)_2||AgNO_3|Ag$

$Zn|ZnCl_2||Pb(NO_3)_2|Pb$

$Pb|Pb(NO_3)_2||NiCl_2|Ni$

2. Which of the following reactions should have the larger emf under standard conditions? Why?

$$CuSo_4(aq) + Pb(s) \rightleftharpoons PbSO_4(s) + Cu(s)$$

$$Cu(NO_3)_2(aq) + Pb(s) \rightleftharpoons Pb(NO_3)_2(aq) + Cu(s).$$

3. Calculate ΔG for the reaction in Example 16.3.

4. Voltages listed in textbooks and handbooks are given as *standard-cell potentials* (voltages). What is meant by a standard cell? Were the cells constructed in this experiment standard cells? Why or why not?

5. As a standard voltaic cell runs, it "discharges" and the cell potential decreases with time. Explain.

6. Using standard potentials given in the appendices, calculate the equilibrium constants for the following reactions:

$$Cu(s) + 2Ag^{+}(aq) \rightleftharpoons Cu^{2+}(aq) + 2Aq(s)$$

$$Zn(s) + Fe^{2+}(aq) \rightleftharpoons Zn^{2+}(aq) + Fe(s)$$

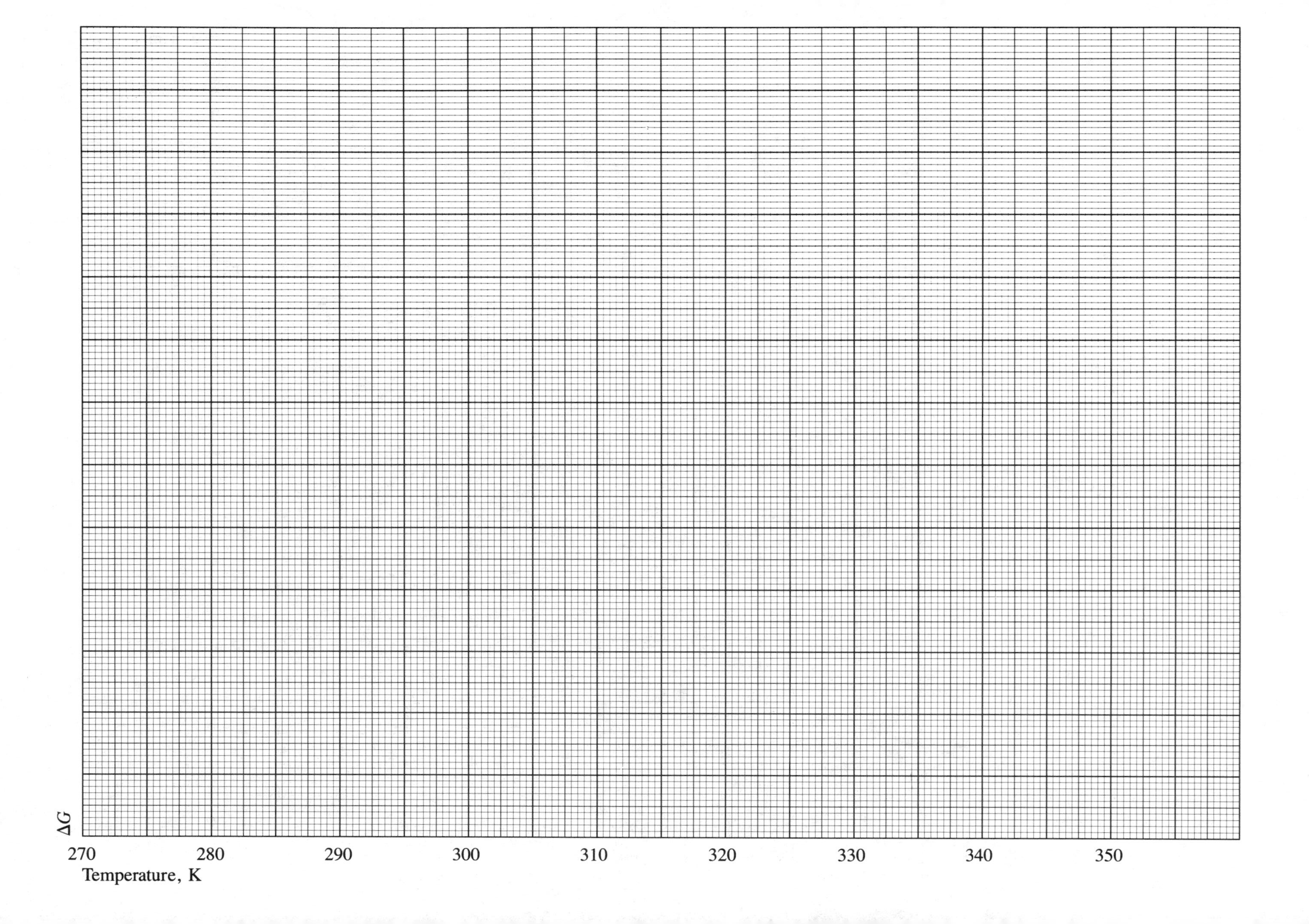

ΔG
270
280
290
300
310
320
330
340
350
Temperature, K

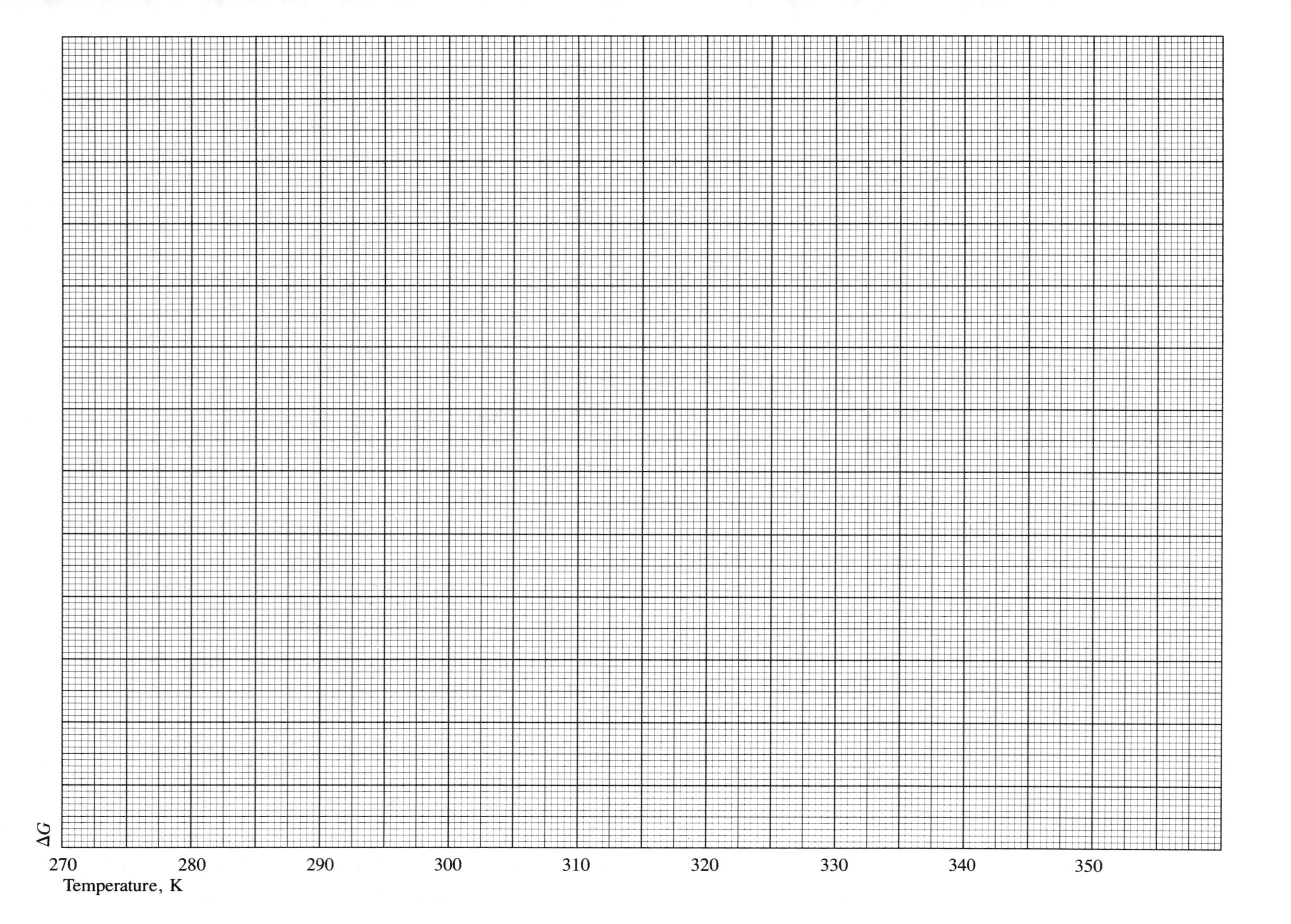
ΔG
270
280
290
300
310
320
330
340
350
Temperature, K

The Chemistry of Oxygen: Basic and Acidic Oxides and the Periodic Table

EXPERIMENT 17

OBJECTIVE

To illustrate the chemistry of oxides and become familiar with acids and bases and the concept of pH.

APPARATUS AND CHEMICALS

deflagrating spoon
glass squares (6)
pneumatic trough
250-mL wide-mouth bottles
6-in. test tube
Bunsen burner and hose
ring stand and clamp
rubber tubing
pH paper
crucible tongs
crucible and lid
clay triangle
150-mL beakers (5)
no. 2 1-hole stopper
400-mL beaker

calcium metal shavings
steel wool
magnesium ribbon, 2-in. piece
red phosphorus
sulfur lumps
charcoal pieces
$KClO_3$ (solid)
MnO_2 (powdered)
H_3BO_3 (solid)
Na_2O_2(solid)
$HClO_4$ (10 percent solution)
dilute HCl (6 *M*)
zinc oxide (solid)
19 *M* NaOH (50 percent by weight)

YOU WILL WORK IN PAIRS IN THIS EXPERIMENT, BUT EVALUATE YOUR DATA INDEPENDENTLY

DISCUSSION

Most elements react with molecular oxygen under various conditions to produce oxides. These reactions are oxidation-reduction reactions and are illustrated by the examples below. The symbol Ae stands for any element; the formula for the oxide depends upon the valence exhibited by the element.

$$Ae + O_2 \longrightarrow AeO_2$$

$$2Ae + O_2 \longrightarrow 2AeO$$

$$4Ae + 3O_2 \longrightarrow 2Ae_2O_3$$

Oxidation is always accompanied by a reduction reaction. In the examples above, the element Ae is oxidized while oxygen (O_2) is reduced.

Oxides of elements may be either ionic or covalent in character. In general, the more ionic oxides are formed from the elements at the left of the periodic table, while the elements at the right form covalent oxides. The ionic oxides that dissolve to any extent in water react with water to form basic solutions; for example,

$$AeO(s) + H_2O(l) \longrightarrow Ae^{2+}(aq) + 2OH^-(aq)$$

or

$$O^{2-}(aq) + H_2O(l) \longrightarrow 2OH^-(aq)$$

The oxide ion reacts with water to form hydroxide. If the water were evaporated from the above solution, we could obtain the base $Ae(OH)_2$. These ionic oxides are called *basic oxides* or *basic anhydrides* (base without water); for example,

$$Ba(OH)_2 - H_2O = BaO \qquad \text{(basic anhydride)}$$

The covalent oxides react with water to form acidic solutions; for example,

$$AeO(g) + H_2O(l) \longrightarrow H_2AeO_2(aq)$$
$$AeO_2(g) + H_2O(l) \longrightarrow H_2AeO_3(aq)$$

In water these acids ionize to various degrees to furnish hydrogen ions:

$$H_2AeO_3(aq) + H_2O(l) \rightleftharpoons H_3O^+(aq) + HAeO_3^-(aq)$$
$$HAeO_3^-(aq) + H_2O(l) \rightleftharpoons H_3O^+(aq) + AeO_3^{2-}(aq)$$

Thus the covalent oxides are termed *acidic oxides* or *acid anhydrides*. We can determine the corresponding formula for the acidic or basic anhydride from the formula of a given acid or base by subtracting water from the formula to eliminate *all* hydrogen atoms. For example,

$$2HNO_2 - H_2O \longrightarrow N_2O_3$$
$$2H_3BO_3 - 3H_2O \longrightarrow B_2O_3$$
$$2NaOH - H_2O \longrightarrow Na_2O$$

To determine whether an oxide is acidic or basic, the water solution of the oxide can be checked with litmus paper. However, not all of the oxides are appreciably soluble. In those cases in which the oxides are not water-soluble, one determines whether the oxides are acidic or basic by noting whether the oxide reacts with bases or acids. Acidic oxides react with bases, and basic oxides with acids.

It is not possible to classify all oxides either as acidic or basic. Some behave as both and are called *amphoteric*. Aluminum oxide is amphoteric, for it reacts both with acids and with strong bases:

$$Al_2O_3(s) + 6H^+(aq) \longrightarrow 2Al^{3+}(aq) + 3H_2O(l)$$
$$Al_2O_3(s) + 3H_2O(l) + 2OH^-(aq) \longrightarrow 2Al(OH)_4^-(aq)$$

Note that the formulas $Ae(OH)_2$ and H_2AeO_2 represent the same chemical composition. By convention, we indicate bases by placing the hydroxyl group at the end of the formula. Acids are written with the acidic hydrogens at the front of the formula.

Although there are several different definitions of acids and bases, a particularly useful one for aqueous solutions was proposed in 1923 by J. N. Brønsted and T. M. Lowry. In this scheme, *an acid is defined as a proton donor* and a *base as a proton acceptor*.

Consider the reaction of a strong acid, such as HCl, with water:

$$HCl(aq) + H_2O(l) \longrightarrow H_3O^+(aq) + Cl^-(aq) \qquad [1]$$

The HCl acts as a proton donor (acid), and the H_2O acts as a proton acceptor (base). Because HCl is a good donor of protons, the reaction essentially goes to completion in dilute solution—that is, all the HCl is present as H_3O^+ and Cl^-. These reactions are then merely proton-transfer reactions in which the stronger base (in this case, H_2O compared with Cl^-) competes for the protons. Recall that oxidation-reduction reactions are electron-transfer reactions in which a competition for electrons is developed. The more electronegative element gains the electrons from the less electronegative element.

A weak acid, such as acetic acid, is a poor donor of protons. Such a weak acid can donate protons to water only to a limited extent, and in the resulting equilibrium the concentrations of the undissociated reactants are much greater than the concentrations of the dissociated products. The reaction of acetic acid with water can be expressed as

$$HC_2H_3O_2(aq) + H_2O(l) \rightleftharpoons H_3O^+(aq) + C_2H_3O_2^-(aq) \qquad [2]$$

This is an equilibrium, and we can write the *equilibrium-constant* (K_c) expression for it. Thus*

$$K_c = \frac{[H_3O^+][C_2H_3O_2^-]}{[HC_2H_3O_2][H_2O]} \qquad [3]$$

Because the concentration of molecular water is large in comparison with the concentration of all other species present, the concentration of water changes relatively little in the course of the reaction and remains essentially constant. Thus the concentration of water can be combined with the equilibrium constant, K_c, and we obtain what we call an *acid-ionization constant* or *acid dissociation constant*, K_a,

$$K_a = K_c[H_2O] = \frac{[H_3O^+][C_2H_3O_2^-]}{[HC_2H_3O_2]} \quad \text{or} \quad K_a = \frac{[H^+][C_2H_3O_2^-]}{[HC_2H_3O_3]} \qquad [4]$$

If the concentration of all the species in Equation [4] are known, the dissociation constant, K_a, of acetic acid can be calculated.

Similarly, the weak base ammonia only partially reacts according to

* Note that both $H^+(aq)$ and $H_3O^+(aq)$ are used to denote the hydrated hydrogen ion.

$$NH_3(aq) + H_2O(l) \rightleftharpoons NH_4^+(aq) + OH^-(aq) \quad [5]$$

for which the equilibrium constant expression is

$$K_c = \frac{[NH_4^+][OH^-]}{[NH_3][H_2O]} \quad [6]$$

and the dissociation-constant expression is

$$K_b = K_c[H_2O] = \frac{[NH_4^+][OH^-]}{[NH_3]}$$

K_b is called the *base-dissociation constant*.

The reaction of HCl, an acid, with NaOH, a base, is also an acid-base reaction. Remembering that an aqueous solution of HCl consists of H_3O^+ and Cl^- ions and that NaOH is a strong electrolyte that dissociates to produce Na^+ and OH^- ions when dissolved in water, we can write

$$H_3O^+(aq) + Cl^-(aq) + Na^+(aq) + OH^-(aq) \longrightarrow 2H_2O(l) + Na^+(aq) + Cl^-(aq)$$

Here the proton is transferred from the H_3O^+ ion to the OH^- ion. Noting that the Na^+ and Cl^- ions appear as both reactants and products (that is, they are spectators and can be canceled out of the equation), we can write the following net ionic equation:

$$H_3O^+(aq) + OH^-(aq) \longrightarrow 2H_2O(l) \quad [7]$$

Equation [7] is the net equation for the reaction of any strong acid with any strong OH^--containing base. The reaction is referred to as *neutralization*.

Water is an amphoteric or amphiprotic substance—that is, it can act as either a Brønsted acid or a Brønsted base. This is evidenced by the reaction of water with itself:

$$H_2O(l) + H_2O(l) \rightleftharpoons H_3O^+(aq) + OH^-(aq) \quad [8]$$

where one molecule of water donates a proton to the other. Since Equation [8] is also an equilibrium, we can write the equilibrium-constant expression for this reaction:

$$K_c = \frac{[H_3O^+][OH^-]}{[H_2O]^2} \quad [9]$$

Again, because the concentration of molecular water is large in aqueous solutions when compared with the concentration of the other species present, its concentration changes very little during the course of the reaction and can be incorportated into the dissociation constant for water, known as K_w:

$$K_w = K_c[H_2O]^2 = [H_3O^+][OH^-] \quad \text{or} \quad K_w = [H^+][OH^-] \quad [10]$$

K_w is often called the ion-product constant for water. At 25°C, $K_w = 1.0 \times 10^{-14}$. This relationship is essentially true for any aqueous solution at 25°C, irrespective of the presence of other ions in solution. Hence, if we know the concentration of H_3O^+ ions in an aqueous solution, we can always use Equation [10] to calculate the concentration of the OH^- ions present in this same solution. In fact, since $K_w = [H_3O^+][OH^-] = 1.0 \times 10^{-14}$, if either $[OH^-]$ or $[H_3O^+]$ is known the other may be calculated. As can be seen from Equation [10], since $K_w = 1.0 \times 10^{-14}$, the value of the H_3O^+ or OH^- concentration can be small but can *never* be equal to zero!

EXAMPLE 17.1

Calculate the $[OH^-]$ concentration in an aqueous solution where $[H_3O^+] = 3.0 \times 10^{-5}\ M$.

Solution:

$$[H_3O^+][OH^-] = 1.0 \times 10^{-14}$$

$$[OH^-] = \frac{1 \times 10^{-14}}{[H_3O^+]} = \frac{1.0 \times 10^{-14}}{3.0 \times 10^{-5}}$$

$$= 3.3 \times 10^{-10}\ M$$

In a neutral solution, one where $[OH^-] = [H_3O^+]$, both concentrations are equal to 1.0×10^{-7}, or 0.00000010. This latter number and others like it are very inconvenient to write. Because of this, the concept of pH was developed as a convenience in expressing the concentration of the hydronium ion in dilute solutions of acids and bases. The pH of a solution is defined as the negative logarithm of the hydronium ion concentration or hydrogen ion concentration:

$$\text{pH} = -\log [H_3O^+] \quad \text{or} \quad \text{pH} = -\log [H^+]. \qquad [11]$$

And, consequently

$$[H_3O^+] = 10^{-\text{pH}}$$

For example, a solution containing 1.0×10^{-3} mol of HCl in 1 L of aqueous solution ($10^{-3}\ M$) contains 0.0010 M $[H_3O^+]$ and has a pH of 3.00. This pH is calculated as follows:

$$\text{pH} = -\log [H_3O^+] = -\log [1.0 \times 10^{-3}] = -[0 - 3.0] = 3.00$$

EXAMPLE 17.2

What is the pH of a 0.033 M HCl solution?

Solution:

$$\text{pH} = -\log [H_3O^+]$$

$$= -\log [3.3 \times 10^{-2}]$$

$$= -(\log 3.3 + \log 10^{-2})$$

$$= -(0.52 - 2.00)$$

$$= 1.48$$

From Equation [8] we can see that in a neutral solution, the hydronium ion concentration must equal the hydroxide ion concentration. These concentrations can be calculated by letting

$$[H_3O^+] = [OH^-] = x$$

Substituting into Equation [10], we have

$$(x)(x) = x^2 = 1.0 \times 10^{-14}$$
$$x = 1.0 \times 10^{-7}\ M$$

Then the pH of a neutral solution can be calculated:

$$pH = -\log (1.0 \times 10^{-7}) = 7.00$$

Since an acid solution is one in which $[H_3O^+] > [OH^-]$, then $[H_3O^+] > 10^{-7}$ and $[OH^-] < 10^{-7}$.* Likewise, since a basic solution is one in which $[OH^-] > [H_3O^+]$, then $[OH^-] > 10^{-7}$ and $[H_3O^+] < 10^{-7}$. These results are summarized in Table 17.1.

Table 17.1 Important Relations in Acidic and Basic Aqueous Solutions

	$[H_3O^+]$	pH	$[OH^-]$
Acidic	$>10^{-7}$	<7	$<10^{-7}$
Neutral	10^{-7}	7	10^{-7}
Basic	$<10^{-7}$	>7	$>10^{-7}$

The hydronium ion concentration can also be calculated from the pH. For example, if a solution has a pH of 4.0 or 5.3, the hydronium ion concentration can be calculated:

$$-\log [H_3O^+] = 4.0$$
$$\log [H_3O^+] = -4.0$$
$$[H_3O^+] = 1.0 \times 10^{-4}\ M$$

or

$$-\log [H_3O^+] = 5.3$$
$$\log [H_3O^+] = -5.3 = -6.0 + 0.7$$
$$[H_3O^+] = 5.0 \times 10^{-6}\ M$$

The pH of aqueous solutions may be determined in essentially two ways. One is by use of an electronic instrument having an electrode that is sensitive to hydronium ion concentration. These instruments are called "pH meters."

* The symbols > and < mean "greater than" and "less than," respectively.

The other way is by the use of organic dyes that possess two different pH-dependent colored forms. These dyes are called *acid-base indicators*.

Acid-base indicators derive their ability to act as pH indicators from the fact that they can exist in two forms. The acid form possesses one color, which by loss of a proton is converted to a differently colored base form. The pertinent equilibrium for the indicator is

$$\underset{\text{(Color A)}}{HIn_A} \rightleftharpoons \underset{\text{(Color B)}}{H^+ + In_B^-} \qquad [12]$$

where H In_A is the acid form of the indicator having color A and In_B^- is the base form of the indicator having color B. As an acid is added to a solution containing an indicator, Le Châtelier's principle tells us that the equilibrium in Equation [12] will shift to the left, the indicator will be present as H In_A, and the solution will have color A. Likewise, if a base is added, the indicator will be present in the form of In_B^-, and the solution will have color B. Table 17.2 lists some common indicators, with their respective colors and pH-transition regions.

Table 17.2 Colors and pH Transition Regions for Some Common Indicators

Indicator	Color change	pH range in which color change takes place
Methyl violet	Yellow-blue	0.1–1.5
Methyl violet	Blue-violet	1.5–3.2
Methyl orange	Red-yellow	3.2–4.4
Methyl red	Red-yellow	4.8–6.0
Bromothymol blue	Yellow-blue	6.0–7.6
Phenol red	Yellow-red	6.6–8.0
Phenolphthalein	Colorless-red	8.2–10.0
Alizarin yellow	Yellow-red	10.1–12.0

The pH paper is paper that has been impregnated with a series of organic dyes similar to those listed in Table 17.2. The paper changes various colors according to the pH of the solution in which it is immersed. Using pH paper, you should be able to estimate the pH of a solution to the nearest pH unit in the region pH 1–11.

In this experiment you will prepare oxygen gas and a number of typical oxides of various elements. You will then dissolve these oxides in water and test the acidity or basicity of the resulting solution. The following key points are to be especially noted:

1. Is there any correlation between the locations of the elements in the periodic table and the acidic and basic properties of their oxides?
2. Is there any correlation between the location of the element in the periodic table and its reactivity toward oxygen?
3. How does a catalyst affect a reaction?

PROCEDURE

A. Preparation of Oxygen

Assemble the apparatus as illustrated in Figure 17.1, but do not yet connect the test tube to the delivery tube. Be certain that the test tube is securely clamped. It *must* NOT be free to wiggle or tilt accidentally so that the contents ($KClO_3$ and MnO_2) come into contact with the rubber stopper, or a SEVERE EXPLOSION may result!

First heat about 0.25 g of potassium chlorate in a clean, dry test tube and compare the vigor and rate of the reaction with the following mixture: carefully mix about 10 g of potassium chlorate with 0.2 g of manganese dioxide. Place about 0.25 g of this mixture in a clean, dry test tube and heat it gently—just enough to melt the potassium chlorate. If it decomposes readily without sparks or combusting, the mixture is safe to use in your apparatus. *(Potassium chlorate is a powerful oxidizing agent and must be kept away from material that can easily be oxidized, or explosions may occur.)* Allow the test tube to cool to room temperature. Then add the remainder of the mixture and connect the delivery tube.

Completely fill six 250-mL wide-mouth bottles with tap water; cover the bottles with a glass square and invert them in the trough as required to collect the oxygen. Gently heat the mixture of potassium chlorate and manganese dioxide so as to maintain a moderate rate of formation of oxygen. Prior to collecting any oxygen, allow some of the gas to escape into the atmosphere. This will sweep air from the generator and allow you to collect pure oxygen. Collect six bottles of oxygen by displacing water and immediately cover the bottles with glass squares as they are filled. Remove the delivery tube from the water before you remove the flame or before the formation of oxygen stops. Why?

B. Preparation of Oxides

Prepare oxides of the elements as directed below by burning them in the oxygen you collected. Label each bottle so that you can identify it and avoid con-

Figure 17.1 Apparatus for oxygen production.

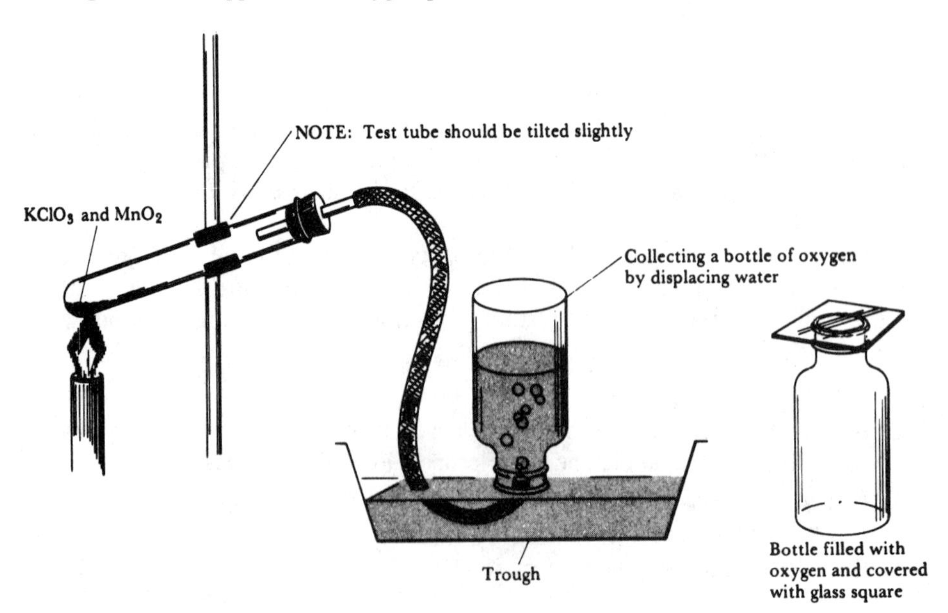

fusion. Keep the bottles covered with your glass square as much as possible during the combustion (see Figure 17.2). Immediately after each combustion, add about 30–50 mL of water; cover the bottle with the glass square; shake the bottle to dissolve the oxide; and set it aside for later tests.

Magnesium Holding a 10-cm piece of magnesium with crucible tongs, ignite it and quickly place it into a bottle of oxygen. TO AVOID INJURY TO YOUR EYES, DO NOT LOOK DIRECTLY AT THE BRILLIANT LIGHT.

Calcium Calcium burns brilliantly but is difficult to ignite. Place a shaving of calcium in a crucible and ignite it in air over a Bunsen burner for 15 min using the maximum temperature possible. After the crucible has cooled, wash out the contents with 50 mL of water into a 150-mL beaker.

Iron Pour a little water into a bottle of oxygen—enough to cover the bottom. The water layer will help prevent cracking of the bottle. Ignite a small, loosely packed wad of steel wool by holding it with tongs in a Bunsen burner flame until it ignites. Quickly thrust it into the bottle.

Carbon Ignite a small piece of charcoal, holding it either with tongs or in a clean deflagrating spoon. Thrust the glowing charcoal into a bottle of oxygen.

Phosphorous and Sulfur DO THESE COMBUSTIONS IN THE HOOD. Clean the deflagrating spoon in the dilute HCl and rinse with water. Then heat the spoon to remove any combustible material before it is used for the combustion of each of these elements. After the spoon has cooled, add a bit (no more in size than half a pea) of sulfur or red phosphorus. DO NOT TOUCH THE PHOSPHORUS WITH YOUR HANDS, SINCE IT WILL CAUSE BURNS. Ignite each with your burner and thrust the spoon into separate bottles of oxygen. After the combustion subsides, heat the spoon to burn off all remaining traces of phosphorus or sulfur; then clean the spoon by dipping it in dilute HCl.

Record your observations on each of the above combustions.

C. Reactions of Oxides with Water

Measure the pH of the water you used to make the above solutions with a strip of pH paper and record your results, then dip a 2-in. strip of pH paper into the

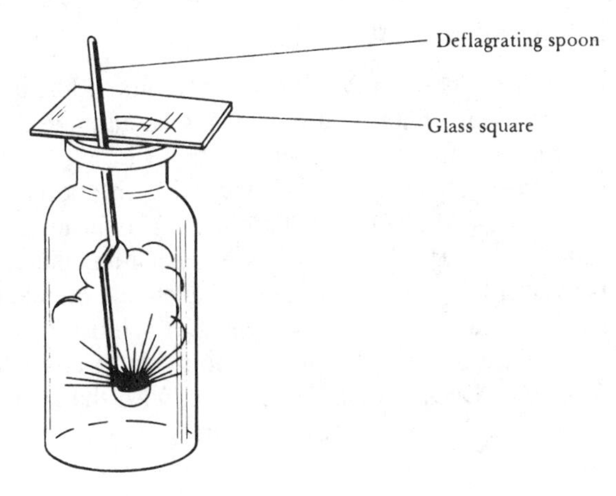

Figure 17.2 Cover bottles with glass square as much as possible during combustion.

solution in each of the above bottles. Estimate the pH to the nearest unit and record your results on your report sheet.

The oxides of elements from groups 1, 3, and 7 of the periodic table are difficult to prepare or are extremely insoluble or unstable. Therefore, you will be provided with either the oxide or a solution of the oxide.

Sodium When sodium burns in air, the peroxide, Na_2O_2, forms rather than the oxide, Na_2O. Because the peroxide is readily available and because it reacts with water similarly to sodium oxide, we shall use it to be indicative of the group 1 elements. In a 6-in. test tube carefully boil a very small amount (no more than the size of half a pea, *since larger quantities react violently!*) of sodium peroxide in 5 mL of water for a few seconds being careful not to point the test tube toward yourself or a neighbor. (DO NOT TOUCH THE Na_2O_2 WITH YOUR HANDS, SINCE IT WILL CAUSE SEVERE BURNS.) Cool the solution and test it with pH paper. Record your results.

Boron When the oxide of boron, B_2O_3, reacts with water, it forms a substance whose composition may be represented either as $B(OH)_3$ or H_3BO_3. Decide which of these formulas is preferred by dissolving a small amount of a substance, which we shall arbitrarily call "boron hydroxide," in about 5 mL of warm water. Cool the solution and test it with pH paper. Record your results. Which formula do you prefer, H_3BO_3 or $B(OH)_3$?

Chlorine The oxides of chlorine—Cl_2O, ClO_2, and Cl_2O_7—are gases and are unstable. An aqueous solution of one of these, labeled $HClO_4$, is available for you to test. Obtain 1 mL of this solution, check its pH, and record your results. DO NOT LET THE $HClO_4$ SOLUTION COME INTO CONTACT WITH ANY ORGANIC MATERIAL, SINCE IT IS A VERY POWERFUL OXIDIZING AGENT AND REACTS VIOLENTLY.

Neutralization Pour the sulfur oxide solution into the calcium oxide solution and test it with pH paper. How does this mixture differ from the individual solutions? Did a chemical reaction occur? What is it? Now mix the phosphorus oxide and magnesium oxide solutions and test the mixture with pH paper. Record your results.

D. Insoluble Oxide

Begin heating about 200 mL of water in a 400-mL beaker. Place about 0.25 g of zinc oxide, ZnO, in a 6-in. test tube and add 5 mL of water. Test the pH of the mixture and record your results.

Place about 0.25 g of zinc oxide in another test tube and *carefully* add 5 mL of 14 *M* NaOH. Be careful not to get any of the 14 *M* NaOH on yourself—if you do, wash it off with water immediately! Place this test tube in the beaker of hot water for several minutes and then allow it to cool. Does the zinc oxide appear to dissolve in the NaOH solution?

Place another 0.25 g of zinc oxide in a test tube and add about 5 mL of 6 *M* HCl. Does the zinc oxide react with the acid? What kind of oxide is zinc oxide—acidic, basic, or amphoteric?

REVIEW QUESTIONS

Before beginning this experiment in the laboratory, you should be able to answer the following questions:

1. Distinguish between ionic and covalent bonding.
2. What is the anhydride of sulfuric acid?
3. Define an acid and a base according to a Lowry-Brønsted definition.
4. How do K_w and K_c differ for the dissociation of water?
5. Why do we use the concept of pH?
6. State when solutions are acidic, neutral, and basic in terms of $[OH^-]$, $[H_3O^+]$, and pH.
7. If the $[H^+]$ in water is 10^{-4}, what is the pH and what is the $[OH^-]$?
8. Complete and balance the following:

$$Li_2O(s) + H_2O(l) \longrightarrow$$
$$N_2O_5(g) + H_2O(l) \longrightarrow$$

9. If the pH of a solution is 9.0, what are the hydrogen- and hydroxide-ion concentrations?
10. What are the hydroxide- and hydrogen-ion concentrations of 0.021 *M* $Ca(OH)_2$?
11. What is the pH of a solution if the indicator phenol red imparts a red color to the solution while phenolpthalein is colorless in the solution? Consult Table 17.2
12. Which solution would have a higher pH, 0.01 *M* HCl or 0.01 *M* NaOH? Which solution is acidic and which is basic?

NOTES AND CALCULATIONS

Name ______________________________ Desk ______

Date ______________ Laboratory Instructor ______________________

REPORT SHEET FOR EXPERIMENT 17

THE CHEMISTRY OF OXYGEN: BASIC AND ACIDIC OXIDES AND THE PERIODIC TABLE

A. Preparation of Oxygen

1. Write the equation for the reaction by which you prepared oxygen.
2. Compare the thermal decomposition of potassium chlorate alone with the decomposition of a mixture of potassium chlorate and manganese dioxide as to following:
   (a) The relative ease or the temperature at which decomposition occurs:

   (b) The amount of oxygen that may be produced from each sample:

3. The term used to refer to substances that alter the rate of a reaction is ______________.

B. Preparation of Oxides

Describe any changes occurring during the reaction of each element with oxygen, and properties of the products formed. Write the equation for each reaction.

	Observations	**Equations**	**Oxidation state of element in the oxide**
Mg			
Ca			
Fe			
C			
P			
S			

C. Reactions of Oxides with Water

pH of water ______________

On the line opposite its periodic group, write the formula for each oxide (or hydroxide) studied in this experiment. Indicate the pH of its solution, the $[H_3O^+]$ and $[OH^-]$ concentrations, and the equations for the formation of the acid or base.

Group	Formula of oxide	pH	$[H_3O^+]$	$[OH^-]$	Equation for reaction
1 Na					
2 Mg 2Ca					
3 B					
4 C					
5 P					
6 S					
7 Cl					
Transition Fe					

Write equations for the chemical reactions that occurred between the aqueous solutions of the oxides of sulfur and calcium, and between the aqueous solutions of the oxides of phosphorus and magnesium. Indicate the acids and bases in each reaction.

D. Insoluble Oxide

Compare the solubility of zinc oxide in water, sodium hydroxide, and hydrochloric acid.

What kind of oxide in zinc oxide?

QUESTIONS

1. Comment on the positions of the elements in the periodic table and on whether their oxides are acid or base producers.

2. Where in the periodic table would you expect most amphoteric oxides to be located?

3. Deduce and write formulas for the anhydrides of the following.

$HClO_2$	________	KOH	________	H_7SbO_6	________
H_2SO_4	________	$Ba(OH)_2$	________	HNO_3	________
H_3AsO_4	________	$Al(OH)_3$	________	H_2CO_3	________

4. If the pH of an aqueous solution is 3.28, what are the hydrogen- and hydroxide-ion concentrations of this solution?

5. If 3.70 g of $Ca(OH)_2$ is dissolved in sufficient water to make 100 mL of solution, what is the hydroxide-ion concentration of this solution?

6. Determine the hydroxide-ion concentration, the pH, and the hydrogen-ion concentrations of solutions containing 15.0 g of the following substances in 1.00 L of solution:

Substance	Molarity of substance	$[OH^-]$	pH	$[H_3O^+]$
CaO				
Na_2O				

(HINT: First write balanced equations for the reactions that occur when the oxides are dissolved in water.)

7. If 50 mL of 0.50 *M* KOH solution is added to 75 mL of 0.20 *M* H_2SO_4 solution, will the solution be neutral, acidic, or basic? Write a balanced chemical equation for the reaction and justify your answer.

8. Complete and balance the following equations and tell whether you predict the oxide to be acidic, basic, or amphoteric.

$Zn + O_2 \rightarrow$

$Ga + O_2 \rightarrow$

$As_4 + O_2 \rightarrow$

$Li + O_2 \rightarrow$

Colligative Properties: Freezing-Point Depression and Molecular Weight

EXPERIMENT 18

OBJECTIVE

To become familiar with colligative properties and to use them to determine the molecular weight of a substance.

APPARATUS AND CHEMICALS

- ring and ring stand
- clamp
- wire gauze
- thermometer
- large test tube
- wire stirrer
- Bunsen burner and hose
- 600-mL beaker
- sulfur, "roll" or precipitated; or unknown solid (2 g)
- naphthalene (50 g)
- 2-hole rubber stopper with slit
- towel
- wide-mouth glass bottle
- weighing paper

WORK IN PAIRS BUT EVALUATE YOUR DATA INDEPENDENTLY.

DISCUSSION

Solutions are homogeneous mixtures that contain two or more substances. The major component is called the *solvent,* and the minor component is called the *solute*. Since the solution is primarily composed of solvent, physical properties of a solution resemble those of the solvent. Some of these physical properties, called *colligative properties,* are independent of the nature of the solute and depend only upon the solute concentration. The colligative properties include vapor-pressure lowering, boiling-point elevation, freezing-point lowering, and osmotic pressure. The *vapor pressure* is just the escaping tendency of the solvent molecules. When the vapor pressure of a solvent is equal to atmospheric pressure, the solvent boils. At this temperature the gaseous and liquid states of the solvent are in dynamic equilibrium, and the rate of molecules going from the liquid to the gaseous state is equal to the rate of molecules going from the gaseous state to the liquid state. It has been found experimentally that the dissolution of a nonvolatile solute (one with very low vapor pressure) in a solvent lowers the vapor pressure of the solvent, which in turn raises the boiling point and lowers the freezing point. This is shown graphically by the phase diagram given in Figure 18.1

You are probably familiar with some common uses of these effects: Antifreeze is used to lower the freezing point and raise the boiling point of coolant (water) in an automobile radiator; and salt is used to melt ice. These effects are expressed quantitatively by the *colligative-property law,* which states that the freezing point and boiling point of a solution differ from those of the pure sol-

Figure 18.1 Phase diagram for a solvent and a solution.

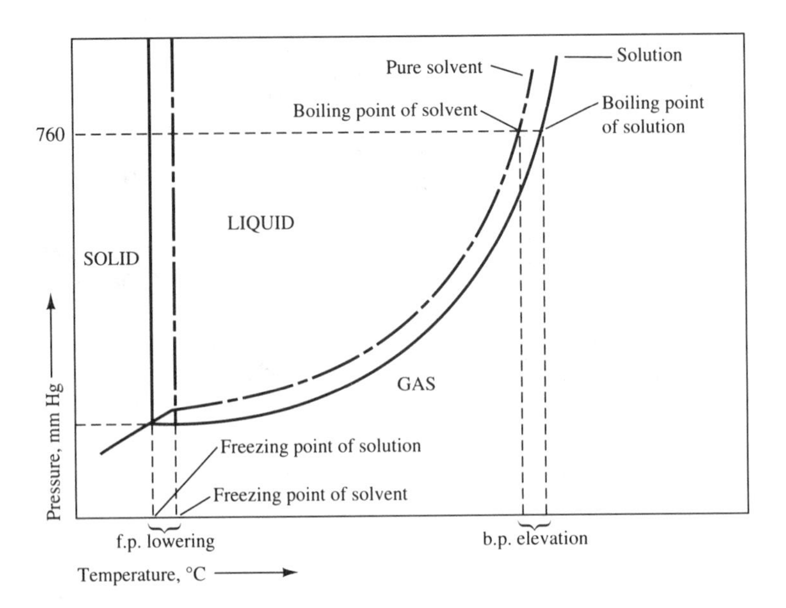

vent by amounts that are directly proportional to the molal concentration of the solute. This relationship is expressed by Equation [1] for the freezing-point lowering and boiling-point elevation:

$$\Delta T = Km \qquad [1]$$

where ΔT is the freezing-point lowering or boiling-point elevation, K is a constant that is specific for each solvent, and m is the molality of the solution (number of moles solute per/1000 g solvent). Some representative constants, boiling points, and freezing points are given in Table 18.1. For naphthalene, the solvent used in this experiment, the molal freezing-point depression constant (K_{fp}) has a value of 6.9°C/m.

Table 18.1 Molal Freezing-Point and Boiling-Point Constants

Solvent	Freezing point (°C)	K_{fp}(°C/m)	Boiling point (°C)	K_{bp}(°C/m)
CH_3COOH (acetic acid)	16.6	3.90	118.1	2.93
C_6H_6 (benzene)	5.4	5.12	80.2	2.53
$CHCl_3$ (chloroform)	−63.5	4.68	61.3	3.63
C_2H_5OH (ethyl alcohol)	−114.1	—	78.4	1.22
H_2O (water)	0.0	1.86	100.0	0.51
$C_{10}H_8$ (naphthalene)	80.6	6.9	218	—

EXAMPLE 18.1

What would be the freezing point of a solution containing 19.5 g of biphenyl ($C_{12}H_{10}$) dissolved in 100 g of naphthalene if the normal freezing point of naphthalene is 80.6°C?

Solution:

$$\text{Moles } C_{12}H_{10} = \frac{19.5 \text{ g}}{154 \text{ g/mol}} = 0.127 \text{ mol}$$

$$\frac{\text{Moles } C_{12}H_{10}}{1000 \text{ g naphthalene}} = \left(\frac{0.127 \text{ mol}}{100 \text{ g}}\right)(1000 \text{ g})$$

$$= 1.27\ m$$

$$\Delta T = (6.9°\text{C}/m)(1.27\ m)$$

$$= 8.76°\text{C} \quad \text{or} \quad 8.8°\text{C}$$

Since the freezing point is lowered, the observed freezing point of this solution will be

$$80.6°\text{C} - 8.8°\text{C} = 71.8°\text{C}$$

Since the molal freezing-point-depression constant is known, it is possible to obtain the molecular weight of a solute by measuring the freezing point of a solution and the weight of both the solute and solvent.

EXAMPLE 18.2

What is the molecular weight of urea if the freezing point of a solution containing 15 g of urea in 100 g of naphthalene is 63.3°C?

Solution: The freezing point of pure naphthalene is 80.6°C. Therefore, the freezing-point lowering (ΔT) is:

$$\Delta T = 80.6°\text{C} - 63.3°\text{C} = 17.3°\text{C}$$

From Equation [1] above,

$$17.3°\text{C} = K_{fp}m$$

$$m = \frac{17.3°\text{C}}{K_{fp}}$$

We know that K_{fp} for naphthalene is 6.9°C/m. Therefore, the molality of this solution is

$$m = \frac{17.3°\text{C}}{6.9°\text{C}/m} = 2.5\ m$$

Remember that molality is the number of moles of solute per 1000 g of solvent. In our solution there are 15 g urea in 100 g of naphthalene, or 150 g of urea in 1000 g of naphthalene. Thus

$$150 \text{ g} = 2.5 \text{ mol}$$

$$1 \text{ mol} = 60 \text{ g}$$

The molecular weight of urea is, therefore, 60 g/mol

In this experiment you will determine the molecular weight of either sulfur or an unknown. You will do this by determining the freezing-point depression of a naphthalene solution having a known concentration of either sulfur or your unknown. The freezing temperature is difficult to ascertain by direct visual ob-

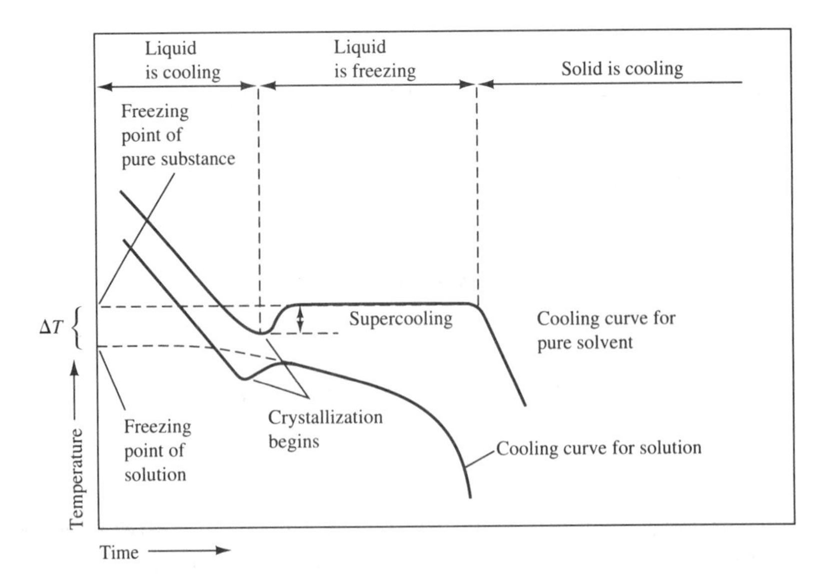

Figure 18.2 Cooling curves for a solvent and a solution.

servation because of a phenomenon called supercooling and also because solidification of solutions usually occurs over a broad temperature range. Temperature-time graphs, called *cooling curves,* reveal freezing temperatures rather clearly. Therefore, you will study the rate at which liquid naphthalene and its solutions cool and will construct a cooling curve similar to the one shown in Figure 18.2.

You will construct cooling curves for both the pure solvent and the solution. Figure 18.2 shows how the freezing point of a solution must be determined by extrapolation of the cooling curve. Extrapolation is necessary because as the solution freezes the solid that is formed is essentially pure solvent and the remaining solution becomes more and more concentrated. Thus its freezing point lowers continuously. Clearly, supercooling produces an ambiguity in the freezing point and should be minimized. Stirring the solution helps to minimize supercooling.

PROCEDURE

A. Cooling Curve for Pure Naphthalene

Weigh a large test tube to the nearest 0.01 g. Add about 15 g of naphthalene and weigh again. The difference in weight is the weight of naphthalene. Assemble the apparatus as shown in Figure 18.3; be certain to use a split two-hole rubber stopper. Carefully insert the thermometer into the hole that has been slit. Bend the stirrer so that the loop encircles the thermometer.

Fill your 600-mL beaker nearly full of water and heat it to about 85°C. Clamp the test tube in the water bath as shown in Figure 18.3. When most of the naphthalene has melted, insert the stopper containing the thermometer and stirrer into the test tube; make certain that the thermometer is not resting on the bottom of or touching the sides of the test tube. When all of the naphthalene has melted, stop heating, remove the beaker of water, and dry the outside of the test tube with a cloth towel. Place the test tube in a wide-mouthed bottle that contains a piece of crumpled paper in the bottom to lessen the chance that impact of the test tube with the bottle will cause the bottle to break. The purpose of the wide-mouth bottle is to minimize drafts. Record temperature readings every 30 s while you are stirring. When the freezing

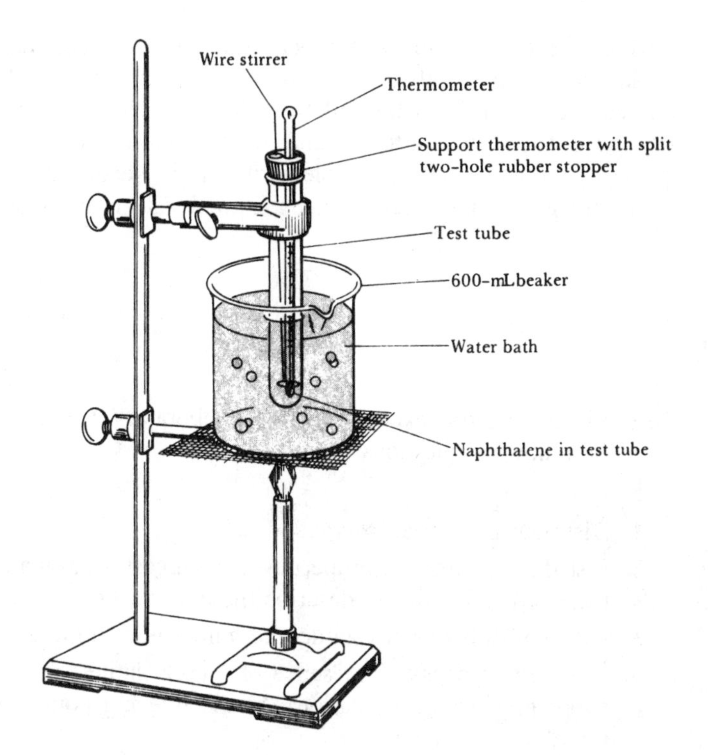

Figure 18.3 Apparatus for determination of cooling curve.

point is reached, crystals will start to form, and the temperature will remain constant. Shortly after this, the naphthalene will solidify to the point where you can no longer stir it.

Your lab instructor will direct you to perform either procedure B or procedure C.

B. Determinations of the Molecular Weight of Sulfur

Using weighing paper, weight to the nearest 0.01 g about 1.2 to 1.5 g of sulfur. CLEAN UP ANY SULFUR SPILLS IN THE BALANCE. Replace the test tube in the water bath and heat until all the naphthalene has melted. Gently remove the stopper, making sure that no naphthalene is lost, and add the sulfur to the test tube. Replace the stopper and stir gently until all the sulfur has dissolved. Remove the water bath, dry the test tube with a towel, and insert the test tube in a wide-mouth glass bottle containing a crumpled piece of paper. Record the temperature every 30 s until all the naphthalene as solidified.

Cleanup To clean out the test tube at the *end* of the experiment, heat the test tube in a water bath until the naphthalene just melts. *Care should be taken not to heat the thermometer beyond its temperature range. Be careful, because naphthalene is flammable.* Remove the stopper and pour the molten naphthalene on a crumpled wad of paper. When the naphthalene has solidified, throw both the paper and solid naphthalene into a waste receptacle. DO NOT POUR LIQUID NAPHTHALENE INTO THE SINK!

C. Determination of the Molecular Weight of an Unknown

Place the test tube in the water bath and heat until all the naphthalene has melted. Using weighing paper, weigh about 2 g of your unknown to the nearest

0.01 g. Gently remove the stopper from the test tube, making sure that no naphthalene is lost, and add your unknown to the test tube. Replace the stopper and stir gently until all the unknown has dissolved. Remove the water bath, dry the test tube with a towel, and insert the test tube in a wide-mouth glass bottle containing a crumpled piece of paper. Record the temperature every 30 s until all the naphthalene has solidified. Clean up as described in Part B above.

REVIEW QUESTIONS

Before beginning this experiment in the laboratory, you should be able to answer the following questions:

1. Distinguish between *solute* and *solvent*.
2. List three colligative properties and suggest a rationale for the choice of the word *colligative* to describe these properties.
3. Distinguish between volatile and nonvolatile substances.
4. What effect does the presence of a nonvolatile solute have upon (a) the vapor pressure of a solution, (b) the freezing point, and (c) the boiling point?
5. What is the molality of a solution that contains 1.5 g urea (MW = 60) in 200 g of benzene?
6. What is supercooling? How can it be minimized?
7. Calculate the freezing point of a solution containing 6.50 g of benzene in 160 g of chloroform.
8. A solution containing 1.00 g of an unknown substance in 12.5 g of naphthalene was found to freeze at 75.4°C. What is the molecular weight of the unknown substance?
9. How many grams of $NaNO_3$ would you add to 250 g of H_2O in order to prepare a solution that is 0.200 molal in $NaNO_3$?
10. Define *molality* and *molarity*.

Name ______________________________ Desk __________

Date ______________ Laboratory Instructor ______________________

REPORT SHEET FOR EXPERIMENT 18

COLLIGATIVE PROPERTIES: FREEZING-POINT DEPRESSION AND MOLECULAR WEIGHT

1. Weight of test tube + naphthalene __________ g
2. Weight of test tube __________ g
3. Weight of naphthalene __________ g
4. Weight of paper + sulfur or unknown __________ g
5. Weight of paper __________ g
6. Weight of sulfur or unknown __________ g

Cooling-curve data

Pure naphthalene		**Naphthalene + sulfur or unknown**	
Temp.	**Time**	**Temp.**	**Time**
________	________	________	________
________	________	________	________
________	________	________	________
________	________	________	________
________	________	________	________
________	________	________	________
________	________	________	________
________	________	________	________
________	________	________	________
________	________	________	________

7. Freezing point of pure naphthalene, from cooling curve __________
8. Freezing point of solution of sulfur or unknown in naphthalene __________ ΔT __________
9. Molality of sulfur or unknown (show calculations) __________

10. Molecular weight of sulfur or unknown (show calculations) ___________

HAND IN YOUR COOLING CURVES WITH YOUR REPORT SHEET.

QUESTIONS

1. What are the major sources of error in this experiment?

2. Suppose your thermometer consistently read a temperature 1.2° lower than the correct temperature throughout the experiment. How would this have affected the molecular weight you found?

3. If the freezing point of the solution had been incorrectly read 0.3° lower than the true freezing point, would the calculated molecular weight of the solute be too high or too low? Explain your answer.

4. Arrange the following aqueous solutions in order of increasing freezing points (lowest to highest temperature): 0.10 *m* glucose, 0.10 *m* $BaCl_2$, 0.20 *m* NaCl, and 0.20 *m* Na_2SO_4.

5. What mass of NaCl is dissolved in 150 g of water in a 0.050 *m* solution?

6. Calculate the molalities of some commercial reagents from the following data:

	HCl	**$HC_2H_3O_2$**	**$NH_3(aq)$**
Molecular weight	36.465	60.05	17.03
Density of solution (g/mL)	1.19	1.05	0.90
Weight %	37.2	99.8	28.0
Molarity	12.1	17.4	14.8

COOLING CURVE FOR PURE NAPHTHALENE

COOLING CURVE FOR SOLUTION OF SULFUR OR UNKNOWN IN NAPHTHALENE

NOTES AND CALCULATIONS

EXPERIMENT 19

Titration of Acids and Bases

OBJECTIVE

To become familiar with the techniques of titration, a volumetric method of analysis; to determine the amount of acid in an unknown.

APPARATUS AND CHEMICALS

500-mL Erlenmeyer flask	pint bottle with rubber stopper
50-mL buret	phenolphthalein solution
buret clamp	potassium acid phthalate (primary standard)
ring stand	250-mL Erlenmeyer flasks (3)
19 *M* NaOH	weighing bottle
balance	unknown acid
wash bottle	
600-mL beaker	

DISCUSSION

One of the most common and familiar reactions in chemistry is the reaction of an acid with a base. This reaction is termed *neutralization,* and the essential feature of this process in aqueous solution is the combination of hydronium ions with hydroxide ions to form water:

$$H_3O^+(aq) + OH^-(aq) \longrightarrow 2H_2O(l)$$

In this experiment you will use this reaction to determine accurately the concentration of a sodium hydroxide solution that you have prepared. The process of determining the concentration of a solution is called *standardization*. Next you will measure the amount of acid in an unknown. To do this, you will accurately measure with a buret the volume of your standard base that is required to exactly neutralize the acid present in the unknown. The technique of accurately measuring the volume of a solution required to react with another reagent is termed *titration*.

An indicator solution is used to determine when an acid has exactly neutralized a base, or vice versa. A suitable indicator changes colors when equivalent amounts of acid and base are present. The color change is termed the *end point* of the titration. Indicators change colors at different pH values. Phenolphthalein, for example, changes color from colorless to pink at a pH of about 9; in slightly more acidic solutions it is colorless, whereas, in more alkaline solutions it is pink. (If you have not done Experiment 17, read its discussion of indicators.)

By definition, one *gram-equivalent weight of an acid* is that mass of the acid in grams that will provide 1 mol of protons (H^+) in a reaction. A *gram-*

equivalent weight of a base is defined as that mass of the base in grams that will provide 1 mol of hydroxide ions in a reaction or that will react with 1 mol of protons. A gram-equivalent weight is often referred to as an *equivalent* (equiv). In acid-base reactions, 1 equiv of an acid will react with 1 equiv of a base. Thus 1 mol of HCl, which is 36.5 g, is 1 equivalent weight of HCl. However, 1 mol of H_2SO_4 reacts with 2 mol of NaOH, and 2 mol of protons are transferred from the acid to the base:

$$H_2SO_4(aq) + 2NaOH(aq) \longrightarrow Na_2SO_4(aq) + 2H_2O(l)$$

Therefore, 1 mol of H_2SO_4 (98.0 g) corresponds to 2 gram-equivalent weights. In other words, 1 gram-equivalent weight of H_2SO_4, or 49.0 g, is equal to $\frac{1}{2}$ mol of the acid (2 equiv H_2SO_4/1 mol H_2SO_4).

For volumetric work, normality is the most convenient method of expressing concentrations. *Normality, N*, is defined as the number of equivalents of solute in a liter of solution (which is also the number of milliequivalents of solute in a milliliter of solution). A useful mathematical relationship for normality is

$$N = \frac{\text{equivalents of solute}}{\text{liter of solution}} = \frac{\text{equiv}}{V_L} = \frac{m}{\text{eqw} \times V_L} \qquad [1]$$

where m is the mass of the solute in grams, eqw is the equivalent weight (grams per equivalent), and V_L is the volume of the solution in liters.

Because equivalent weight and molecular weight are related, normality and molarity are related. In general,

$$\text{Equivalent weight} = \frac{\text{molecular weight}}{a}$$

where a is the number of moles of acidic hydrogen (H^+) per mole of acid that react, or the moles of OH^- per mole of base that react. Thus the normality of a solution and its molarity are related:

$$N = a \times M$$

Normality is always equal to or greater than molarity.

It should be obvious that a 1 *M* HCl solution is also 1 *N*, because 1 gram-equivalent weight of HCl is exactly 1 mol of HCl. However, a 1 *M* H_2SO_4 solution is 2 *N* because 1 mol of H_2SO_4 is equal to 2 equiv of H_2SO_4.

EXAMPLE 19.1

What is the normality of a solution that contains 2.45 g of H_2SO_4 in 0.250 L of solution?

Solution:

$$N = \frac{2.45\ \text{g}}{(49\ \text{g/equiv})(0.250\ \text{L})} = 0.20\ N$$

*Time may be saved in this experiment if the students are provided with an approximate 0.1 1*N* NaOH solution.

In this experiment your solution of NaOH* will be standardized by titrating it against a very pure sample of potassium hydrogen phthalate, $KHC_8H_4O_4$, of known weight. Potassium hydrogen phthalate (hereafter abbreviated as KHP) has only one replaceable acid hydrogen. Its structure is shown below. It is a monoprotic acid with the acidic hydrogen bonded to oxygen and has a molecular weight of 204.2; hence 1 equiv of KHP weighs 204.2 g.

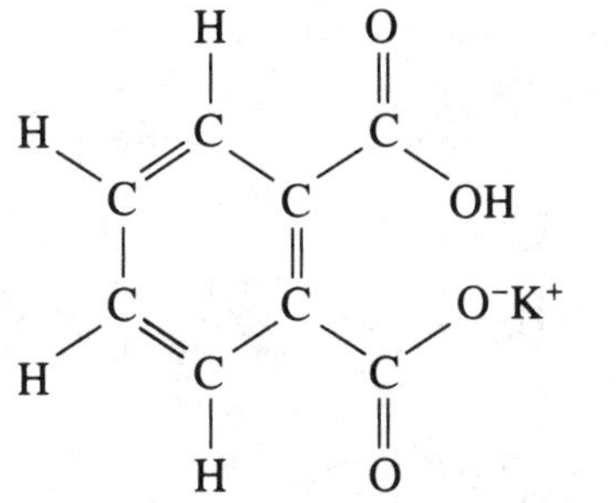

KHP

In the titration of a base against KHP, an equal number of equivalents of base and acid are present at the *equivalence point*. In other words, at the equivalence point

$$\text{Equivalents NaOH} = \text{equivalents KHP} \qquad [2]$$

Equation [2] can be expressed in terms of Equation [1] by rearranging the equation as follows:

$$N \times V_L = \frac{m}{\text{eqw}} = \text{number of equivalents}$$

so that

$$N_{\text{base}} \times V_{\text{base}} = \frac{m \text{ of KHP}}{\text{eqw of KHP}} \qquad [3]$$

If one measures the volume (in liters) of base required to neutralize a known weight (in grams) of KHP, it is possible to calculate the normality (N) of the base because the equivalent weight of KHP is known.

Because volumes are measured in milliliters, a more convenient form of Equation [3] is

$$N \times \text{mL} = \frac{m(10^3 \text{ mL/L})}{\text{eqw}} = \text{milliequivalents (meq)} \qquad [4]$$

EXAMPLE 19.2

What is the normality of an NaOH solution if 35.75 mL of it is required to neutralize 1.070 g KHP?

Solution:

$$N \times 35.75 \text{ mL} = \frac{(1.070\text{g})(10^3\text{mL/L})}{(204.2 \text{ g/equiv})}$$

$$N = \frac{(1.070 \text{ g})(10^3 \text{ mL/L})}{(35.75 \text{ mL})(204.2 \text{ g/equiv})}$$

$$= 0.1466\ N \text{ NaOH}$$

Once the normality of the NaOH solution is accurately known, the base can be used to determine the amount of KHP or any other acid present in a known weight of an impure sample. The percentage of KHP in an impure sample is

$$\% \text{ KHP} = \frac{\text{g KHP}}{\text{weight of sample}} \times 100$$

In this experiment we use an acid-base indicator, phenolphthalien, to signal the *end point* in the titration. The end point is the point at which the indicator changes color. We choose an indicator such that its color change occurs as closely as possible to the *equivalence point*. See experiment 17 for more discussion of indicators. Recall that the equivalence point is the point at which the number of equivalents of reactants are equal.

EXAMPLE 19.3

What is the percentage of KHP in a sample if 2.537 g requires 32.77 mL of 0.1466 N NaOH to neutralize it?

Solution: The actual number of grams of KHP in the 2.537-g sample must first be determined. Remember that at the equivalence point the number of equivalents of base equals the number of equivalents of KHP present. Let the left-hand member of Equation [4] refer to the base, and the right-hand member to the KHP:

$$\text{meq of base} = \text{meq of acid}$$

$$0.1466\ N \times 32.77 \text{ mL} = \frac{\text{grams} \times 10^3 \text{ mL/L}}{204.2 \text{ g/equiv}}$$

Solving for grams,

$$\text{Grams} = \frac{0.1466\ N \times 32.77 \text{ mL} \times 204.2 \text{ g/equiv}}{10^3 \text{ mL/L}}$$

$$= 0.9810 \text{ g}$$

Thus

$$\% \text{ KHP} = \frac{0.9810 \text{ g}}{2.537 \text{ g}} \times 100 = 38.67\%$$

PROCEDURE

Preparation of Approximately 0.100 N Sodium Hydroxide Heat 500 mL of distilled water to boiling in a 600-mL flask,* *and after cooling under the*

* the water is boiled to remove carbon dioxide, which would react with the sodium hydroxide and change its normality. (Recall the results of Experiment 17: $H_2O + CO_2 \longrightarrow$?)

water tap, transfer to a 1-pt bottle fitted with a rubber stopper.[†] Add 3 mL of stock solution of carbonate-free sodium hydroxide (approximately 19 *M*) and shake vigorously for at least 1 min.

Preparation of a Buret for Use Clean a 50-mL buret with soap solution and a buret brush and thoroughly rinse with tap water. Then rinse with at least five 10-mL portions of distilled water. The water must run freely from the buret without leaving any drops adhering to the sides. Make sure that the buret does not leak and that the stopcock turns freely.

Reading a Buret All liquids, when placed in a buret, form a curved meniscus at their upper surfaces. In the case of water or water solutions, this meniscus is concave (see Figure 19.1), and the most accurate buret readings are obtained by observing the position of the lowest point on the meniscus on the graduated scales.

To avoid parallax errors when taking readings, the eye must be on a level with the meniscus. Wrap a strip of paper around the buret and hold the top edges of the strip evenly together. Adjust the strip so that the front and back edges are in line with the lowest part of the meniscus and take the reading by estimating to the nearest tenth of a marked division (0.01 mL). A simple way of doing this for repeated readings on a buret is illustrated in Figure 19.1.

A. Standardization of Sodium Hydroxide Solution

Prepare about 400–450 mL of CO_2-free water by boiling for about 5 min. Weigh from a weighing bottle (your lab instructor will show you how to use a weighing bottle if you don't already know) triplicate samples of between 0.4 and 0.6 g each of pure potassium acid phthalate into 250-mL Erlenmeyer flasks; accurately weigh to four significant figures.* Do not weigh the flasks. Record the weights and label the three flasks in order to distinguish among them. Add to each sample about 100 mL of distilled water that has been freed

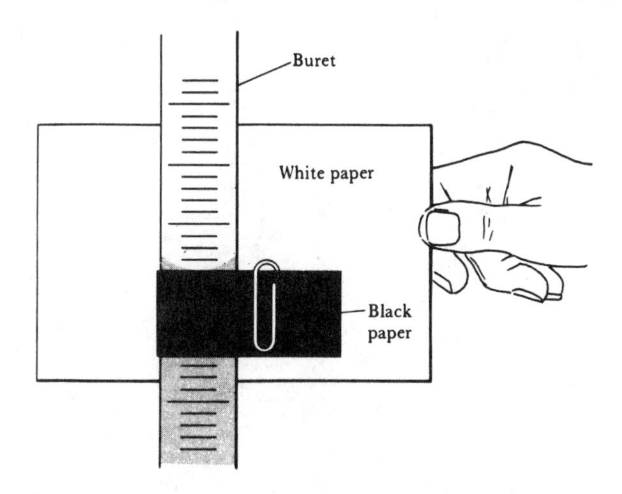

Figure 19.1 Reading a buret.

[†] A rubber stopper should be used for a bottle containing sodium hydroxide solution. A strongly alkaline solution tends to cement a glass stopper so firmly that it is difficult to remove.

* In cases where the weight of a sample is larger than 1 g, it is necessary to weigh only to the nearest milligram to obtain four significant figures. Buret readings can be read only to the nearest 0.01 mL, and for readings greater than 10 mL this represents four significant figures.

from carbon dioxide by boiling, and warm gently with swirling until the salt is completely dissolved. Add to each flask two drops of phenolphthalein indicator solution.

Rinse the previously cleaned buret with at least four 5-mL portions of the approximately 0.100 N sodium hydroxide solution that you have prepared. Discard each portion. *Do not return any of the washings to the bottle*. Completely fill the buret with the solution and remove the air from the tip by running out some of the liquid into an empty beaker. Make sure that the lower part of the meniscus is at the zero mark or slightly lower. Allow the buret to stand for at least 30 s before reading the exact position of the meniscus. Remove any hanging drop from the buret tip by touching it to the side of the beaker used for the washings. Record the initial buret reading.

Slowly add the sodium hydroxide solution to one of your flasks of potassium hydrogen phthalate solution while gently swirling the contents of the flask, as illustrated in Figure 19.2. As the sodium hydroxide solution is added, a pink color appears where the drops of the base come in contact with the solution. This coloration disappears with swirling. As the end point is approached, the color disappears more slowly, at which time the sodium hydroxide should be added drop by drop. It is most important that the flask be swirled constantly throughout the entire titration. The end point is reached when one drop of the sodium hydroxide solution turns the entire solution in the flask from colorless to pink. The solution should remain pink when it is swirled. Allow the titrated solution to stand for at least 1 min so the buret will drain properly. Remove any hanging drop from the buret tip by touching it to the side of the flask and wash down the sides of the flask with a stream of water from the wash bottle. Record the buret reading. Repeat this procedure with the other two samples.

From the data you obtain in the three titrations, calculate the normality of the sodium hydroxide solution to four significant figures as in Example 19.2.

The three determinations should agree within 1.0 percent. If they do not, the standardization should be repeated until agreement is reached. The average

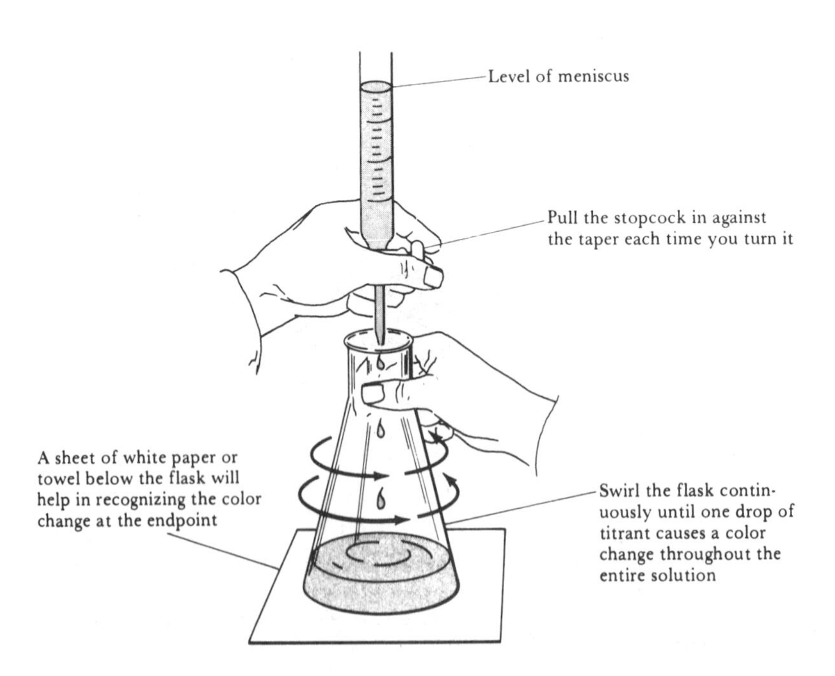

Figure 19.2 Titration procedure.

of the three acceptable determinations is taken as the normality of the sodium hydroxide. Calculate the standard deviation of your results. (Standard deviation is discussed in Experiment 8.) *Save your standardized solution for the unknown determination.*

B. Analysis of an Unknown Acid

Calculate the approximate weight of unknown that should be taken to require about 20 mL of your standardized sodium hydroxide, assuming that your unknown sample is 75 percent KHP.

Weigh by difference from a weighing bottle triplicate portions of the sample to four significant figures and place them in three separate 250-mL flasks. The sample size should be about the amount determined by the above computation. Dissolve the sample in 100 mL of CO_2-free distilled water (prepared by boiling) and add two drops of phenolphthalein indicator solution. Titrate with your standard sodium hydroxide solution to the faintest visible shade of pink (not red) as described above in the standardization procedure. Calculate the percentage of potassium hydrogen phthalate, $KHC_8H_4O_4$, in the samples as in Example 19.3. For good results the three determinations should agree within 1.0 percent. Your answers should have four significant figures. Compute the standard deviation of your results.

Test your results by computing the average deviation from the mean. If one result is noticeably different from the others, perform an additional titration. If any result is more than two standard deviations away from the mean, discard it and titrate another sample.

REVIEW QUESTIONS

Before beginning this experiment in the laboratory, you should be able to answer the following questions:

1. Define standardization and state how you would go about doing it.
2. Define titration.
3. Define normality and molarity and distinguish between them.
4. Why do you weigh by difference?
5. What are equivalence points and end points and how do they differ?
6. What is parallax and why should you avoid it?
7. Why is it necessary to rid the distilled water of CO_2?
8. What is the normality of a solution that contains 1.89 g $H_2C_2O_4 \cdot 2H_2O$ in 100 mL of solution?
9. If 50.0 mL of NaOH solution is required to react completely with 1.24 g KHP, what are the molarity and normality of the NaOH solution?
10. In the titration of an impure sample of KHP, it was found that 29.4 mL of 0.100 *M* NaOH was required to react completely with 0.745 g of sample. What is the percentage of KHP in this sample?

NOTES AND CALCULATIONS

Reactions in Aqueous Solution: Metathesis Reactions and Net Ionic Equations

EXPERIMENT 20

OBJECTIVE

To become familiar with writing equations for metathesis reactions, including net ionic equations.

APPARATUS AND CHEMICALS

small test tubes (12)
evaporating dish
100-mL beaker (2)
600-mL beaker
thermometer
ring stand and ring
Bunsen burner and hose
funnel (2)
filter paper
magnifying glass
sodium nitrate
potassium chloride
0.1 *M* sodium acetate
0.1 *M* lead nitrate
0.1 *M* potassium chloride
short-stem funnel
funnel support
0.1 *M* copper (II) sulfate
0.2 *M* sodium carbonate
0.1 *M* barium chloride
0.1 *M* trisodium phosphate
1.0 *M* sulfuric acid
1.0 *M* hydrochloric acid
0.1 *M* silver nitrate
0.1 *M* nickel chloride
0.1 *M* cadmium chloride
0.1 *M* sodium sulfide
0.1 *M* sodium hydroxide
0.1 *M* ammonium chloride
0.1 *M* sodium nitrate
ice

ALL SOLUTIONS SHOULD BE PROVIDED IN DROPPER BOTTLES

DISCUSSION

In Experiment 4 we briefly encountered metathesis, or double decomposition, reactions. We will now examine these reactions in more detail, recognizing the ionic character of the species in solution. You may recall that metathesis reactions have the general form

$$AB + CD \longrightarrow AD + CB \qquad [1]$$

This kind of reaction is fairly common, especially in aqueous solution, where the cations and anions of the substances involved exchange partners. The reaction of barium chloride with silver nitrate is a typical example:

$$BaCl_2(aq) + 2AgNO_3(aq) \longrightarrow Ba(NO_3)_2(aq) + 2AgCl(s) \qquad [2]$$

This form of the equation for this reaction is referred to as the *molecular equa-*

tion. Because we know that the salts $BaCl_2$, $AgNO_3$, and $Ba(NO_3)_2$ are strong electrolytes and are completely dissociated in solution, we can more realistically write the equation as follows:

$$Ba^{2+}(aq) + 2Cl^-(aq) + 2Ag^+(aq) + 2NO_3^-(aq) \longrightarrow Ba^{2+}(aq) + 2NO_3^-(aq) + 2AgCl(s) \quad [3]$$

This form is known as the *ionic equation*. Reaction [2] occurs because the insoluble substance AgCl precipitates out of solution. The other product, barium nitrate, is soluble in water and remains in solution. We see that Ba^{2+} and NO_3^- ions appear on both sides of the equation and thus do not enter into the reaction. Such ions are called *spectator ions*. If we eliminate or omit them from both sides, we obtain the *net ionic equation*

$$Ag^+(aq) + Cl^-(aq) \longrightarrow AgCl(s) \quad [4]$$

This equation focuses our attention on the salient feature of the reaction: the formation of the precipitate AgCl. It tells us that solutions of any soluble Ag^+ salt and any soluble Cl^- salt, when mixed, will form insoluble AgCl. When writing net ionic equations, remember that only *strong electrolytes* are written in the ionic form. Solids, gases, nonelectrolytes, and weak electrolytes are written in the molecular form. Frequently the symbol (aq) is omitted from ionic equations. The symbols (g) for gas and (s) for solid should not be omitted. Thus equation [4] can be written

$$Ag^+ + Cl^- \longrightarrow AgCl(s) \quad [5]$$

Consider mixing solutions of KCl and $NaNO_3$. The ionic equation for the reaction is

$$K^+(aq) + Cl^-(aq) + Na^+(aq) + NO_3^-(aq) \longrightarrow K^+(aq) + NO_3^-(aq) + Na^+(aq) + Cl^-(aq) \quad [6]$$

Because all the compounds are water-soluble and are strong electrolytes, they have been written in the ionic form. They completely dissolve in water. If we eliminate spectator ions from the equation, nothing remains. Hence, there is no reaction:

$$K^+(aq) + Cl^-(aq) + Na^+(aq) + NO_3^-(aq) \longrightarrow \text{no reaction} \quad [7]$$

Metathesis reactions occur when a precipitate, a gas, a weak electrolyte, or a nonelectrolyte is formed. The following equations are further illustrations of such processes.

Formation of a Gas

Molecular equation: $2HCl(aq) + Na_2S(aq) \longrightarrow 2NaCl(aq) + H_2S(g)$

Ionic equation: $2H^+(aq) + 2Cl^-(aq) + 2Na^+(aq) + S^{2-}(aq) \longrightarrow 2Na^+(aq) + 2Cl^-(aq) + H_2S(g)$

Net ionic equation: $2H^+(aq) + S^{2-}(aq) \longrightarrow H_2S(g)$

or

$$2H^+ + S^{2-} \longrightarrow H_2S(g)$$

Formation of Weak Electrolyte

Molecular equation: $HNO_3(aq) + NaOH(aq) \longrightarrow H_2O(l) + NaNO_3(aq)$

Ionic equation: $H^+(aq) + NO_3^-(aq) + Na^+(aq) + OH^-(aq) \longrightarrow H_2O(l) + Na^+(aq) + NO_3^-(aq)$

Net ionic equation: $H^+(aq) + OH^-(aq) \longrightarrow H_2O(l)$

In order to decide if a reaction occurs, we need to be able to determine whether or not a precipitate, a gas, a nonelectrolyte or a weak electrolyte will be formed. The following brief discussion is intended to aid you in this regard. Table 20.1 summarizes solubility rules and should be consulted while performing this experiment.

The common gases are CO_2, SO_2, H_2S, and NH_3. Carbon dioxide and sulfur dioxide may be regarded as resulting from the decomposition of their corresponding weak acids, which are initially formed when carbonate and sulfite salts are treated with acid:

$$H_2CO_3(aq) \longrightarrow H_2O(l) + CO_2(g)$$

and

$$H_2SO_3(aq) \longrightarrow H_2O(l) + SO_2(g)$$

Table 20.1 Solubility Rules

Water-soluble salts	
Na^+, K^+, NH_4^+	All sodium, potassium, and ammonium salts are soluble
NO_3^-, ClO_3^-, $C_2H_3O_2^-$	All nitrates, chlorates, and acetates are soluble.
Cl^-	All chlorides are soluble except AgCl, Hg_2Cl_2, and $PbCl_2$.[a]
Br^-	All bromides are soluble except AgBr, Hg_2Br_2, $PbBr_2$,[a] and $HgBr_2$.[a]
I^-	All iodides are soluble except AgI, Hg_2I_2, PbI_2, and HgI_2.
SO_4^{2-}	All sulfates are soluble except $CaSO_4$,[a] $SrSO_4$, $BaSO_4$, Hg_2SO_4, $PbSO_4$, and Ag_2SO_4.
Water-insoluble salts	
CO_3^{2-}, SO_3^{2-}, PO_4^{3-}, CrO_4^{2-}	All carbonates, sulfites, phosphates, and chromates are insoluble except those of alkali metals and NH_4^+.
OH^-	All hydroxides are insoluble except those of alkali metals and $Ca(OH)_2$,[a] $Sr(OH)_2$,[a] and $Ba(OH)_2$.
S^{2-}	All sulfides are insoluble except those of the alkali metals, alkaline earths, and NH_4^+

[a] Slightly soluble.

Ammonium salts form NH_3 when they are treated with strong bases:

$$NH_4^+(aq) + OH^- \longrightarrow NH_3(g) + H_2O(l)$$

Which are the weak electrolytes? The easiest way of answering this question is to identify all of the strong electrolytes, and if the substance does not fall in that category it is a weak electrolyte. Note, water is a nonelectrolyte. Strong electrolytes are summarized in Table 20.2.

Table 20.2 Strong Electrolytes

Salts	All common soluble salts.
Acids	$HClO_4$, HCl, HBr, HI, HNO_3, and H_2SO_4 are strong electrolytes. All others are weak.
Bases	Alkali metal hydroxides, $Ca(OH)_2$, $Sr(OH)_2$, and $Ba(OH)_2$ are strong electrolytes. All others are weak.

In the first part of this experiment, you will study some metathesis reactions. In some instances it will be very evident that a reaction has occurred, while in others it will not be so apparent. In the doubtful case, use the guidelines above to decide whether or not a reaction has taken place. You will be given the names of the compounds to use but not their formulas. This is being done deliberately to give practice in writing formulas from names.

In the second part of this experiment, you will study the effect of temperature on solubility. The effect that temperature has on solubility varies from salt to salt. We conclude that mixing solutions of KCl and $NaNO_3$ resulted in no reaction (see Equation [6] and [7]). What would happen if we cooled such a mixture? The solution would eventually become saturated with respect to one of the salts, and crystals of that salt would begin to appear as its solubility was exceeded. Examination of Equation [6] reveals that crystals of any of the following salts could appear initially: KNO_3, KCl, $NaNO_3$, or NaCl.

Consequently, if a solution containing Na^+, K^+, Cl^-, and NO_3^- ions is evaporated at a given temperature, the solution becomes more and more concentrated and will eventually become saturated with respect to one of the four compounds. If evaporation is continued, that compound will crystallize out, removing its ions from solution. The other ions will remain in solution and increase in concentration. Before coming to lab you are to plot the solubilities of the four salts given in Table 20.3 on the graph on your report sheet.

Table 20.3 Molar Solubilities of NaCl, $NaNO_3$, KCl, and KNO_3 (mol/L)

Compound	0°C	20°C	40°C	60°C	80°C	100°C
NaCl	5.4	5.4	5.5	5.5	5.5	5.6
$NaNO_3$	6.7	7.6	8.5	9.4	10.4	11.3
KCl	3.4	4.0	4.6	5.1	5.5	5.8
KNO_3	1.3	3.2	5.2	7.0	9.0	11.0

PROCEDURE

A. Metathesis Reactions

The report sheet lists 16 pairs of chemicals that are to be mixed. Use about 1 mL of 0.1 *M* solutions of each of the reagents (use 1.0 *M* hydrochloric acid and 1.0 *M* sulfuric acid) to be combined as indicated on the report sheet. Mix the solutions in small test tubes and record your observations on the report sheet. If there is no reaction, write N.R. (The reactions need not be carried out in the order listed. Congestion at the reagent shelf can be avoided if everyone does not start with reagents for reaction 1.)

B. Solubility, Temperature, and Crystallization

Place 8.5 g of sodium nitrate and 7.5 g of potassium chloride in a 100-mL beaker and add 25 mL of water. Warm the mixture, stirring, until the solids completely dissolve. Assuming a volume of 25 mL for the solution, calculate the molarity of the solution with respect to $NaNO_3$, KCl, NaCl, and KNO_3, and record these molarities (1).

Cool the solution to about 10°C by dipping the beaker in ice water in a 600-mL beaker and stirring the solution carefully with a thermometer, being careful not to break it. (SHOULD THE THERMOMETER BREAK, IMMEDIATELY CONSULT YOUR INSTRUCTOR.) When no more crystals form, at approximately 10°C, filter the cold solution quickly and allow the filtrate to drain thoroughly into an evaporating dish. Dry the crystals between two dry pieces of filter paper. Examine the crystals with a magnifying glass (or fill a Florence flask with water and look at the crystals through it.) Describe the shape of the crystals, that is, needles, cubes, plates, rhombs, and so forth (2). Based upon your solubility graph, which compound crystallized out of solution (3)?

Evaporate the filtrate to about half of its volume using a Bunsen burner and ring stand. A second crop of crystals should form. Record the temperature (4) and rapidly filter the hot solution, collecting the filtrate in a clean 100-mL beaker. Dry the second batch of crystals between two pieces of filter paper and examine their shape. Compare their shape with the first batch of crystals (5). Based upon your solubility graph, what is this substance (6)?

Finally, cool the filtrate to 10° while stirring carefully with a thermometer to obtain a third crop of crystals. Carefully observe their shapes and compare them with those of the first and second batches (7). What compound is the third batch of crystals (8)?

REVIEW QUESTIONS

Before beginning this experiment in the laboratory, you should be able to answer the following questions:

1. Write molecular, ionic, and net ionic equations for the reactions that occur, if any, when solutions of the following substances are mixed:
  (a) nitric acid and barium carbonate;
  (b) zinc chloride and lead nitrate;
  (c) acetic acid and sodium hydroxide;

(d) calcium nitrate and sodium carbonate;

(e) ammonium chloride and potassium hydroxide.

2. Which of the following are not water-soluble:
$Ba(NO_3)_2$, $FeCl_3$, $CuCO_3$, $CuSO_4$, ZnS, $ZnSO_4$?

3. Write equations for the decomposition of H_2CO_3 and H_2SO_3.

4. At what temperature (from your graph) do KNO_3 and NaCl have the same molar solubility?

5. Which of the following are strong electrolytes: $BaCl_2$, $AgNO_3$, HCl, HNO_3, $HC_2H_3O_2$?

6. Which of the following are weak electrolytes: HNO_3, HF, HCl, $NH_3(aq)$, NaOH?

7. For each of the following water-soluble compounds, indicate the ions formed in an aqueous solution: NaI, K_2SO_4, NaCN, $Ba(OH)_2$, $(NH_4)_2SO_4$.

8. Write a balanced chemical equation showing how you could prepare each of the following salts from an acid-base reaction: $NaNO_3$, KCl, $BaSO_4$.

Name ______________________________ Desk ____________

Date ______________ Laboratory Instructor ______________________________

REPORT SHEET FOR EXPERIMENT 20

REACTIONS IN AQUEOUS SOLUTIONS: METATHESIS REACTIONS AND NET IONIC EQUATIONS

A. Metathesis Reactions

1. Copper (II) sulfate + sodium carbonate
Observations ______________________________
Molecular equation ______________________________
Ionic equation ______________________________
Net ionic equation ______________________________

2. Copper(II) sulfate + barium chloride
Observations ______________________________
Molecular equation ______________________________
Ionic equation ______________________________
Net ionic equation ______________________________

3. Copper(II) sulfate + trisodium phosphate
Observations ______________________________
Molecular equation ______________________________
Ionic equation ______________________________
Net ionic equation ______________________________

4. Sodium carbonate + sulfuric acid
Observations ______________________________
Molecular equation ______________________________
Ionic equation ______________________________
Net ionic equation ______________________________

5. Sodium carbonate + hydrochloric acid
Observations ______________________________
Molecular equation ______________________________
Ionic equation ______________________________
Net ionic equation ______________________________

6. Cadmium chloride + sodium sulfide
Observations ______________________________
Molecular equation ______________________________

Ionic equation ______

Net ionic equation ______

7. Cadmium chloride + sodium hydroxide

Observations ______

Molecular equation ______

Ionic equation ______

Net ionic equation ______

8. Nickel chloride + silver nitrate

Observations ______

Molecular equation ______

Ionic equation ______

Net ionic equation ______

9. Nickel chloride + sodium carbonate

Observations ______

Molecular equation ______

Ionic equation ______

Net ionic equation ______

10. Hydrochloric acid + sodium hydroxide

Observations ______

Molecular equation ______

Ionic equation ______

Net ionic equation ______

11. Ammonium chloride + sodium hydroxide

Observations ______

Molecular equation ______

Ionic equation ______

Net ionic equation ______

12. Sodium acetate + hydrochloric acid

Observations ______

Molecular equation ______

Ionic equation ______

Net ionic equation ______

13. Sodium sulfide + hydrochloric acid

Observations ______

Molecular equation ______

Ionic equation ______

Net ionic equation ______

14. Lead nitrate + sodium sulfide

Observations ______

Molecular equation ______

Ionic equation ______

Net ionic equation ______

15. Lead nitrate + sulfuric acid

Observations ______________________________

Molecular equation ______________________________

Ionic equation ______________________________

Net ionic equation ______________________________

16. Potassium chloride + sodium nitrate

Observations ______________________________

Molecular equation ______________________________

Ionic equation ______________________________

Net ionic equation ______________________________

B. Solubility, Temperature and Crystallization

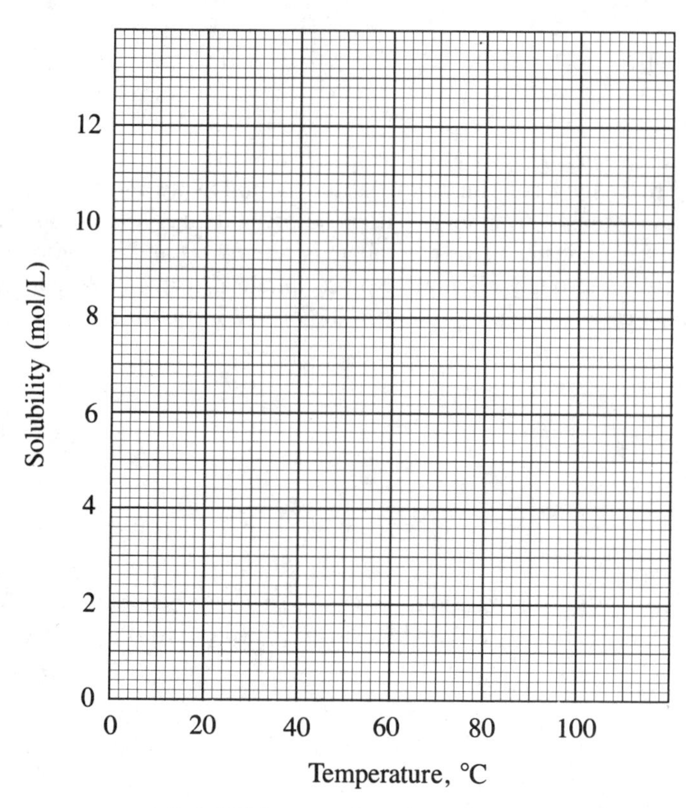

Solubilities as a function of temperature

1. Molarities

_______ M $NaNO_3$, _______ M KCl, _______ M NaCl, _______ M KNO_3

2. Crystal shape ______________________________

3. Identity of crystals ______________________________

4. Temperature of filtrate ______________________________

5. Crystal shape of second batch ______________________________

6. Identity of second batch of crystals ______________________

7. Crystal shape of third batch ______________________

8. Identity of third batch of crystals ______________________

QUESTIONS

1. Which of the following reactions are metathesis reactions?
   (a) $2KClO_3 \rightarrow 2KCl + 3O_2$
   (b) $Cu(NO_3)_2 + Zn \rightarrow Cu + Zn(NO_3)_2$
   (c) $BaCO_3 + 2HCl \rightarrow BaCl_2 + H_2O + CO_2$
   (d) $Na_2CO_3 + CuSO_4 \rightarrow Na_2SO_4 + CuCO_3$.

2. How many grams of each of the following substances will dissolve in 100 mL of cold water? Consult a handbook.

   $Ce(IO_3)_4$, $RaSO_4$, $Pb(NO_3)_2$, $(NH_4)_2SeO_4$

3. Suppose you have a solution that might contain any or all of the following cations: Cu^{2+}, Ag^+, Ba^{2+} and Mn^{2+}. Addition of HCl causes a precipitate to form. After the precipitate is filtered off, H_2SO_4 is added to the supernate and another precipitate forms. This precipitate is filtered off and a solution of NaOH is added to the supernatant liquid until it is strongly alkaline. No precipitate is formed. Which ions are present in each of the precipitates? Which cations are not present in the original solution?

4. Write balanced net ionic equations for the reactions, if any, that occur between (a) $FeS(s)$ and $HBr(aq)$; (b) $K_2CO_3(aq)$ and $CuCl_2(aq)$; (c) $Fe(NO_3)_2(aq)$ and $HCl(aq)$; (d) $Bi(OH)_3(s)$ and $HNO_3(aq)$.

Determination of the Dissociation Constant of a Weak Acid

EXPERIMENT 21

OBJECTIVE

To become familiar with the operation of a pH meter and quantitative equilibrium constants.

APPARATUS AND CHEMICALS

- pH meter with electrodes
- potassium acid phthalate (KHP)
- sodium hydroxide solution, about 0.1 *M*
- unknown solution of a weak acid (100 mL of approx. 0.1 *M* solution)
- weighing bottle
- 25-mL pipet
- buret, buret clamp, and ring stand
- 150-mL beaker
- 250-mL beakers (3)
- standard buffer solution
- phenolphthalein indicator solution

DISCUSSION

General Theory

According to the Brønsted-Lowry acid-base theory, the strength of an acid is related to its ability to donate protons. All acid-base reactions are then competitions between bases of various strengths for these protons. For example, the strong acid HCl reacts with water according to Equation [1]:

$$HCl(aq) + H_2O(l) \longrightarrow H_3O^+(aq) + Cl^-(aq) \qquad [1]$$

This acid is a strong acid and is completely dissociated—in other words, 100 percent dissociated—in dilute aqueous solution. Consequently, the $[H_3O^+]$ concentration of 0.1 *M* HCl is 0.1 *M*.

By contrast, acetic acid, $HC_2H_3O_2$ (abbreviated HOAc), is a weak acid and is only slightly dissociated, as shown in Equation [2]:

$$H_2O(l) + HOAc(aq) \rightleftharpoons H_3O^+(aq) + OAc^-(aq) \qquad [2]$$

Its acid dissociation constant, as shown by Equation [3], is therefore small:

$$K_a = \frac{[H_3O^+][OAc^-]}{[HOAc]} = 1.8 \times 10^{-5} \qquad [3]$$

Acetic acid only partially dissociates in aqueous solution, and an appreciable quantity of undissociated acetic acid remains in solution.

For the general weak acid HA the dissociation reaction and dissociation constant expression are

$$HA(aq) + H_2O(l) \rightleftharpoons H_3O^+(aq) + A^-(aq) \qquad [4]$$

$$K_a = \frac{[H_3O^+][A^-]}{[HA]} \qquad [5]$$

Recall that pH is defined as

$$-\log [H_3O^+] = pH \qquad [6]$$

Solving Equation [5] for $[H_3O^+]$ and substituting this quantity into Equation [6] yields

$$[H_3O^+] = K_a\frac{[HA]}{[A^-]} \qquad [7]$$

$$-\log [H_3O^+] = -\log K_a - \log \frac{[HA]}{[A^-]} \qquad [8]$$

$$pH = pK_a - \log \frac{[HA]}{[A^-]} \qquad [9]$$

$$\text{where } pK_a = -\log K_a$$

If we titrate the weak acid HA with a base, there will be a point in the titration at which the number of equivalents of base is just one-half the number of equivalents of acid present.* This is the point at which 50 percent of the acid has been titrated to produce A^- and 50 percent remains as HA. At this point $[HA] = [A^-]$, the ratio $[HA]/[A^-] = 1$, and $\log [HA]/[A^-] = 0$. Hence, at this point in a titration, that is, at one-half the equivalence point, Equation [9] becomes

$$pH = pK_a \qquad [10]$$

By titrating a weak acid with a strong base and recording the pH versus the volume of base added, we can determine the ionization constant of the weak acid. From the resultant titration curve we obtain the ionization constant, as explained in the following paragraph.

From the titration curve (Figure 21.1), we see that at the point denoted as half the equivalence point, where $[HA] = [A^-]$, the pH is 4.3. Thus, from Equation [10], at this point $pH = pK_a$, or

$$pK_a = 4.3$$

$$-\log K_a = 4.3$$

$$\log K_a = -4.3$$

$$K_a = 5.01 \times 10^{-5}$$

(Your instructor will show you a graphical method for locating the equivalence point on your titration curves.)

---

*It may be advantageous to review Experiment 19 for the terms *equivalents* and *equivalence point*.

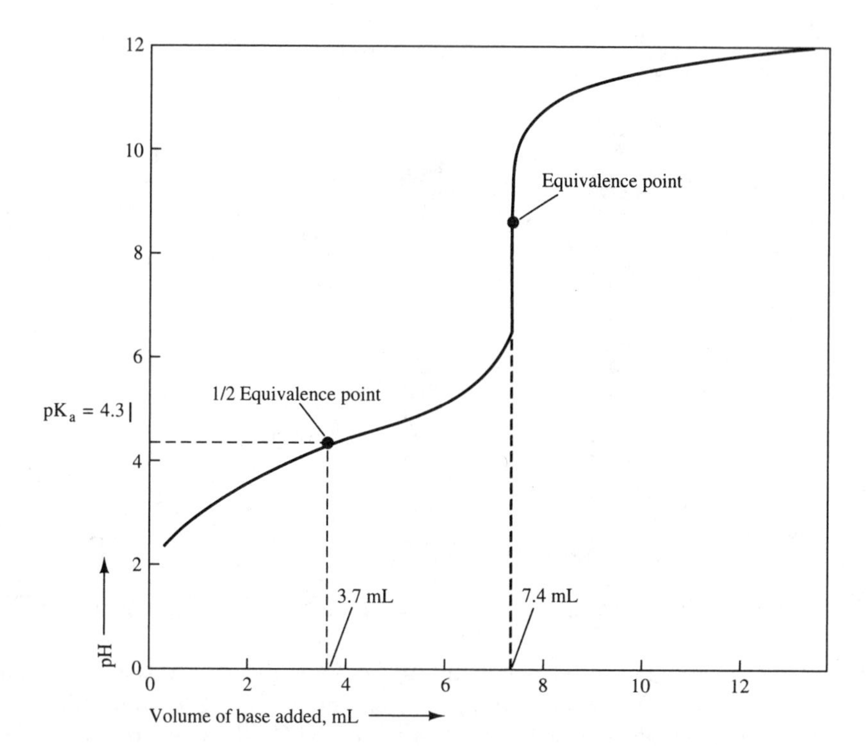

Figure 21.1 Exemplary titration curve for the titration of the weak acid HA with a base.

Operation of the pH Meter

In order to measure the pH during the course of the titration, we shall use an electronic instrument called a pH meter. This device consists of a meter and two electrodes, as illustrated in Figure 21.2.

The main variations among different pH meters involve the positions of the control knobs and the types of electrodes and electrode-mounting devices. The measurement of pH requires two electrodes: a sensing electrode that is sensitive to H_3O^+ concentrations, and a reference electrode. This is because the pH meter is really just a voltmeter that measures the electrical potential of

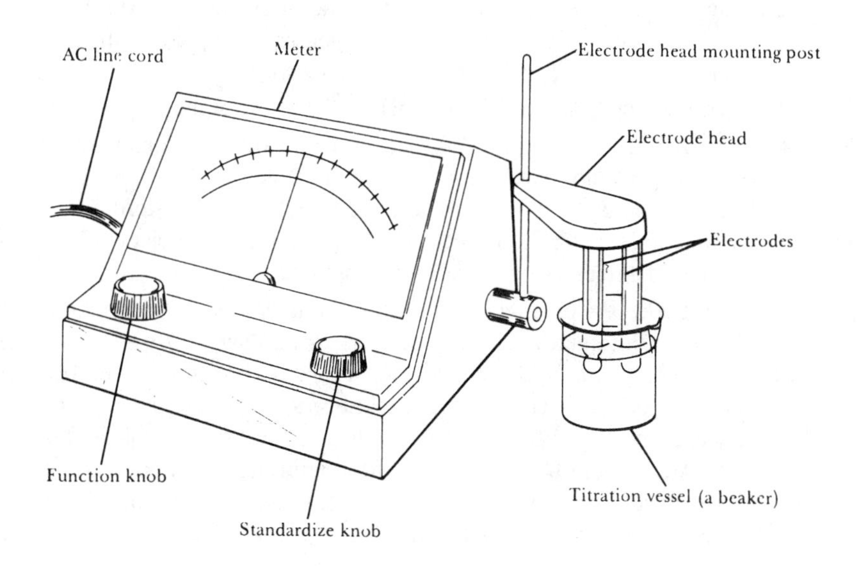

Figure 21.2 A pH meter.

Figure 21.3

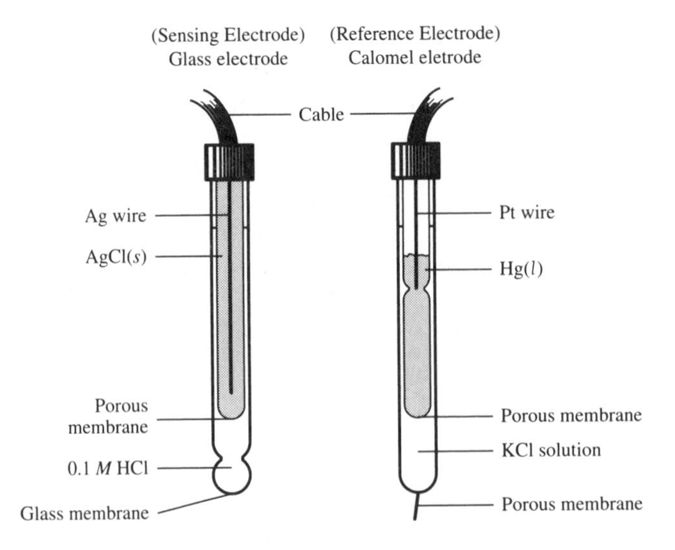

a solution. Typical sensing and reference electrodes are illustrated in Figure 21.3.

The reference electrode is just an electrode that develops a known potential that is essentially independent of the contents of the solution into which it is placed. The glass electrode is sensitive to the H_3O^+ concentration of the solution into which it is placed; its potential is a function of $[H_3O^+]$. It operates by transport of H_3O^+ ions through the glass membrane. This can be more precisely described, but for our purposes here it is sufficient for you to understand that two electrodes are required. These two electrodes are sometimes combined into an electrode called a combination electrode, which appears to be a single electrode. The combination electrode, however, does contain both a reference and a sensing electrode.

Preliminary Operations with the pH Meter

1. Obtain a buffer solution of known pH.
2. Plug in the pH meter to line current and allow at least 10 min for warm-up. It should be left plugged in until you are completely finished with it. *This does not apply to battery-operated meters.*
3. Turn the function knob on the pH meter to the standby position.
4. *Prepare the electrodes.* Make certain that the solution in the reference electrode extends well above the internal electrode. If it does not, ask your instructor to fill it with saturated KCl solution. Remove the rubber tip and slide down the rubber collar on the reference electrode. Rinse the outside of the electrodes well with distilled water.
5. *Standardize the* pH *meter*. Carefully immerse the electrodes in the buffer solution contained in a small beaker. *Remember that the glass electrode is very fragile; it breaks easily!* Don't touch the bottom of the beaker with the electrodes!! Turn the function knob to "read" or "pH." Turn the standardize knob until the pH meter indicates the exact pH of the buffer solution. Wait 5 s to be certain that the reading remains constant. *Once you have standardized the* pH *meter, don't readjust the standardize knob.* Turn the function knob to standby. Carefully lift electrodes from the

buffer and rinse them with distilled water. The pH meter is now ready to use to measure pH.

RECORD ALL DATA DIRECTLY ONTO THE REPORT SHEETS.

PROCEDURE

A. Standardization of NaOH

Prepare approximately 500 mL of 0.1 *M* NaOH as described in Experiment 19. Standardize your NaOH solution as described in Experiment 19 by titrating against KHP, using phenolphthalein indicator. *Save the remaining standardized solution for this experiment and the following one.*

B. Determination of pK_a of Unknown Acid

With the aid of a pipet bulb, pipet a 25-mL aliquot of your unknown acid solution into a 250-mL beaker and carefully immerse the previously rinsed electrodes in this solution. Measure the pH of this solution by turning the function knob to "read" or "pH." Record the pH on the report sheet. Begin your titration by adding 1 mL of your standardized base from a buret and record the volume of titrant and pH. Repeat with successive additions of 1 mL of base until you approach the end point; then add 0.1-mL increments of base and record the pH and milliliters of NaOH added. When the pH no longer changes upon addition of NaOH, your titration is completed. From these data plot a titration curve of pH versus mL titrant added. Repeat the titration with two more 25-mL aliquots of your unknown acid and plot the titration curves. From these curves calculate the ionization constant. Time may be saved if the first titration is run with larger-volume increments of the titrant to locate an approximate equivalence point; then the second and third titrations may be run with the small increments indicated above.

C. Concentration of Unknown Acid

Using the volume of base at the equivalence point, its normality, and the fact that you used 25.0 mL of acid, calculate the concentration of the unknown acid, record this on the report sheet. Turn the function knob to standby, rinse the electrodes with distilled water, and wipe them with a clean dry tissue.

REVIEW QUESTIONS

Before beginning this experiment in the laboratory, you should be able to answer the following questions:

1. Define Brønsted-Lowry acids and bases.
2. Differentiate between the dissociation constant and equilibrium constant for the dissociation of a weak acid, HA, in aqueous solution.
3. Why isn't the pH at the equivalence point always equal to 7 in a neutralization titration? When would it be 7?
4. What is the pK_a of an acid whose K_a is 3.6×10^{-6}?

5. Why must two electrodes be used to make an electrical measurement such as pH?
6. What is a buffer solution?
7. The pH at one-half the equivalence point in an acid-base titration was found to be 5.32. What is the value of K_a for this unknown acid?
8. If 30.15 mL of 0.0995 M NaOH is required to neutralize 0.216 g of an unknown acid, HA, what is the molecular weight of the unknown acid?
9. If K_a is 1.85×10^{-5} for acetic acid, calculate the pH at one-half the equivalence point and at the equivalence point for a titration of 50 mL of 0.100 M acetic acid with 0.100 M NaOH.

Name ____________________ Desk __________

Date __________ Laboratory Instructor ____________________

REPORT SHEET FOR EXPERIMENT 21

DETERMINATION OF THE DISSOCIATION CONSTANT OF A WEAK ACID

A. Standardization of NaOH

	Trial 1	Trial 2	Trial 3
Weight of bottle + KHP	______	______	______
Weight of bottle	______	______	______
Weight of KHP used	______	______	______
Final buret reading	______	______	______
Initial buret reading	______	______	______
mL of NaOH used	______	______	______
Normality of NaOH	______	______	______

Average normality (show calculations and standard deviation) ______

Standard deviation (see Experiment 8) ______

C. Concentration of Unknown Acid

	Trial 1	Trial 2	Trial 3
Volume of unknown acid	______	______	______
Normality of NaOH from above	______	______	______
mL of NaOH at eq. point	______	______	______
Normality of unknown acid (see Experiment 19)	______	______	______

Average normality (show calculations) ____________ Standard deviation

B. Determination of pK_a of Unknown Acid

First determination		**Second determination**		**Third determination**	
mL NaOH	**pH**	**mL NaOH**	**pH**	**mL NaOH**	**pH**

First determination		Second determination		Third determination	
mL NaOH	pH	mL NaOH	pH	mL NaOH	pH
________	________	________	________	________	________
________	________	________	________	________	________
________	________	________	________	________	________

Volume at eq. point ________ ________ ________

Volume at $\frac{1}{2}$ eq. point ________ ________ ________

pK_a ________ pK_a ________ pK_a ________

K_a ________ K_a ________ K_a ________

Average K_a (show calculations) ________ Standard deviation of K_a ________

QUESTIONS

1. What are the largest sources of error in this experiment?

2. What is the pH of the solution obtained by mixing 30.00 mL of 0.250 *M* HCl and 30.00 mL of 0.125 *M* NaOH?

3. What is the pH of a solution that is 0.50 *M* in sodium acetate and 0.75 *M* in acetic acid? (K_a for acetic acid is 1.85×10^{-5}.)

4. Calculate the pH of a solution prepared by mixing 15.0 mL of 0.50 *M* NaOH and 30.0 mL of 0.50 *M* benzoic acid solution. (Benzoic acid is monoprotic; its dissociation constant is 6.5×10^{-5}.)

5. K_a for hypochlorous acid, HClO, is 3.0×10^{-8}. Calculate the pH after 10.0, 20.0, 30.0, and 40.0 mL of 0.100 *M* NaOH have been added to 40.0 mL of 0.100 *M* HOCl.

ration curve

mL NaOH added ⟶

pH ⟶

0 1 2 3 4 5 6 7 8 9 10 11 12 13

Titration curve

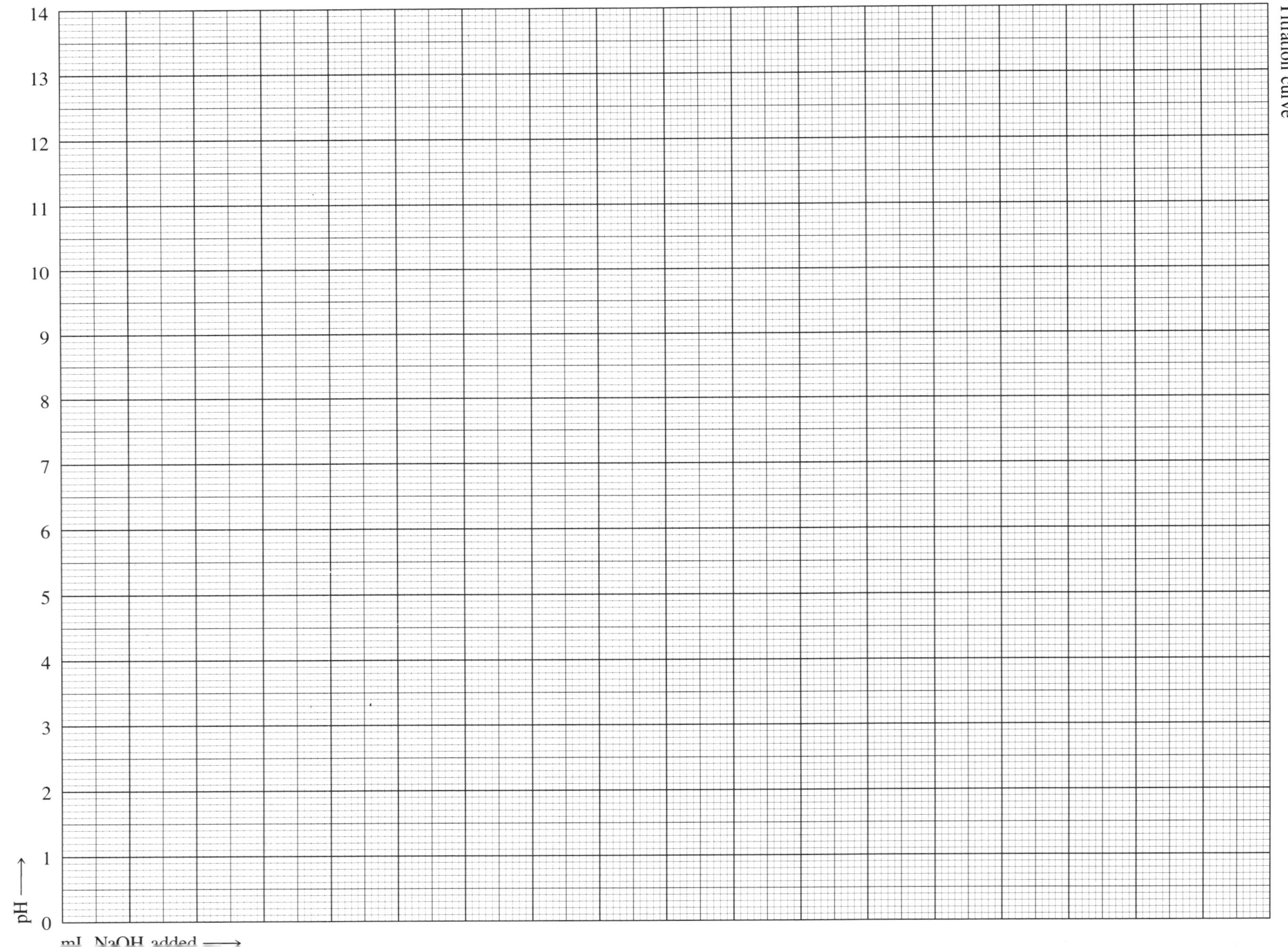

Titration curve

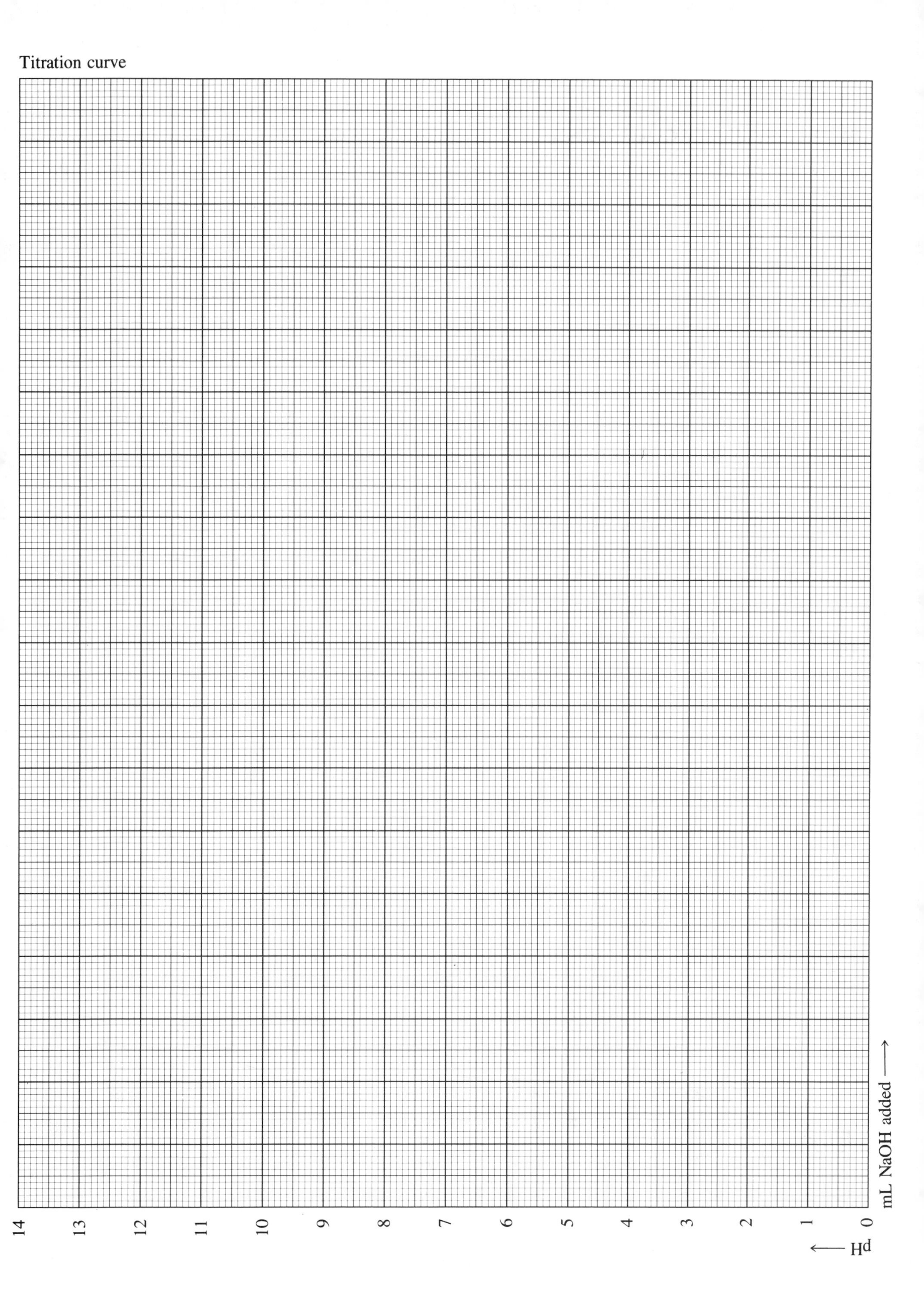

NOTES AND CALCULATIONS

Titration Curves of Polyprotic Acids

EXPERIMENT 22

OBJECTIVE

To become familiar with consecutive equilibria, normality, and molarity.

APPARATUS AND CHEMICALS

pH meter with electrodes
potassium acid phthalate
sodium hydroxide solution
weighing bottle
buret, buret clamp, and ring stand
25-mL pipet
standard buffer solution
unknown solution of a polyprotic acid (100 mL of approx. 0.1 M solution)
phenolphthalein indicator solution
150-mL beaker
250-mL beakers (3)

DISCUSSION

In Experiment 21 the dissociation of a monobasic weak acid, HA, was discussed, and its dissociation constant, K_a, was determined by titration. In this experiment a similar approach will be taken to determine the dissociation constants of an unknown polyprotic acid, H_nA. Consider the triprotic acid H_3PO_4. It undergoes the following dissociations in aqueous solution:

$$H_3PO_4 + H_2O \rightleftharpoons H_2PO_4^- + H_3O^+ \qquad K_{a1} = \frac{[H_2PO_4^-][H_3O^+]}{[H_3PO_4]} \quad [1]$$

$$H_2PO_4^- + H_2O \rightleftharpoons HPO_4^{2-} + H_3O^+ \qquad K_{a2} = \frac{[HPO_4^{2-}][H_3O^+]}{[H_2PO_4^-]} \quad [2]$$

$$HPO_4^{2-} + H_2O \rightleftharpoons PO_4^{3-} + H_3O^+ \qquad K_{a3} = \frac{[PO_4^{3-}][H_3O^+]}{[HPO_4^{2-}]} \quad [3]$$

The acid H_3PO_4 possesses three dissociable protons, and for this reason it is termed a triprotic acid. If you were to perform a titration of H_3PO_4 with NaOH in the same manner as you did in Experiment 21 with the acid HA, the following reactions would occur in turn:

$$H_3PO_4 + NaOH \rightleftharpoons NaH_2PO_4 + H_2O \quad [4]$$

$$NaH_2PO_4 + NaOH \rightleftharpoons Na_2HPO_4 + H_2O \quad [5]$$

$$Na_2HPO_4 + NaOH \rightleftharpoons Na_3PO_4 + H_2O \quad [6]$$

The resultant titration curve, when plotted as pH versus milliliters of NaOH added, would be similar to that shown in Figure 22.1. At the point at

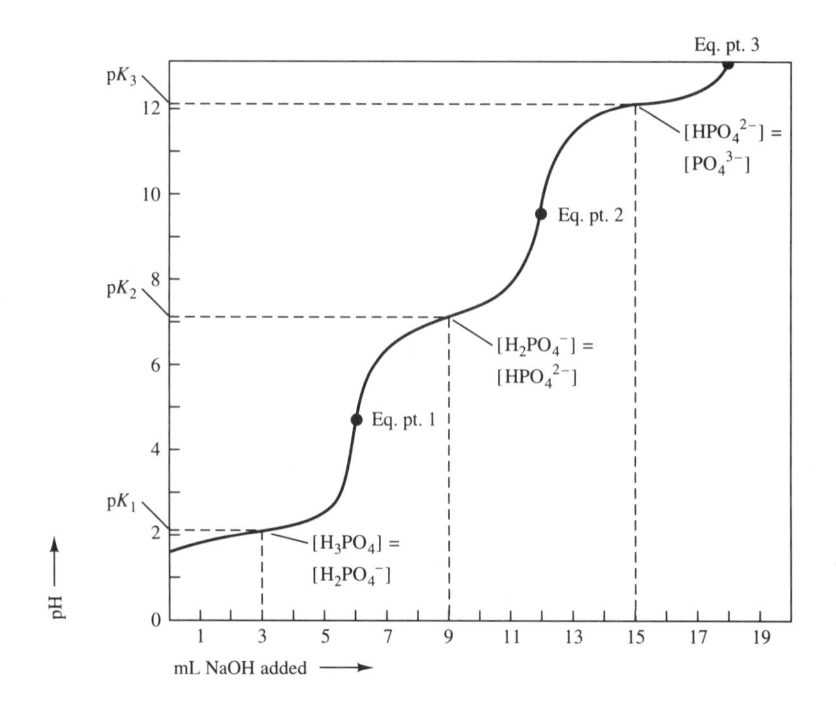

Figure 22.1 Titration curve for titration of H_3PO_4 with NaOH.

which one-half of the protons in the first dissociation step of H_3PO_4 have been titrated with NaOH, the H_3PO_4 concentration is equal to the $H_2PO_4^-$ concentration. Substituting $[H_3PO_4] = [H_2PO_4^-]$ into Equation [1] yields $K_{a1} = [H_3O^+]$, or $pH = pK_{a1}$ at this point.

Similarly, at one-half the second equivalence point, one-half of the $H_2PO_4^-$ has been neutralized and $[H_2PO_4^-] = [HPO_4^{2-}]$. Substituting this into Equation [2] yields $K_{a2} = [H_3O^+]$, or $pK_{a2} = pH$ at this point.

In the same manner, at one-half the third equivalence point, $[HPO_4^{2-}] = [PO_4^{3-}]$. Substituting this into Equation [3], we obtain the expression $K_{a3} = [H_3O^+]$, or $pK_{a3} = pH$.

The same type of result is obtained for any polyprotic acid. If a titration of the acid is performed with a pH meter, the dissociation constants may be obtained from titration curves as long as the dissociation constants exceed the ion product of water, which you should recall is 10^{-14} for the reaction

$$2H_2O \rightleftharpoons H_3O^+ + OH^-$$

In practice, if the acidity of the acid being studied approaches that of water, as in the case for the third proton of H_3PO_4 for which K_{a3} is 4.8×10^{-13}, it is difficult to determine the dissociation constant in this manner. Thus for H_3PO_4, both K_{a1} and K_{a2} are readily obtained in this way, but K_{a3} is not.

In this experiment you will determine the dissociation constants K_{a1} and K_{a2} of a diprotic acid and both its normality and molarity.

Recall that the definitions for normality and molarity are

$$\text{Normality } (N) = \frac{\text{number of equivalents solute}}{\text{liter solution}}$$

$$\text{Molarity } (M) = \frac{\text{number of moles solute}}{\text{liter solution}}$$

For acid-base reactions, an equivalent of acid is defined as the amount of acid that contains 1 mol of dissociable protons. Consequently,

HCl has one equivalent/mole;	a 1 molar solution is 1 normal
H_2SO_4 has two equivalents/mole;	a 1 molar solution is 2 normal
H_3PO_4 has three equivalents/mole;	a 1 molar solution is 3 normal

(Normality is always equal to or greater than molarity.)

PROCEDURE

RECORD ALL DATA DIRECTLY ONTO THE REPORT SHEETS.

1. Standardize the pH meter as described in Experiment 20.
2. Prepare 200 mL of approximately 0.1 *M* NaOH solution and standardize it as described in Experiment 19 or use the standardized NaOH that remains from Experiment 20.
3. Titrate three separate 25-mL aliquots of the unknown acid in three separate 250-mL beakers and plot the titration curves.
4. Using the relations given above, and the relevant equations from Experiment 21, determine the pK_a and K_a values for your unknown acid.
5. Determine the normality and molarity of your unknown acid solution and the volumes necessary to reach the equivalence points.

(HINT: You can save time if you do your first titration rapidly so that you know the approximate volumes of the equivalence points; then you can do the next two titrations with large-volume increments away from the equivalence points and small-volume increments near the equivalence points.)

REVIEW QUESTIONS

Before beginning this experiment in the laboratory, you should be able to answer the following questions:

1. What is a polyprotic acid?
2. If 20.2 mL of 0.122 *N* NaOH is required to reach the first equivalence point of a solution of citric acid ($H_3C_6H_5O_7$), how many mL of NaOH are required to completely neutralize this solution?
3. What is the molarity of a 6.2 *N* $H_2C_2O_4$ solution?
4. How many moles and how many equivalents of H_3O^+ are present in 50 mL of a 0.3 *N* solution of H_2SO_4?
5. Why is it necessary to standardize a pH meter?
6. If the pH at one-half the first and second equivalence points of a dibasic acid is 3.52 and 6.31, respectively, what are the values for pK_{a1} and pK_{a2}? From pK_{a1} and pK_{a2} calculate the K_{a1} and K_{a2}.
7. Derive the relationship between pH and pK_a at one-half the equivalence point for the titration of a weak acid with a strong base.

8. Could K_b for a weak base be determined in the same way that K_a for a weak acid is determined in this experiment?
9. What is the relationship of the successive equivalence-point volumes in the titration of a polyprotic acid?
10. The titration of a weak acid with NaOH showed two equivalence points. Using only this information determine the relationship between the molarity and normality of this acid solution.

Name ______________________________ Desk ____________

Date ____________ Laboratory Instructor ______________________________

REPORT SHEET FOR EXPERIMENT 22

TITRATION CURVES OF POLYPROTIC ACIDS

A. Standardization of NaOH

	Trial 1	Trail 2	Trial 3
Weight of bottle + KHP	________	________	________
Weight of bottle	________	________	________
Weight of KHP used	________	________	________
Final buret reading	________	________	________
Initial buret reading	________	________	________
mL of NaOH used	________	________	________
Normality of NaOH	________	________	________

Average normality (show calculations) ________

Standard deviation (show calculations) ________

B. Concentration of Unknown Acid

	Trial 1	Trial 2	Trail 3
Volume of unknown acid	25 mL	25 mL	25 mL
Normality of NaOH from above	________	________	________
mL of NaOH at eq. point 1	________	________	________
mL of NaOH at eq. point 2	________	________	________
Normality of unknown acid	________	________	________

Average normality ________ Standard deviation ________

	Trial 1	Trial 2	Trial 3
Molarity of unknown acid	________	________	________

Average molarity (show calculations) ________

Standard deviation (show calculations) ____________

C. Determination of pK_a values of Unknown Acid

First determination		Second determination		Third determination	
mL NaOH	pH	mL NaOH	pH	mL NaOH	pH
0		0		0	

pK_{a1}	________	pK_{a1}	________	pK_{a1}	________
K_{a1}	________	K_{a1}	________	K_{a1}	________
Average K_{a1}	________			Standard deviation	________
pK_{a2}	________	pK_{a2}	________	pK_{a2}	________
K_{a2}	________	K_{a2}	________	K_{a2}	________
Average K_{a2}	________			Standard deviation	________

D. pH of Unknown Acid

Using the measured K_1 and the concentration, calculate the pH of your unknown acid solution and compare it with that measured.

Calculated pH ________ Calculated pH ________ Calculated pH ________

Measured pH ________ Measured pH ________ Measured pH ________

Average: Calculated pH ________ Measured pH ________

E. Identity of Unknown Acid

Consult a reference book such as the *Handbook of Chemistry and Physics,* and using tables of acid dissociation constants contained therein, identify your unknown acid ________________________

Titration curve

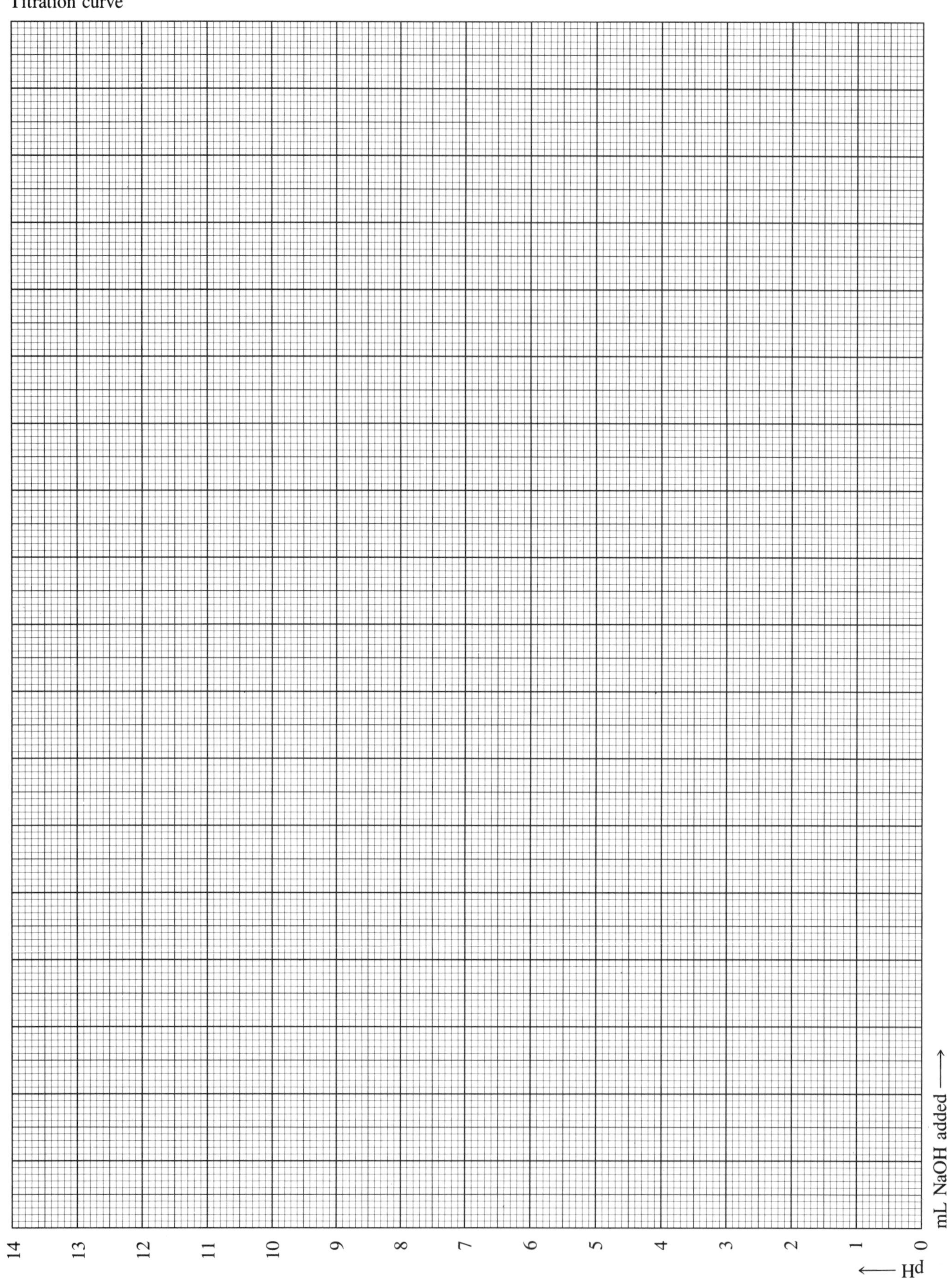

Titration curve

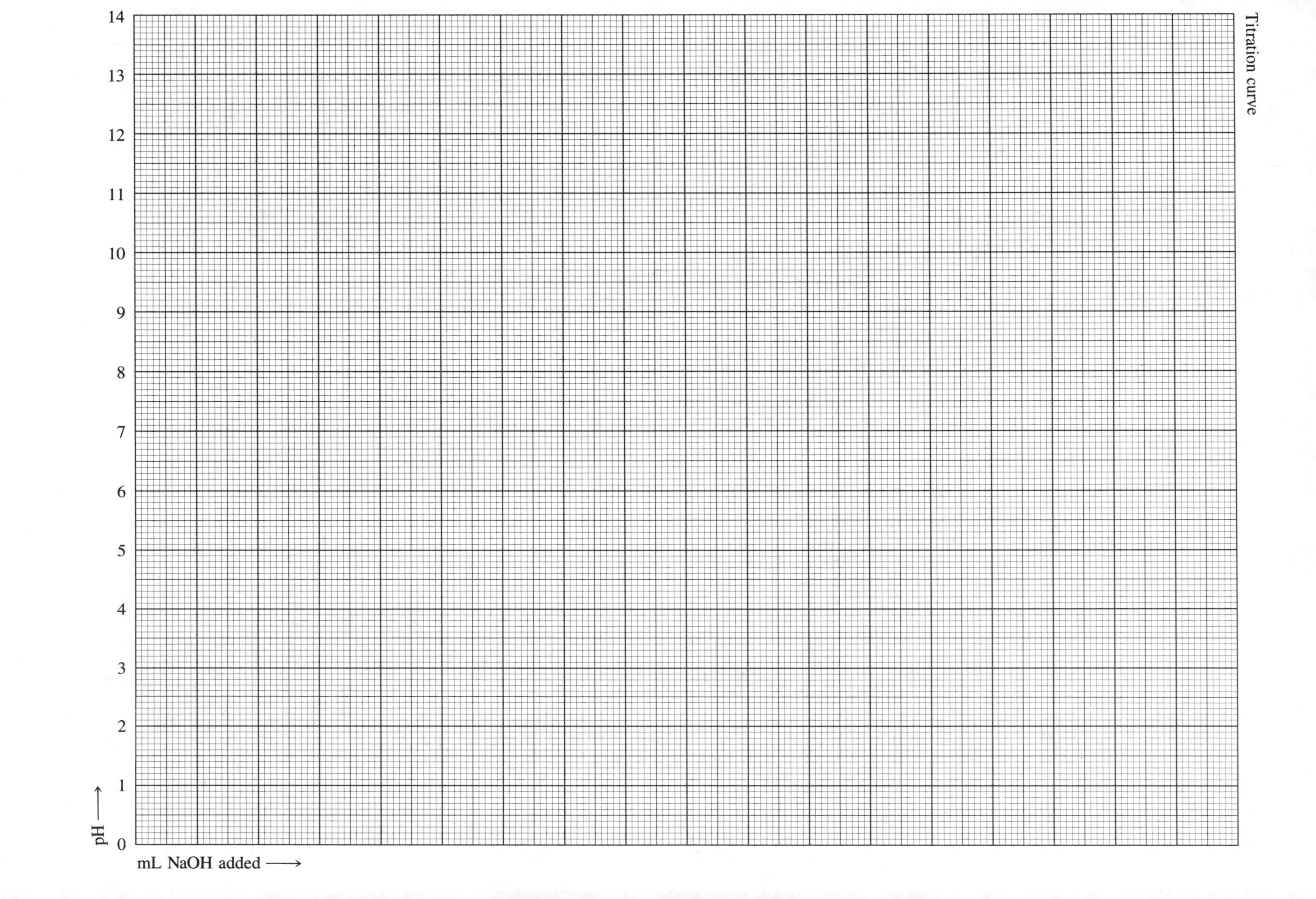

Titration curve

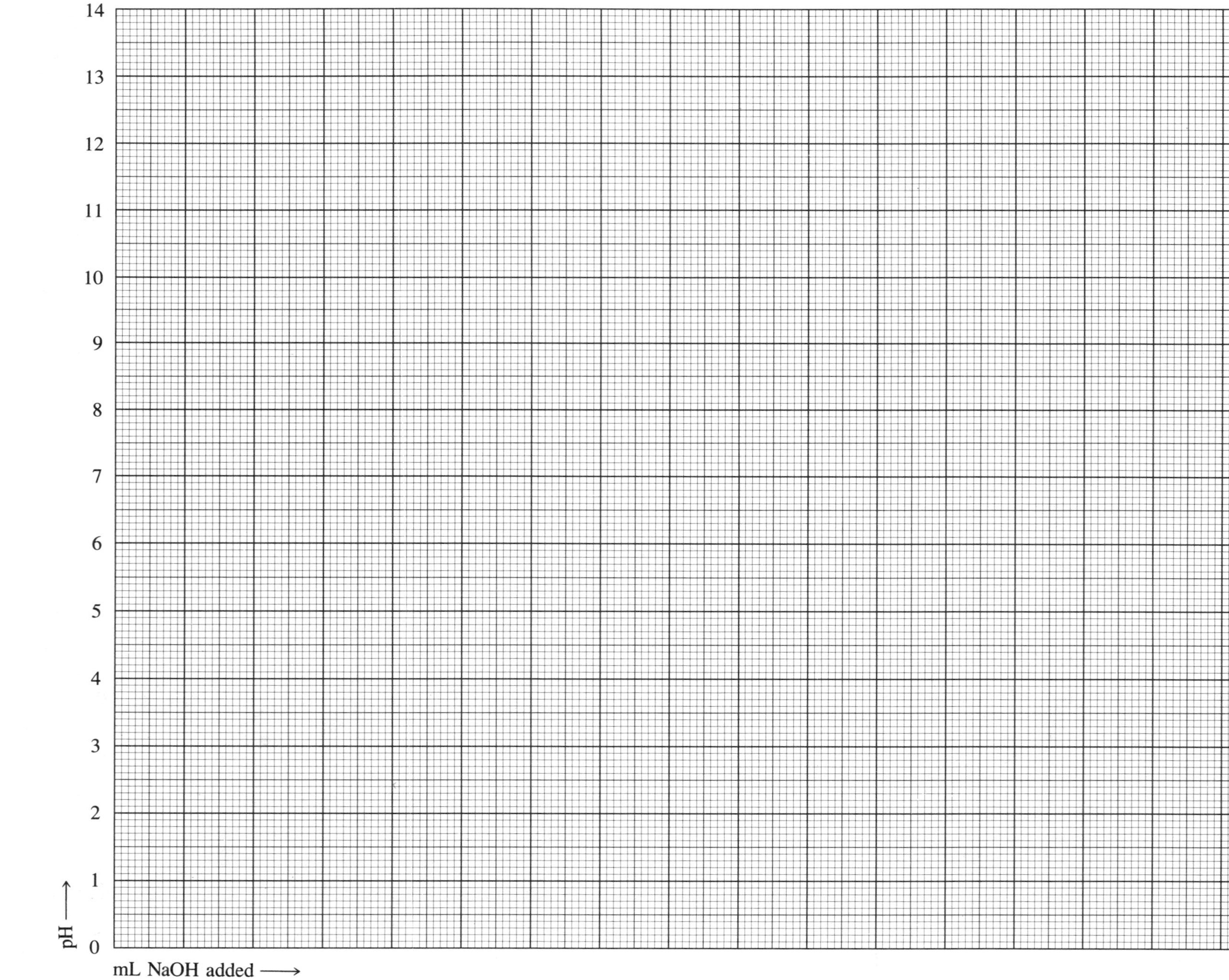

Acid-Base Properties of Salt Solutions: Hydrolysis

EXPERIMENT 23

OBJECTIVE

To learn about the concept of hydrolysis and gain familiarity with acid-base indicators.

APPARATUS AND CHEMICALS

500-mL Erlenmeyer flask
Bunsen burner
ring stand and iron ring
wire gauze
dropping bottles of methyl orange, methyl red, bromothymol blue, phenol red, phenolphthalein, and alizarin yellow-R

0.1 *M* solutions of NaCl, $NaC_2H_3O_2$ (sodium acetate), copper nitrate, ammonium chloride, zinc chloride, potassium aluminum sulfate, and sodium carbonate
test tubes (6)
test-tube rack
10-mL graduated cylinder

DISCUSSION

We expect solutions of substances such as HCl and HNO_2 to be acidic and solutions of NaOH and NH_3 to be basic. However, we may be somewhat surprised at first to discover that aqueous solutions of some salts such as sodium nitrite, $NaNO_2$, and potassium acetate, $KC_2H_3O_2$, are basic while others such as NH_4Cl and $FeCl_3$ are acidic. Recall that salts are the products formed in neutralization reactions of acids and bases. For example, when NaOH and HNO_2 (nitrous acid) react, the salt $NaNO_2$ is formed:

$$NaOH(aq) + HNO_2(aq) \longrightarrow NaNO_2(aq) + H_2O(l)$$

Most salts are strong electrolytes and exist as ions in aqueous solutions. Many ions react with water to produce acidic or basic solutions. The reactions of ions with water are frequently called *hydrolysis reactions*. We will see that anions such as CN^- and $C_2H_3O_2^-$ that are conjugate bases of weak acids react with water to form OH^- ions. Cations such as NH_4^+ and Fe^{3+} come from weak bases and react with water to form H^+ ions.

Hydrolysis of Anions

Let us consider the behavior of anions first. Anions of weak acids react with proton sources. When placed in water these anions react to some extent with water to accept protons and generate OH^- ions and thus cause the solution pH to be greater than 7. Recall that proton acceptors are Brønsted bases. Thus the anions of weak acids are basic in two senses: They are proton acceptors, and their aqueous solutions have pH's above 7. The nitrite ion, for example, reacts with water to increase the concentration of OH^- ions:

$$NO_2^-(aq) + H_2O(l) \rightleftharpoons HNO_2(aq) + OH^-(aq)$$

This reaction of the nitrite ion is similar to that of weak bases such as NH_3 with water:

$$NH_3(aq) + H_2O(l) \rightleftharpoons NH_4^+(aq) + OH^-(aq)$$

Thus both NH_3 and NO_2^- are bases and as such have a basicity or base-dissociation constant, K_b, associated with their corresponding equilibria.

According to the Brønsted theory, the nitrite ion is the conjugate base of nitrous acid. Let's consider the conjugate acid-base pair HNO_2 and NO_2^- and their behavior in water:

$$HNO_2 \rightleftharpoons H^+ + NO_2^- \qquad K_a = \frac{[H^+][NO_2^-]}{[HNO_2]}$$

$$NO_2^- + H_2O \rightleftharpoons HNO_2 + OH^- \qquad K_b = \frac{[HNO_2][OH^-]}{[NO_2^-]}$$

Multiplication of these dissociation constants yields:

$$K_a \times K_b = \left(\frac{[H^+][NO_2^-]}{[HNO_2]}\right)\left(\frac{[HNO_2][OH^-]}{[NO_2^-]}\right) = [H^+][OH^-] = K_w$$

where K_w is the ion-product constant of water.

Thus the product of the acid-dissociation constant for an acid and the base-dissociation constant for its conjugate base is the ion-product constant for water:

$$K_a \times K_b = K_w = 1.0 \times 10^{-14} \qquad [1]$$

Knowing the K_a for a weak acid, we can easily find the K_b for the anion of the acid:

$$K_b = \frac{K_w}{K_a} \qquad [2]$$

By consulting a table of acid-dissociation constants, we can find that K_a for nitrous acid is 4.5×10^{-4}. Using this value, we can readily determine K_b for NO_2^-:

$$K_b = \frac{1.0 \times 10^{-14}}{4.5 \times 10^{-4}} = 2.2 \times 10^{-11}$$

We further note that the stronger the acid, that is, the larger the K_a, the weaker its conjugate base. Similarly, the weaker the acid (the smaller the K_a), the stronger the conjugate base.

Anions derived from *strong acids,* such as Cl^- from HCl, do not react with water to affect the pH. Nor do Br^-, I^-, NO_3^-, SO_4^{2-}, and ClO_4^- affect the pH, for the same reason. They are spectator ions in the acid-base sense and can be described as neutral ions. Similarly, cations from strong bases, such as Na^+ from NaOH or K^+ from KOH, do not react with water to affect the pH. Hydrolysis of an ion occurs only when it can form a molecule or ion that is a weak electrolyte in the reaction with water. Strong acids and bases do not exist as molecules in dilute water solutions.

EXAMPLE 23.1

What is the pH of a 0.10 *M* NaClO solution?

$$K_a \text{ for HClO is } 3.0 \times 10^{-8}.$$

Solution: The salt NaClO exists as Na^+ and ClO^-. The Na^+ ions are spectator ions, but ClO^- ions undergo hydrolysis to form the weak acid HClO. Let x equal the equilibrium concentration of HClO (and OH^-):

$$\underset{(0.10 - x)M}{ClO^-(aq)} + H_2O(l) \rightleftharpoons \underset{xM}{HClO(aq)} + \underset{xM}{OH^-(aq)}$$

The value of K_b for the reaction is $(1.0 \times 10^{-14})/(3.0 \times 10^{-8}) = 3.3 \times 10^{-7}$. Because K_b is so small, we can neglect x in comparison with 0.10 and thus $0.10 - x \simeq 0.10$.

$$\frac{[HClO][OH^-]}{[ClO^-]} = K_b$$

$$\frac{x^2}{0.10} = 3.3 \times 10^{-7}$$

$$x^2 = 3.3 \times 10^{-8}$$

$$x = 1.8 \times 10^{-4}\ M$$

$$\text{pOH} = 3.74$$

$$\text{and pH} = 14 - 3.74 = 10.26$$

Anions with ionizable protons such as HCO_3^-, $H_2PO_4^-$, and HPO_4^{2-} may be either acidic or basic, depending on the relative values of K_a and K_b for the ion. We will not consider such ions in this experiment.

Hydrolysis of Cations

Cations that are derived from weak bases react with water to increase the hydrogen-ion concentration; they form acidic solutions. The ammonium ion is derived from the weak base NH_3 and reacts with water as follows:

$$NH_4^+(aq) + H_2O(l) \rightleftharpoons H_3O^+(aq) + NH_3(aq)$$

This reaction is completely analogous to the dissociation of any other weak acid, such as acetic acid or nitrous acid. We can represent this acid-dissociation of NH_4^+ more simply:

$$NH_4^+(aq) \rightleftharpoons NH_3(aq) + H^+(aq)$$

Here too the acid-dissociation constant is related to the K_b of NH_3, which is the conjugate base of NH_4^+:

$$NH_3(aq) + H_2O(l) \rightleftharpoons NH_4^+(aq) + OH^-(aq)$$

Knowing the value of K_b for NH_3, we can readily calculate the acid dissociation constant from Equation [1]:

$$K_a = \frac{K_w}{K_b} \qquad [3]$$

Cations of the alkali metals (Group 1A) and the larger alkaline earth ions, Ca^{2+}, Sr^{2+}, and Ba^{2+}, do not react with water, because they come from strong bases. Thus these ions have no influence on the pH of aqueous solutions. They are merely spectator ions in acid-base reactions. Consequently, they are described as being neutral in the acid-base sense. The cations of most other metals do hydrolyze to produce acidic solutions. Metal cations are coordinated with water molecules, and it is the hydrated ion that serves as the proton donor. The following equations illustrate this behavior for the iron (III) ion:

$$Fe(H_2O)_6^{3+}(aq) + H_2O(l) \rightleftharpoons Fe(OH)(H_2O)_5^{2+}(aq) + H_3O^+(aq) \qquad [4]$$

We frequently omit the coordinated water molecules from such equations. For example, Equation [4] may be written as

$$Fe^{3+}(aq) + H_2O(l) \rightleftharpoons Fe(OH)^{2+}(aq) + H^+(aq) \qquad [5]$$

Additional hydrolysis reactions can occur to form $Fe(OH)_2^+$ and even lead to the precipitation of $Fe(OH)_3$. The equilibria for such cations are often complex, and not all species have been identified. However, equations such as [4] and [5] serve to illustrate the acidic character of dipositive and tripositive ions and account for most of the H^+ in these solutions.

Summary of Hydrolysis Behavior

Whether a solution of a salt will be acidic, neutral, or basic can be predicted on the basis of the strenghts of the acid and base from which the salt was formed.

1. *Salt of a strong acid and a strong base*. Examples: NaCl, KBr, and $Ba(NO_3)_2$. Neither the cation nor anion hydrolyzes, and the solution has a pH of 7.
2. *Salt of a strong acid and a weak base:* Examples: NH_4Br, $ZnCl_2$, and $Al(NO_3)_3$. The cation hydrolyzes, forming H^+ ions, and the solution has a pH less than 7.
3. *Salt of a weak acid and a strong base:* Examples: $NaNO_2$, $KC_2H_3O_2$, and $Ca(OCl)_2$. The anion hydrolyzes, forming OH^- ions, and the solution has a pH greater than 7.
4. *Salt of a weak acid and a weak base:* Examples: NH_4F, $NH_4C_2H_3O_2$, and $Zn(NO_2)_2$. Both ions hydrolyze. The pH of the solution is determined by the relative extent to which each ion hydrolyzes.

In this experiment, we will test the pH of water and of several aqueous salt solutions to determine whether these solutions are acidic, basic, or neutral. In each case, the salt solution will be 0.1 *M*. Knowing the concentration of the salt solution and the measured pH of each solution allows us to calculate K_a or K_b for the ion that hydrolyzes. Example 23.2 illustrates such calculations.

EXAMPLE 23.2

Calculate K_b for OBr^- if a 0.10 *M* solution of NaOBr has a pH of 10.85.

Solution: The spectator ion is Na^+. Alkali metal ions do not react with water and have no influence on pH. The ion OBr^- is the anion of a weak acid and thus reacts with water to produce OH^- ions:

$$OBr^- + H_2O \rightleftharpoons HOBr + OH^-$$

and the corresponding expression for the base dissociation constant is:

$$K_b = \frac{[HOBr][OH^-]}{[OBr^-]} \qquad [6]$$

If the pH is 10.85, then

$$pOH = 14.00 - 10.85 = 3.15$$

and

$$[OH^-] = \text{antilog}\ (-3.15) = 7.1 \times 10^{-4}\ M$$

The concentration of HOBr that is formed along with OH^- must also be 7.1×10^{-4} *M*. The concentration of OBr^- that has not hydrolyzed is:

$$[OBr^-] = 0.10\ M - 0.00071\ M \simeq 0.10\ M$$

Substituting these values into Equation [6] for K_b yields:

$$K_b = \frac{[7.1 \times 10^{-4}][7.1 \times 10^{-4}]}{[0.10]}$$

$$= 5.0 \times 10^{-6}$$

The behavior of indicators was discussed in Experiment 17. We will use a set of indicators to determine the pH of various salt solutions. The dark areas in Figure 23.1 denote the transition ranges for the indicators you will use.

We will generally find that the solutions that we test will be more acidic than we would predict them to be. A major reason for this increased acidity is the occurrence of CO_2 dissolved in the solutions. CO_2 reacts with water to generate H^+:

$$CO_2(g) + H_2O(l) \rightleftharpoons H_2CO_3(aq) \rightleftharpoons H^+(aq) + HCO_3^-(aq)$$

The solubility of CO_2 is greatest in basic solutions, intermediate in neutral ones, and least in acidic ones. Even distilled water will therefore be somewhat acidic, unless it is boiled to remove the dissolved CO_2.

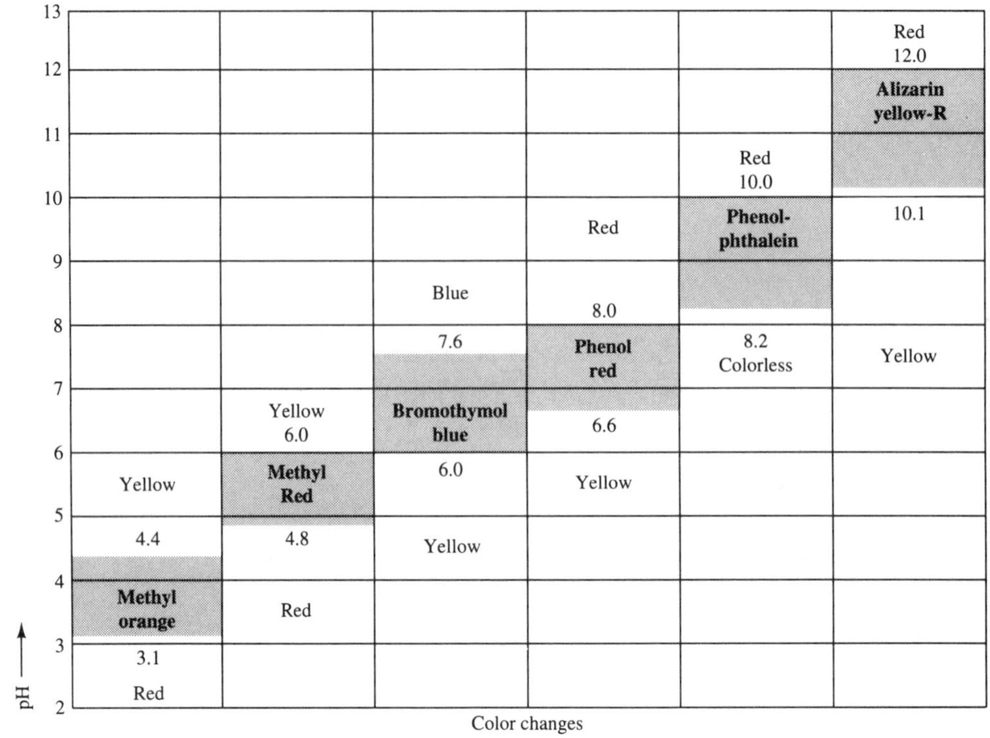

pH

13
12
11
10
9
8
7
6
5
4
3
2

Yellow
4.4
Methyl orange
3.1
Red

Yellow
6.0
Methyl Red
4.8
Red

Blue
7.6
Bromothymol blue
6.0
Yellow

Red
8.0
Phenol red
6.6
Yellow

Red
10.0
Phenol-phthalein
8.2
Colorless

Red
12.0
Alizarin yellow-R
10.1
Yellow

Color changes

Figure 23.1 The color behavior of indicators.

PROCEDURE

Boil approximately 450 mL of distilled water for about 10 min to expel dissolved carbon dioxide. Allow the water to cool to room temperature. While the water is boiling and subsequently cooling, add about 5 mL of unboiled distilled water to each of six test tubes. Add 3 drops of a different indicator to each of these six test tubes (one indicator per tube) and record the colors on the report sheet. From these colors and the data given in Figure 23.1 determine the pH of the unboiled water to the nearest pH unit. (Remember that we would expect its pH to be below 7 because of dissolved CO_2). Empty the contents of the test tubes and rinse the test tubes three times with about 3 mL of boiled distilled water. Then pour about 5 mL of the boiled distilled water into each of the six test tubes and add 3 drops of each of the indicators (one indicator per tube) to each tube. Record the colors and determine the pH. Empty the contents of the test tubes and rinse each tube three times with about 3 mL of boiled distilled water.

Repeat the same procedure to determine the pH of each of the following solutions that are 0.1 *M*: NaCl, $NaC_2H_3O_2$, $Cu(NO_3)_2$, NH_4Cl, $ZnCl_2$, $KAl(SO_4)_2$, and Na_2CO_3. Use 5 mL of each of these solutions per test tube. Do not forget to rinse the test tubes with boiled distilled water when you go from one solution to the next.

From the pH values that you determined, calculate the hydrogen- and hydroxide-ion concentrations for each solution. Complete the tables on the report sheets and calculate the K_a or K_b as appropriate.

REVIEW QUESTIONS

Before beginning this experiment in the laboratory, you should be able to answer the following questions:

1. Define Brønsted acids and bases.
2. Which of the following ions will react with water in a hydrolysis reaction: Na^+, Ca^{2+}, Cu^{2+}, Zn^{2+}, F^-, SO_3^{2-}, Br^-?
3. For those ions in question 2 that undergo hydrolysis, write net ionic equations for the hydrolysis reaction.
4. The K_a for HCN is 4.9×10^{-10}. What is the value of K_b for CN^-?
5. What are the conjugate base and conjugate acid of $H_2PO_4^-$?
6. From what acid and what base were the following salts made: $CaSO_4$, NH_4Br, and $BaCl_2$?
7. Define the term *salt*.
8. Tell whether 0.1 *M* solutions of the following salts would be acidic, neutral, or basic: $BaCl_2$, $CuSO_4$, $(NH_4)_2SO_4$, $ZnCl_2$, NaCN.
9. If the pH of a solution is 9, what are the hydrogen- and hydroxide-ion concentrations?
10. The pH of a 0.1 *M* MCl (M^+ is an unknown cation) was found to be 4.6. Write a net ionic equation for the hydrolysis of M^+ and its corresponding equilibrium expression K_b. Calculate the value of K_b.

NOTES AND CALCULATIONS

Name ______________________________ Desk ____________

Date ______________ Laboratory Instructor ______________________________

REPORT SHEET FOR EXPERIMENT 23

ACID-BASE PROPERTIES OF SALT SOLUTIONS: HYDROLYSIS

(See tables on pages 260 to 262)

QUESTIONS

1. Using the K_a's for $HC_2H_3O_2$ and HCO_3^- (from Appendix G), calculate the K_b's for the $C_2H_3O_2^-$ and CO_3^{2-} ions. Compare these values with those calculated from your measured pH's.

2. Using K_b for NH_3 (from Appendix H), calculate K_a for the NH_4^+ ion. Compare this value with that calculated from your measured pH's.

3. How should the pH of a 0.1 M solution of $NaC_2H_3O_2$ compare with that of a 0.1 M solution of $KC_2H_3O_2$? Explain briefly.

4. What is the greatest source of error in this experiment? How could you minimize this source of error?

Solution	Indicator Color*								
	Methyl orange	Methyl red	Bromo-thymol blue	Phenol red	Phenol-phtha-lein	Alizarin yellow-R	pH	$[H^+]$	$[OH^+]$
H_2O (unboiled)									
H_2O (boiled)									
NaCl									
$NaC_2H_3O_2$									
$Cu(NO_3)_2$									
NH_4Cl									
$ZnCl_2$									
$KAl(SO_4)_2$									
Na_2CO_3									

*color key: org = orange; ppl = purple; -- = colorless.

Solution	Ion Expected to Hydrolyze (If Any)	Spectator Ion(s) (If Any)
0.1 *M* $NaCl$	____________	____________
0.1 *M* $NaC_2H_3O_2$	____________	____________
0.1 *M* Na_2CO_3	____________	____________
0.1 *M* NH_4Cl	____________	____________
0.1 *M* $ZnCl_2$	____________	____________
0.1 *M* $Cu(NO_3)_2$	____________	____________
0.1 *M* $KAl(SO_4)_2$	____________	____________

CALCULATIONS

Solution	Net-Ionic Equation for Hydrolysis	Expression for Equilibrium Constant (K_a or K_b)	Value of K_a or K_b
$NaC_2H_3O_2$			
Na_2CO_3			
NH_4Cl			
$ZnCl_2$			
$Cu(NO_3)_2$			
$KAl(SO_4)_2$			

Determination of the Solubility-Product Constant for a Sparingly Soluble Salt

EXPERIMENT 24

OBJECTIVE

To become familiar with equilibria involving sparingly soluble substances by determining the value of the solubility-product constant for a sparingly soluble salt.

APPARATUS AND CHEMICALS

buret	0.0024 M K_2CrO_4
ring stand and buret clamp	0.004 M $AgNO_3$
centrifuge	0.25 M $NaNO_3$
75-mm test tubes (3)	100-mL volumetric flasks (4)
spectrophotometer and cuvettes	no. 1 corks (3)
	5-mL pipets (2)

DISCUSSION

Inorganic substances may be broadly classified into three different categories: acids, bases, and salts. According to the Brønsted-Lowry theory (as discussed in Experiment 21), acids are proton donors, and bases are proton acceptors. When an acid reacts with a base in aqueous solution, the products are a salt and water, as illustrated by the reaction of H_2SO_4 and $Ba(OH)_2$:

$$H_2SO_4(aq) + Ba(OH)_2(aq) \rightleftharpoons BaSO_4(s) + 2H_2O(l) \qquad [1]$$

With but a few exceptions, nearly all common salts are strong electrolytes. The solubilities of salts span a broad spectrum, ranging from slightly or sparingly soluble to very soluble. This experiment is concerned with heterogeneous equilibria of slightly soluble salts. In order for a true equilibrium to exist between a solid and solution, the solution must be saturated. Barium sulfate is a slightly soluble salt, and in a saturated solution this equilibrium may be represented as follows:

$$BaSO_4(s) \rightleftharpoons Ba^{2+}(aq) + SO_4^{2-}(aq) \qquad [2]$$

The equilibrium constant for Equation [2] is

$$K_c = \frac{[Ba^{2+}][SO_4^{2-}]}{[BaSO_4]} \qquad [3]$$

The terms in the numerator refer to the molar concentration of ions in solution. The term in the denominator refers to the "concentration" of solid $BaSO_4$. Because the concentration of a pure solid is a constant, $[BaSO_4]$ can be combined with K_c to give a new equilibrium constant, K_{sp}, which is called the solubility-product constant.

$$K_{sp} = K_c[BaSO_4] = [Ba^{2+}][SO_4^{2-}]$$

At a given temperature, the value of K_{sp} is a constant. The solubility product for a sparingly soluble salt can easily be calculated by determining the solubility of the substance in water. Suppose, for example, we determined that 2.42×10^{-4} g of $BaSO_4$ dissolves in 100 mL of water. The molar solubility of this solution (that is, the molarity of the solution) is

$$\left(\frac{2.42 \times 10^{-4} \text{ g } BaSO_4}{100 \text{ mL}}\right)\left(\frac{1000 \text{ mL}}{\text{liter}}\right)\left(\frac{1 \text{ mol } BaSO_4}{233.4 \text{ g } BaSO_4}\right) = 1.04 \times 10^{-5}\ M$$

We see from Equation [2] that for each mole of $BaSO_4$ that dissolves, 1 mol of Ba^{2+} and 1 mol of SO_4^{2-} are formed. It follows, therefore, that

$$\begin{aligned}\text{Solubility of } BaSO_4 \text{ in moles/liter} &= [Ba^{2+}] \\ &= [SO_4^{2-}] \\ &= 1.04 \times 10^{-5}\ M\end{aligned}$$

and

$$\begin{aligned}K_{sp} &= [Ba^{2+}][SO_4^{2-}] \\ &= [1.04 \times 10^{-5}][1.04 \times 10^{-5}] \\ &= 1.08 \times 10^{-10}\end{aligned}$$

In a saturated solution the product of the molar concentrations of Ba^{2+} and SO_4^{2-} cannot exceed 1.08×10^{-10}. If the ion product $[Ba^{2+}][SO_4^{2-}]$ exceeds 1.08×10^{-10}, precipitation of $BaSO_4$ would occur until this product is reduced to the value of K_{sp}. Or if, for example, a solution of Na_2SO_4 is added to a solution of $Ba(NO_3)_2$, $BaSO_4$ would precipitate if the ion product $[Ba^{2+}][SO_4^{2-}]$ is greater than K_{sp}.

Similarly, if we determined that the solubility of Ag_2CO_3 were 3.49×10^{-3} g/100 mL, we could calculate the solubility-product constant for Ag_2CO_3 as follows. The solubility equilibrium involved is

$$Ag_2CO_3(s) \rightleftharpoons 2\ Ag^+(aq) + CO_3^{2-}(aq) \qquad [4]$$

and the corresponding solubility-product expression is

$$K_{sp} = [Ag^+]^2[CO_3^{2-}]$$

The rule for writing the solubility-product expression states that K_{sp} is equal to the product of the concentration of the ions involved in the equilibrium, each raised to the power of its coefficient in the equilibrium equation.

The solubility of Ag_2CO_3 in moles per liter is

$$\left(\frac{3.49 \times 10^{-3} \text{ g } Ag_2CO_3}{100 \text{ mL}}\right)\left(\frac{1000 \text{ mL}}{\text{liter}}\right)\left(\frac{1 \text{ mol } Ag_2CO_3}{278.5 \text{ g } Ag_2CO_3}\right) = 1.27 \times 10^{-4}\ M$$

whence

$$[CO_3^{2-}] = 1.27 \times 10^{-4}\ M \qquad \text{(from Equation [4])}$$

and

$$\begin{aligned} [Ag^+] &= 2(1.27 \times 10^{-4}) \\ &= 2.54 \times 10^{-4}\ M \qquad \text{(from Equation [4])} \\ K_{sp} &= [Ag^+]^2[CO_3^{2-}] \\ &= [2.54 \times 10^{-4}]^2[1.27 \times 10^{-4}] \\ &= 8.19 \times 10^{-12} \end{aligned}$$

In order to determine the solubility-product constant for a sparingly soluble substance, we need only determine the concentration of one of the ions, because the concentration of the other ion is related to the first ion's concentration by a simple stoichiometric relationship. Any method that we could use to determine accurately the concentration would be suitable. In this experiment you will determine the solubility-product constant for Ag_2CrO_4. This substance contains the yellow chromate ion, CrO_4^{2-}. You will determine the concentration of the chromate ion spectrophotometrically at 375 nm. Consult Experiment 30 for directions on the principles involved and the use of a spectrophotometer.

In order to determine the solubility of Ag_2CrO_4, you will first prepare it by the reaction of $AgNO_3$ with K_2CrO_4:

$$2AgNO_3(aq) + K_2CrO_4(aq) \rightleftharpoons Ag_2CrO_4(s) + 2KNO_3(aq)$$

If a solution of $AgNO_3$ is added to a solution of K_2CrO_4, precipitation will occur if the ion product $[Ag^+]^2[CrO_4^{2-}]$ numerically exceeds the value of K_{sp}; if not, no precipitation will occur.

EXAMPLE 24.1

If the K_{sp} for PbI_2 is 7.1×10^{-9}, will precipitation of PbI_2 occur when 10 mL of $1.0 \times 10^{-4}\ M$ $Pb(NO_3)_2$ is mixed with 10 mL of $1.0 \times 10^{-3}\ M$ KI?

Solution:

$$PbI_2(s) \rightleftharpoons Pb^{2+}(aq) + 21^-(aq)$$
$$K_{sp} = [Pb^{2+}][I^-]^2 = 7.1 \times 10^{-9}$$

Precipitation will occur if $[Pb^{2+}][I^-]^2 > 7.1 \times 10^{-9}$.

$$\begin{aligned} [Pb^{2+}] &= \left(\frac{10 \text{ mL}}{20 \text{ mL}}\right)(1.0 \times 10^{-4}\ M) \\ &= 5.0 \times 10^{-5}\ M \end{aligned}$$

$$[I^-] = \left(\frac{10\text{ mL}}{20\text{ mL}}\right)(1.0 \times 10^{-3}\ M)$$
$$= 5.0 \times 10^{-4}\ M$$
$$[Pb^{2+}][I^-]^2 = [5.0 \times 10^{-5}][5.0 \times 10^{-4}]^2$$
$$= 125 \times 10^{-13}$$
$$= 1.3 \times 10^{-11}$$

Since $1.3 \times 10^{-11} < 7.1 \times 10^{-9}$, no precipitation will occur. However, if 10 mL of 1.0×10^{-2} M $Pb(NO_3)$ is added to 10 mL of 2.0×10^{-2} M KI, then

$$[Pb^{2+}] = \left(\frac{10\text{ mL}}{20\text{ mL}}\right)(1.0 \times 10^{-3}\ M)$$
$$= 5.0 \times 10^{-3}\ M$$
$$[I^-] = \left(\frac{10\text{ mL}}{20\text{ mL}}\right)(2.0 \times 10^{-3})$$
$$= 1.0 \times 10^{-2}\ M$$

and

$$[Pb^{2+}][I^-]^2 = [5.0 \times 10^{-3}][1.0 \times 10^{-2}]^2 = 5.0 \times 10^{-7}$$

Because $5.0 \times 10^{-7} > 7.1 \times 10^{-9}$, precipitation of PbI_2 will occur in this solution.

PROCEDURE

A. Preparation of a Calibration Curve

WORK IN GROUPS OF FOUR TO OBTAIN YOUR CALIBRATION CURVE, BUT EVALUATE YOUR DATA INDIVIDUALLY. See Experiment 30 for instructions on the use of the spectrophotometer. Using a buret, add 1, 5, 10, and 15 mL of standardized 0.0024 M K_2CrO_4 to each of four clean, dry 100-mL volumetric flasks and dilute to the 100 mL mark with 0.25 M $NaNO_3$. Calculate the CrO_4^{2-} concentration in each of these solutions. Measure the absorbance of these solutions at 375 nm and plot the absorbance versus concentration to construct your calibration curve.

B. Determination of the Solubility-Product Constant

Accurately prepare three separate solutions in separate 150-mm test tubes by adding 5 mL of 0.004 M $AgNO_3$ to 5 mL of 0.0024 M K_2CrO_4.

Stopper each test tube. Shake the solutions thoroughly at periodic intervals for about 15 min to establish equilibrium between the solid phase and the ions in solution. Transfer approximately 3 mL of each solution along with most of the insoluble Ag_2CrO_4 to 75-mm test tubes and centrifuge. Discard the supernatant liquid and retain the precipitate. To each of the test tubes add 2 mL 0.25 M $NaNO_3$. Shake each test tube thoroughly and centrifuge again. Discard the supernatant liquid, then add 2 mL of 0.25 M $NaNO_3$ to each of the test tubes and shake the test tubes vigorously and periodically for about 15 min to establish an equilibrium betwen the solid and the solution. There must be some solid Ag_2CrO_4 remaining in these test tubes. If there is not, start over again. After shaking the test tubes for about 15 min, centrifuge the mixtures. Transfer the clear, pale yellow supernatant liquid from each of the three test tubes to a clean, dry cuvette. Measure and record the absorbance of the three solutions.

Using your calibration curve, calculate the molar concentration of CrO_4^{2-} in each solution.

Note on Calculations

You are determining the K_{sp} of Ag_2CrO_4 in this experiment. The equilibrium reaction for the dissolution of Ag_2CrO_4 is

$$Ag_2CrO_4(s) \rightleftharpoons 2Ag^+(aq) + CrO_4^{2-}(aq)$$

for which $K_{sp} = [Ag^+]^2[CrO_4^{2-}]$.

You should note that at equilibrium $[Ag^+] = 2[CrO_4^{2-}]$; hence, having determined the concentration of chromate ions, you know the silver-ion concentration.

REVIEW QUESTIONS

Before beginning this experiment in the laboratory, you should be able to answer the following questions:

1. Write the solubility equilibrium and the solubility-constant expression for the slightly soluble salt CaF_2.
2. Calculate the number of moles of Ag^+ in 5 mL of 0.004 *M* $AgNO_3$ and the number of moles of CrO_4^{2-} in 5 mL of 0.0024 *M* K_2CrO_4.
3. If 10 mL of 0.004 *M* $AgNO_3$ is added to 10 mL of 0.0024 *M* K_2CrO_4, is either Ag^+ or CrO_4^{2-} in stoichiometric excess? If so, which is in excess?
4. The K_{sp} for $BaCrO_4$ is 1.095×10^{-5}. Will $BaCrO_4$ precipitate upon mixing 10 mL of 1×10^{-4} *M* $Ba(NO_3)_2$ with 10 mL of 1×10^{-4} *M* K_2CrO_4?
5. The K_{sp} for $BaCO_3$ is 5.1×10^{-9}. How many grams of $BaCO_3$ will dissolve in 100 mL of water?
6. Distinguish between the equilibrium-constant expression and K_{sp} for the dissolution of a sparingly soluble salt.
7. List as many experimental techniques as you can that may be used to determine K_{sp} for a sparingly soluble salt.
8. Why must some solid remain in contact with a solution of a sparingly soluble salt in order to ensure equilibrium?
9. In general, when will a sparingly soluble salt precipitate from solution?
10. Sparingly soluble bases and salts, such as $Fe(OH)_2$ and $FeCO_3$, are more soluble in acidic than in neutral solutions. Why?

NOTES AND CALCULATIONS

Name ______________________________ Desk __________

Date ______________ Laboratory Instructor ______________________

REPORT SHEET FOR EXPERIMENT 24

DETERMINATION OF THE SOLUBILITY-PRODUCT CONSTANT FOR A SPARINGLY SOLUBLE SALT

A. Calibration Curve

Initial $[CrO_4^{2-}]$ ________

	Volume of 0.0024 M K_2CrO_4	Total volume	$[CrO_4^{2-}]$	Absorbance
1.				
2.				
3.				
4.				

Molar absorption coefficient for $[CrO_4^{2-}]$

1. ________ **2.** ________ **3.** ________ **4.** ________

Average molar absorption coefficient ________

Standard deviation (show calculations) ________

B. Determination of K_{sp}

	Absorbance	$[CrO_4^{2-}]$	$[Ag^+]$	K_{sp} of $Ag_2(CrO_4)$
1.				
2.				
3.				

Average K_{sp} (show calculations) __________

Standard deviation ___________
(Show calculations)

QUESTIONS

1. If the standard solutions had unknowingly been made up to be 0.0024 *M* $AgNO_3$ and 0.0040 *M* K_2CrO_4, would this have affected your results? How?

2. If your cuvette had been dirty, how would this have affected the value of K_{sp}?

3. Using your determined value of K_{sp}, calculate how many milligrams of Ag_2CrO_4 will dissolve in 10.0 mL of H_2O.

4. The experimental procedure for this experiment has you add 5 mL of 0.004 *M* $AgNO_3$ to 5 mL of 0.0024 *M* K_2CrO_4. Is either of these reagents in excess, and if so, which one?

5. Use your experimentally determined value of K_{sp} and show, by calculation, that Ag_2CrO_4 should precipitate when 5 mL of 0.004 M $AgNO_3$ are added to 5 mL of 0.0024 M K_2CrO_4.

6. Look up the accepted value of K_{sp} for Ag_2CrO_4 in the back of your textbook. Calculate the percentage error in your experimentally determined value for K_{sp}.

7. Although Ag_2CrO_4 is insoluble in water, it is soluble in dilute HNO_3. Explain, using chemical equations.

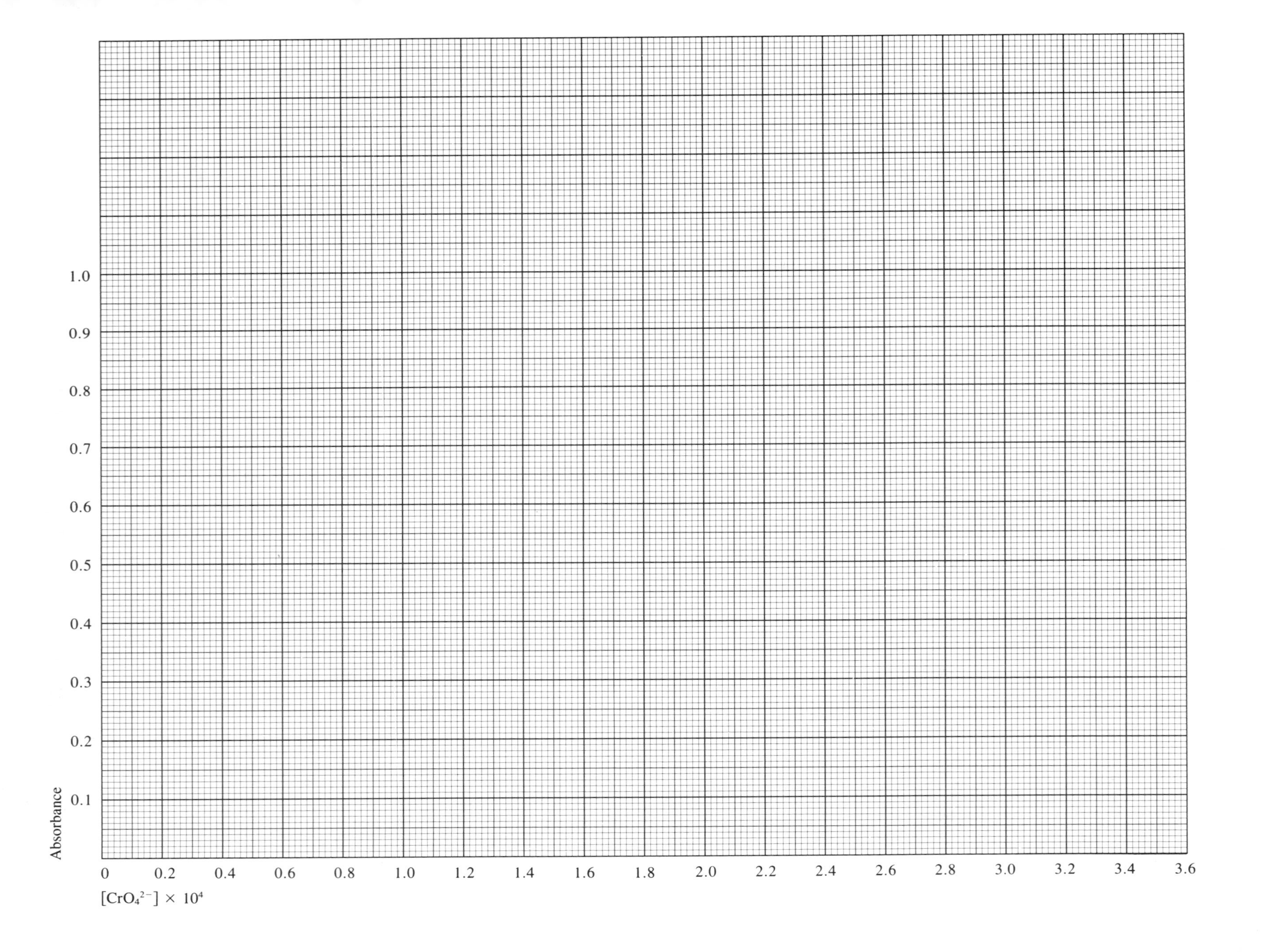
Absorbance
1.0
0.9
0.8
0.7
0.6
0.5
0.4
0.3
0.2
0.1
0
0.2
0.4
0.6
0.8
1.0
1.2
1.4
1.6
1.8
2.0
2.2
2.4
2.6
2.8
3.0
3.2
3.4
3.6
[CrO4 2−] × 10 4

Heat of Neutralization

EXPERIMENT

25

OBJECTIVE

To measure, using a calorimeter, the energy changes accompanying neutralization reactions.

APPARATUS AND CHEMICALS

Styrofoam cups (2)	1 *M* HCl
Bunsen burner	1 *M* NaOH
ring stand and ring	1 *M* acetic acid ($HC_2H_3O_2$)
wire gauze	50-mL graduated cylinder
thermometers (2)	cardboard square with hole in center
250-mL beaker	split one-hole rubber stopper
400-mL beaker	

WORK IN PAIRS.

DISCUSSION

Every chemical change is accompanied by a change in energy, usually in the form of heat. The energy change of a reaction that occurs at constant pressure is termed the *heat of reaction* or the *enthalpy change*. The symbol ΔH (the symbol Δ means "change in") is used to denote the enthalpy change. If heat is evolved, the reaction is *exothermic* ($\Delta H < 0$); and if heat is absorbed, the reaction is *endothermic* ($\Delta H > 0$). In this experiment you will measure the heat of neutralization (or the enthalpy of neutralization) when an acid and a base react to form water.

This quantity of heat is measured experimentally by allowing the reaction to take place in a thermally insulated vessel called a *calorimeter*. The heat liberated in the neutralization will cause an increase in the temperature of the solution and of the calorimeter. If the calorimeter were perfect, no heat would be radiated to the laboratory. The calorimeter you will use in this experiment is shown in Figure 25.1.

Because we are concerned with the heat of the reaction and because some heat is absorbed by the calorimeter itself, we must know the amount of heat absorbed by the calorimeter. This requires that we determine the heat capacity of the calorimeter. By "heat capacity of the calorimeter" we mean the amount of heat (that is, the number of joules) required to raise its temperature 1 kelvin, which is the same as 1°C. In this experiment, the temperature of the calorimeter and its contents is measured before and after the reaction. The change in the enthalpy, ΔH, is equal to the product of the temperature change, ΔT, times the heat capacity of the calorimeter and its contents:

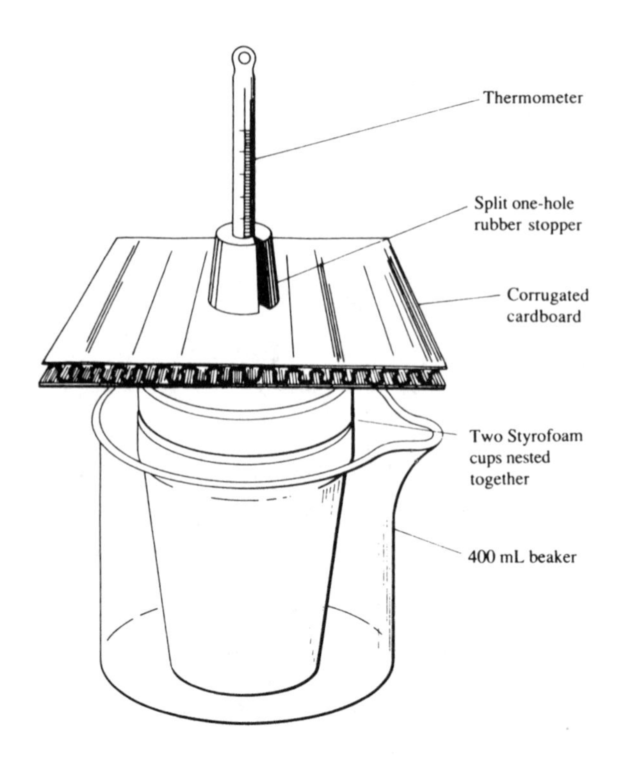

Figure 25.1 A simple calorimeter.

$$\Delta H = \Delta T(\text{heat capacity of calorimeter} + \text{heat capacity of contents}) \quad [1]$$

Note the *numerical difference* on the Celsius scale is the same as the *numerical difference* on the kelvin scale where ΔT is the difference between the final and initial temperatures.

The heat capacity of the calorimeter is determined by measuring the temperature change that occurs when a known amount of hot water is added to a known amount of cold water in the calorimeter. The heat lost by the warm water is equal to the heat gained by the cold water and the calorimeter. (We assume no heat is lost to the laboratory.) For example, if T_1 equals the temperature of a calorimeter and 50 mL of cooler water, if T_2 equals the temperature of 50 mL of warmer water added to it, and if T_f equals the temperature after mixing, then the heat lost by the warmer water is

$$\text{Heat lost by warmer water} = (T_2 - T_f) \times 50 \text{ g} \times 4.18 \text{ J/K-g} \quad [2]$$

The specific heat of water is 4.184 J/K-g, and the density of water is 1.00 g/mL. The heat gained by the cooler water is

$$\text{Heat gained by cooler water} = (T_f - T_1) \times 50 \text{ g} \times 4.18 \text{ J/K-g} \quad [3]$$

The heat lost to the calorimeter is the difference between heat lost by the warmer water and that gained by the cooler water:

$$(\text{heat lost by warmer water}) - (\text{heat gained by cooler water}) = \text{heat lost to calorimeter}$$

Substituting Equations [2] and [3] we have

$$[(T_2 - T_f) \times 50 \text{ g} \times 4.18 \text{ J/K-g}] - [(T_f - T_1) \times 50 \text{ g} \times 4.18 \text{ J/K-g}] = (T_f - T_1) \times \text{heat capacity of calorimeter} \quad [4]$$

Note that the heat lost to the colorimeter equals its temperature change times its heat capacity. Thus by measuring T_1, T_2, and T_f, the heat capacity of the calorimeter can be calculated from Equation [4]. This is illustrated in Example 25.1.

EXAMPLE 25.1

Given the following data, calculate the heat lost by the warmer water, the heat lost to the cooler water, the heat lost to the calorimeter, and the heat capacity of the calorimeter:

Temperature of 50 mL warmer water: 37.9°C = T_2
Temperature of 50 mL cooler water: 20.9°C = T_1
Temperature after mixing: 29.1°C = T_f

Solution: The heat lost by the warmer water, where ΔT = 37.9°C − 29.1°C, is

$$8.8 \text{ K} \times 50 \text{ g} \times 4.18 \text{ J/K-g} = 1840 \text{ J}$$

The heat gained by the cooler water, where ΔT = 29.1°C − 20.9°C, is

$$8.2 \text{ K} \times 50 \text{ g} \times 4.18 \text{ J/K-g} = 1710 \text{ J}$$

The heat lost to the calorimeter is

$$1840 \text{ J} - 1710 \text{ J} = 130 \text{ J}$$

The heat capacity of the calorimeter is, therefore,

$$130\text{K}/8.2 \text{ K} = 16 \text{ J/K}$$

Once the heat capacity of the calorimeter is determined, Equation [1] can be used to determine the ΔH for the neutralization reaction. Example 25.2 illustrates such a calculation.

EXAMPLE 25.2

Given the following data, calculate the heat gained by the solution, the heat gained by the calorimeter, and the heat of reaction:

Temperature of 50 mL of acid before mixing: 21.0°C
Temperature of 50 mL of base before mixing: 21.0°C
Temperature of 100 mL of solution after mixing: 27.5°C

Assume that the density of these solutions is 1.00 g/mL.

Solution: The heat gained by the solution, where ΔT = 27.5°C − 21.O°C, is

$$6.5 \text{ K} \times 100 \text{ g} \times 4.18 \text{ J/K-g} = 2720 \text{ J}$$

The heat gained by the calorimeter, where ΔT = 27.5°C − 21.0°C, is

$$6.5 \text{ K} \times 16 \text{ J/K} = 104 \text{ J}$$

The heat of reaction is therefore

$$2720 \text{ J} + 104 \text{ J} = 2820 \text{ J}$$

or

$$2.82 \text{ kJ}$$

PROCEDURE

A. Heat Capacity of Calorimeter

Construct a calorimeter similar to the one shown in Figure 25.1 by nesting two Styrofoam cups together. Use a cork borer to make a hole in the lid just big enough to admit the thermometer and slip the thermometer into a split one-hole rubber stopper to prevent the thermometer from entering too deeply into the calorimeter. The thermometer should not touch the bottom of the cup. Rest the entire apparatus in a 400-mL beaker to provide stability.

Place exactly 50 mL of tap water in the calorimeter cup and replace the cover and thermometer. Allow 5 to 10 min for the system to reach thermal equilibrium; then record the temperature to the nearest 0.1°C.

Place exactly 50 mL of water in a 250-mL beaker and heat the water with a low flame until the temperature is approximately 15–20°C above room temperature. Do not heat to boiling, or appreciable water wil be lost, leading to an erroneous result. Allow the hot water to stand for a minute or two; quickly record its temperature to the nearest 0.1°C and pour it as completely as possible into the calorimeter. Replace the lid with the thermometer and carefully stir the water with the thermometer. Observe the temperature for the next 3 min and record the temperature every 15 s. Plot the temperature as a function of time, as shown in Figure 25.2. Determine ΔT from your curve and then do the calculations indicated on the report sheet.

B. Heat of Neutralization of HCl–NaOH

Dry the calorimeter and the thermometer with a towel. Carefully measure 50 mL of 1.0 *M* NaOH and add it to the calorimeter. Place the lid on the calorimeter but leave the thermometer out. Measure out exactly 50 mL of 1.0 *M* HCl into a dry beaker. Allow it to stand near the calorimeter for 3 to 4 min. Measure the temperature of the acid; rinse the thermometer with tap

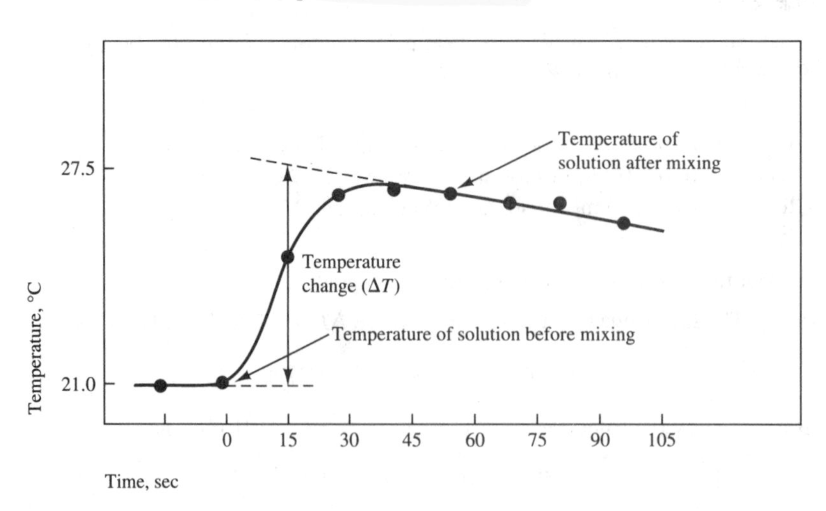

Figure 25.2 Temperature as a function of time.

water and wipe dry. Insert the thermometer into the calorimeter and measure the temperature of the NaOH solution.

The temperatures of the NaOH and the HCl should not differ by more than 0.5°C. If the difference is greater than 0.5°C, adjust the temperature of the HCl by *either* warming it by holding the beaker in your hands or cooling the outside of the beaker with tap water until the temperature of the HCl is within 0.5°C of that of the NaOH.

Record the temperature of the NaOH solution. Lift the lid and carefully add the 1.0 *M* HCl all at once. Be careful not to splash any on the upper sides of the cup. Stir the solution gently with the thermometer and record the temperature as a function of time every 15 s for the next 3 min. Construct a temperature-versus-time curve and determine ΔT. Calculate the heat of neutralization per mole of water formed. You may assume that the NaCl solution has the same density and specific heat as water.

C. Heat of Neutralization of $HC_2H_3O_2$–NaOH

Follow the same procedure as in Part B but substitute 1.0 *M* $HC_2H_3O_2$ for 1.0 *M* HCl. Calculate the heat of neutralization per mole of water formed.

REVIEW QUESTIONS

Before beginning this experiment in the laboratory, you should be able to answer the following questions:

1. Define endothermic and exothermic reactions in terms of the sign of ΔH.
2. A 425 mL sample of water was cooled from 50.0°C to 10°C. How much heat was lost?
3. Define heat capacity.
4. How many joules are required to change the temperature of 80.0 g of water from 23.3°C to 38.8°C?
5. Define specific heat
6. Calculate the final temperatures when 50 mL of water at 40°C are added to 25 mL of water at 20°C.
7. Describe how you could determine the heat capacity of a metal by using the apparatus and techniques in this experiment.
8. A piece of metal weighing 5.10 g at a temperature of 48.6°C was placed into 20.00 mL of water in a calorimeter at 22.1°C, and the final equilibrium temperature was found to be 28.2°C. What is the heat capacity of the metal?
9. If the heat capacity of methanol is 2.51 J/K-g, how many joules are necessary to raise the temperature of 50.0 g of methanol from 18°C to 33°C?
10. When a 3.25 g sample of solid sodium hydroxide was dissolved in a calorimeter in 100.0 g of water, the temperature rose from 23.9°C to 32.0°C. Calculate ΔH (in kJ/mol NaOH) for the solution process:

$$NaOH(s) \longrightarrow Na^+(aq) + OH(aq)$$

Assume it's a perfect calorimeter and that the specific heat of the solution is the same as that of pure water.

NOTES AND CALCULATIONS

Rates of Chemical Reactions I: A Clock Reaction

EXPERIMENT

26

OBJECTIVE

To measure the effect of concentration upon the rate of the reaction of peroxydisulfate ion with iodine ion; to determine the order of the reaction with respect to the reactant concentrations; and to obtain the rate law for the chemical reaction.

APPARATUS AND CHEMICALS

- burets (2)
- 1-mL pipets (2)
- clock or watch with second hand
- 125-mL Florence flask
- 400-mL beaker
- 500-mL Florence flask
- test tube
- pipet bulb
- 250-mL Erlenmeyer flasks (4)
- buret clamp
- ring stand
- 25-mL pipet
- 50-mL pipet
- thermometer
- 0.2 *M* KI (200 mL)
- 0.4 *M* $Na_2S_2O_3$ (100 mL) (freshly prepared)
- 1 percent starch solution, boiled
- 0.2 *M* KNO_3 (300 mL)
- 0.1 *M* solution of Na_2EDTA
- 0.2 *M* $(NH_4)_2S_2O_8$ (200 mL) (prepared from fresh solid)

WORK IN PAIRS BUT EVALUATE YOUR DATA INDIVIDUALLY.

DISCUSSION

Factors Affecting Rates of Reactions

On the basis of the experiments you've performed, you probably have already noticed that reactions occur at varying speeds. There is an entire spectrum of speeds of reactions, ranging from very slow to extremely fast. For example, the rusting of iron is reasonably slow, whereas the decomposition of TNT is extremely fast. The branch of chemistry that is concerned with the rates of reactions is called *chemical kinetics*. Experiments show that rates of homogeneous reactions in solution depend upon:

1. The nature of the reactants;
2. The concentration of the reactants;
3. The temperature; and
4. Catalysis.

Before a reaction can occur, the reactants must come into direct contact via collisions of the reacting particles. However, even then, the reacting particles (ions or molecules) must collide with sufficient energy to result in a reaction; if they do not, their collisions are ineffective and analogous to collisions of billiard balls. With these considerations in mind, we can qualitatively explain how the various factors influence the rates of reactions.

Concentration Changing the concentration of a solution alters the number of particles per unit volume. The more particles present in a given volume, the greater the probability of their colliding. Hence, increasing the concentration of a solution increases the number of collisions per unit time and therefore the rate of reaction.

Temperature Since temperature is a measure of the average kinetic energy, an increase in temperature increases the kinetic energy of the particles. An increase in kinetic energy increases the velocity of the particles and therefore the number of collisions between them in a *given period of time*. Thus, the rate of reaction increases. Also, an increase in kinetic energy results in a greater proportion of the collisions having the required energy for reaction. As a rule of thumb, for each 10° increase in temperature, the rate of reaction doubles.

Catalyst Catalysts, in some cases, are believed to increase reaction rates by bringing particles into close juxtaposition in the correct geometrical arrangement for reaction to occur. In other instances, catalysts offer an alternative route to the reaction, one that requires less energetic collisions between reactant particles. If less energy is required for a successful collision, a larger percentage of the collisions will have the requisite energy, and the reaction will occur faster. Actually, the catalyst may take an active part in the reaction, but at the end of the reaction, the catalyst can be recovered chemically unchanged.

Order of Reaction Defined

Let's examine now precisely what is meant by the expression *rate of reaction*. Consider the hypothetical reaction

$$A + B \longrightarrow C + D \qquad [1]$$

The rate of this reaction may be measured by observing the rate of disappearance of either of the reactants A and B, or the rate of appearance of either of the products C and D. In practice, then, one measures the change of concentration with time of either A, B, C, or D. Which species you choose to observe is a matter of convenience. For example, if A, B, and D are colorless and C is colored, you could conveniently measure the rate of appearance of C by observing an increase in the intensity of the color of the solution as a function of time. Mathematically, the rate of reaction may be expressed as follows:

$$\text{Rate of disappearance of A} = \frac{\text{change in concentration of A}}{\text{time required for change}} = \frac{\Delta[A]}{\Delta t}$$

$$\text{Rate of appearance of C} = \frac{\text{change in concentration of C}}{\text{time required for change}} = \frac{\Delta[C]}{\Delta t}$$

In general, the rate of the reaction will depend upon the concentration of the reactants. Thus, the rate of our hypothetical reaction may be expressed as

$$\text{Rate} = k[A]^x[B]^y \qquad [2]$$

where [A] and [B] are the molar concentrations of A and B, x and y are the powers to which the respective concentrations must be raised to describe the rate, and k is the *specific rate constant*. One of the objectives of chemical kinetics is to determine the rate law. Stated slightly differently, one goal of measuring the rate of the reaction is to determine the numerical values of x and y. Suppose that we found $x = 2$ and $y = 1$ for this reaction. Then

$$\text{Rate} = k[A]^2[B] \qquad [3]$$

would be the rate law. It should be evident from Equation [3] that doubling the concentration of B (keeping [A] the same) would cause the reaction rate to double. On the other hand, doubling the concentration of A (keeping [B] the same) would case the rate to increase by a factor of 4, because the rate of the reaction is proportional to the *square* of the concentration of A. The powers to which the concentrations in the rate law are raised are termed the *order of the reaction*. In this case, the reaction is said to be second order in A and first order in B. The *overall order* of the reaction is the sum of the exponents, 2 + 1 = 3, or a third-order reaction. It is possible to determine the order of the reaction by noting the effects of changing reagent concentrations on the rate of the reaction.

It should be emphasized that k, the specific rate constant, has a definite value that is independent of the concentration. It is characteristic of a given reaction and depends only on temperature. Once the rate is known, the value of k can be calculated.

Reaction of Peroxydisulfate Ion with Iodine Ion

In this experiment you will measure the rate of the reaction

$$S_2O_8^{2-} + 2I^- \longrightarrow I_2 + 2SO_4^{2-} \qquad [4]$$

and you will determine the rate law by measuring the amount of peroxy disulfate, $S_2O_8^{2-}$, that reacts as a function of time. The rate law to be determined is of the form

$$\text{Rate of disappearance of } S_2O_8^{2-} = k[S_2O_8^{2-}]^x[I^-]^y \qquad [5]$$

or

$$\frac{\Delta[S_2O_8^{2-}]}{\Delta t} = k[S_2O_8^{2-}]^x[I^-]^y$$

Your goal will be to determine the values of x and y as well as the specific rate constant, k.

You will add to the solution a small amount of another reagent (sodium thiosulfate, $Na_2S_2O_3$), which will cause a change in the color of the solution.

The amount is such that the color change will occur when 2×10^{-4} mol of $S_2O_8^{2-}$ has reacted. For reasons to be explained shortly, the solution will turn blue-black when 2×10^{-4} mol of $S_2O_8^{2-}$ has reacted. You will quickly add another portion of $Na_2S_2O_3$ after the appearance of the color, and the blue-black color will disappear. When the blue-black color reappears the second time, *another* 2×10^{-4} mol of $S_2O_8^{2-}$ has reacted, making a total of $2(2 \times 10^{-4})$ mol of $S_2O_8^{2-}$ that has reacted. You will repeat this procedure several times, keeping *careful* note of the time for the appearance of the blue-black colors. By graphing the amount of $S_2O_8^{2-}$ consumed versus time, you will be able to determine the rate of the reaction. By changing the initial concentrations of $S_2O_8^{2-}$ and I^- and observing the effects upon the rate of the reaction, you will determine the order of the reaction with respect to $S_2O_8^{2-}$ and I^-.

The blue-black color that will appear in the reaction is due to the presence of a starch-iodine complex that is formed from iodine, I_2, and starch in the solution. The color therefore will not appear until a detectable amount of I_2 is formed according to Equation [4]. The thiosulfate that is added to the solution reacts *extremely rapidly* with the iodine, as follows:

$$I_2 + 2S_2O_3^{2-} \longrightarrow 2I^- + S_4O_6^{2-} \qquad [6]$$

Consequently, until the same amount of $S_2O_3^{2-}$ that is added is *all* consumed, there will not be a sufficient amount of I_2 in the solution to yield the blue-black color. You will add 4×10^{-4} mol of $S_2O_3^{2-}$ each time (these equal portions are termed *aliquots*), and from the stoichiometry of Equations [4] and [6] you can verify that when this quantity of $S_2O_3^{2-}$ has reacted, 2×10^{-4} mol of $S_2O_8^{2-}$ has reacted. Note also that although iodine, I^-, is consumed according to Equation [4], it is rapidly regenerated according to Equation [6] and therefore its concentration does not change during a given experiment.

Graphical Determination of Rate

The more rapidly the 2×10^{-4} mol of $S_2O_8^{2-}$ is consumed, the faster is the reaction. To determine the rate of the reaction, a plot of moles of $S_2O_8^{2-}$ that have reacted versus the time required for the reaction is made, as shown in Figure 26.1. The best straight line passing through the origin is drawn, and the slope is determined. The slope, $\Delta S_2O_8^{2-}/\Delta t$, corresponds to the moles of $S_2O_8^{2-}$ that have been consumed per second and is proportional to the rate. Since the rate corresponds to the change in the concentration of $S_2O_8^{2-}$ per second, dividing the slope by the volume of the solution yields the rate of disappearance of $S_2O_8^{2-}$, that is, $\Delta[S_2O_8^{2-}]\Delta t$. If the total volume of the solution in this example were 75 mL, the rate would be as follows:

$$\frac{4.5 \times 10^{-5}\ \text{mol/s}}{0.075\ \text{L}} = 6.0 \times 10^{-4}\ \text{mol/L-s}$$

If we obtain a rate of 6.0×10^{-4} mol/L-s when $[S_2O_8^{2-}] = 2.0\ M$ and $[I^-] = 2.0\ M$, and a rate of 3.0×10^{-4} mol/L-s when $[S_2O_8^{2-}] = 1.0\ M$ and $[I^-] = 2.0\ M$, we then know that doubling the concentration of $S_2O_8^{2-}$ doubles the rate of the reaction and the reaction is first order in $S_2O_8^{2-}$. By varying the initial concentrations of $S_2O_8^{2-}$ and I^-, you can, via the above analysis, determine the order of the reaction with respect to both of these species.

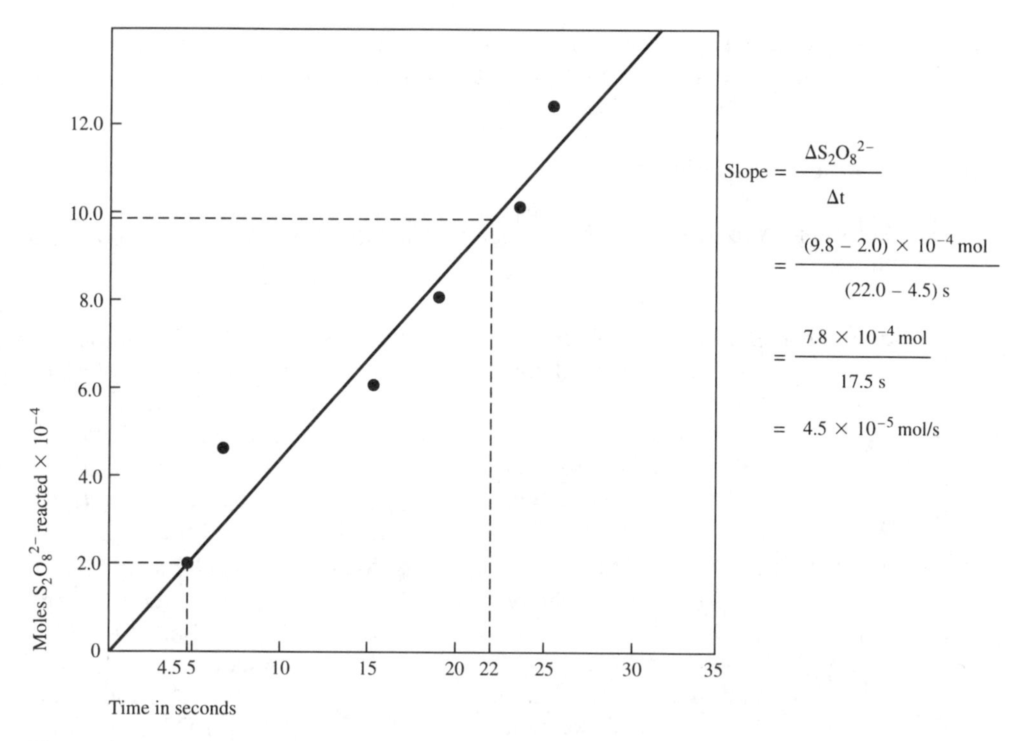

Figure 26.1 Graphical determination of rate.

Helpful Comments

1. According to the procedure of this experiment, the solution will turn blue-black when exactly 2×10^{-4} mol of $S_2O_8^{2-}$ has reacted.
2. The purpose of the KNO_3 solution in this reaction is to keep the *reaction medium* the same in each run in terms of the concentration of ions; it does not enter into the reaction in any way.
3. The reaction studied in this experiment is catalyzed by metal ions. The purpose of the drop of the EDTA solution is to minimize the effects of trace quantities of metal ion impurities that would cause spurious effects on the reaction.
4. You will perform a few preliminary experiments to become acquainted with the observations in this experiment so that you will know what to expect in the reactions.
5. The initial concentrations of the reactants have been provided for you on the report sheet.

PROCEDURE

A. Preliminary Experiments

1. Dilute 5 mL of 0.2 *M* KI solution with 10 mL of distilled water in a test tube, add 3 drops of starch solution and mix thoroughly, and then add 5 mL of 0.2 *M* $(NH_4)_2S_2O_8$ solution. Mix. Wait a while and observe color changes.

2. Repeat the procedure in (1), but when the solution changes color add 4 drops of 0.4 *M* $Na_2S_2O_3$, mix the solution, and note the effect that the addition of $Na_2S_2O_3$ has on the color.

B. Kinetics Experiment

Solution Preparation Prepare four reaction solutions as follows (one at a time):

Solution 1: 25.0 mL KI solution
1.0 mL starch solution
1.0 mL $Na_2S_2O_3$ solution
48.0 mL KNO_3 solution
1 drop EDTA solution
Total volume = 75.0 mL

Solution 2: 25.0 mL KI solution
1.0 mL starch solution
1.0 mL $Na_2S_2O_3$ solution
23.0 mL KNO_3 solution
1 drop EDTA solution
Total volume = 50.0 mL

Solution 3: 50.0 mL KI solution
1.0 mL starch solution
1.0 mL $Na_2S_2O_3$ solution
23.0 mL KNO_3 solution
1 drop EDTA solution
Total volume = 75.0 mL

Solution 4: 12.5 mL KI solution
1.0 mL starch solution
1.0 mL $Na_2S_2O_3$ solution
35.5 mL KNO_3 solution
1 drop EDTA solution
Total volume = 50.0 mL

Equipment Setup Set up two burets held by a clamp on a ring stand as shown in Figure 26.2. Use these burets to measure accurately the volumes of the KI and KNO_3 solutions. Use two separate 1-mL pipets for measuring the volumes of the $Na_2S_2O_3$ and starch solutions and use 25-mL and 50-mL pipets to measure the volumes of the $(NH_4)_2S_2O_8$ solutions. Each solution must be freshly prepared to begin the rate study—that is, *prepare solutions 1, 2, 3, and 4 one at a time as you make your measurements.*

Rate Measurements Prepare solution 1 in a 250-mL Erlenmeyer flask that has been scrupulously cleaned and dried. Pipet 25.0 mL of $(NH_4)_2S_2O_8$ solution into a clean, dry 100-mL beaker. *Be ready* to begin timing the reaction when the solutions are mixed (READ AHEAD). The reaction starts the moment the solutions are mixed! BE PREPARED! ZERO TIME! Quickly pour the 25.0 mL of $(NH_4)_2S_2O_8$ solution into solution 1 and swirl vigorously; note the time you begin mixing to the nearest second. At the instant when the blue-black color appears, 2×10^{-4} mol of $S_2O_8^{2-}$ has reacted. IMMEDIATELY (be prepared!) add a 1-mL aliquot of $Na_2S_2O_3$ solution from the pipet and swirl the solution; the color will disappear.

Record the time for the reappearance of the blue-black color. Add another 1-mL aliquot of $Na_2S_2O_3$ solution and note the time for the reappearance of the color. The time interval being measured is that between the appearance of the blue-black color. For good results, these aliquots of $Na_2S_2O_3$ must be measured as quickly, accurately, and reproducibly as possible. Continue this procedure until you have added seven (7) aliquots to solution 1.

You are finished with solution 1 when you have recorded all your times on the report sheet. (The time intervals are cumulative.)

Solutions 2, 3, and 4 should be treated in exactly the same manner except

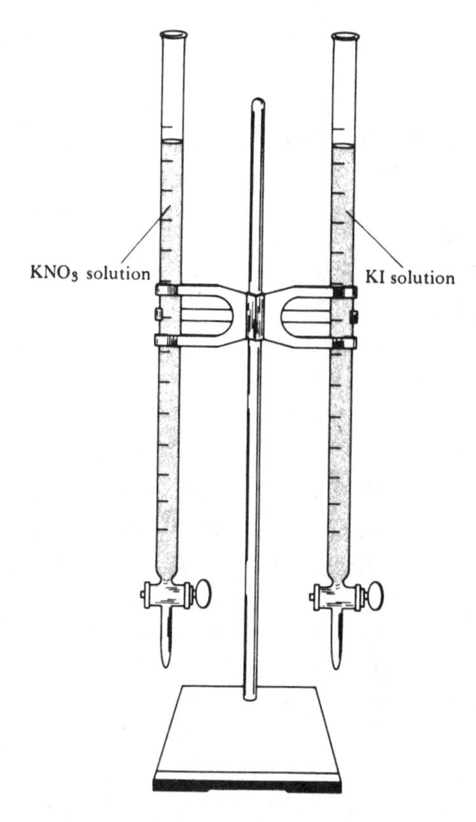

Figure 26.2

that 50.0-mL portions of $(NH_4)_2S_2O_8$ solutions should be added to solutions 2 and 4 and 25 mL of $(NH_4)_2S_2O_8$ solution should be added to solution 3. (CAUTION: *Be on guard—solution 3 will react much more rapidly than solution 1.*) In each of these reactions the final total solution volume is 100 mL.

Timesaving Hint to Instructors

Time may be saved in this experiment by setting up several burets on side tables in the laboratory filled with $(NH_4)_2S_2O_8$, KNO_3, KI, and $Na_2S_2O_3$ solutions. The students may obtain the solution from these burets and refill the burets from bottles of stock solutions also kept there. If the students fill each of seven clean, dry test tubes with 1 mL of $Na_2S_2O_3$ solution, they then can add these aliquots to their reactions at the appearance of the blue color without loss of time.

Calculations

Tabulate on the data sheet for each aliquot of $Na_2S_2O_3$ added to each of the four solutions:

1. The time interval from the start of the reaction (addition of $S_2O_8^{2-}$) to the appearance of color for the first aliquot of $S_2O_3^{2-}$ and the time interval from the preceding color appearance for each succeeding aliquot (column 2);
2. The cumulative time from the start of the reaction to each appearance of color (column 3);
3. The corresponding numbers of moles $S_2O_8^{2-}$ consumed (column 4).

For each solution, plot on the graph paper provided the moles of $S_2O_8^{2-}$ consumed (as the ordinate, vertical axis) versus time in seconds (as the abscissa, horizontal axis), using the data in columns 3 and 4. Calculate the slope of each plot, and from these calculations answer the questions on your report sheet.

REVIEW QUESTIONS

Before beginning this experiment in the laboratory, you should be able to answer the following questions:

1. What factors influence the rate of a chemical reaction?
2. What is the general form of a rate law?
3. What is the order of reaction with respect to A and B for a reaction that obeys the rate law rate $= k[A]^2[B]^3$?
4. Write the chemical equations involved in this experiment and show that the rate of disappearance of $[S_2O_8^{2-}]$ is proportional to the rate of appearance of the blue color of the starch-iodine complex.
5. It is found for the reaction A + C → C that doubling the concentration of either A or B quadruples the rate of the reaction. Write the rate law for this reaction.
6. If 2×10^{-4} mol of $S_2O_8^{2-}$ in 50 mL of solution is consumed in 188 s, what is the rate of consumption of $S_2O_8^{2-}$?
7. Why are chemists concerned with the rates of chemical reactions? What possible practical value does this type of information have?
8. Suppose you were dissolving a metal such as zinc with hydrochloric acid. How would the particle size of the zinc affect the rate of its dissolution?
9. Assuming that a chemical reaction doubles in rate for each 10° temperature increase, by what factor would the rate increase if the temperature were increased 40°C?
10. A reaction between the substances A and B has been found to give the following data:

$$3A + 2B \longrightarrow 2C + D$$

[A] (mol/L)		**[B] (mol/L)**		**Rate of appearance of C (mol/L-h)**
1.0×10^{-2}	______	1.0	______	0.3×10^{-6}
1.0×10^{-2}	______	3.0	______	8.1×10^{-6}
[A] (mol/L)		**[B] (mol/L)**		**Rate of appearance of C (mol/h)**
2.0×10^{-2}	______	3.0	______	3.24×10^{-5}
2.0×10^{-2}	______	1.0	______	1.20×10^{-6}
3.0×10^{-2}	______	3.0	______	7.30×10^{-5}

Using the above data, determine the order of the reaction with respect to A and B and the rate law and calculate the specific rate constant.

Name ______________________________ Desk __________

Date ______________ Laboratory Instructor ______________________

REPORT SHEET FOR EXPERIMENT 26

RATES OF CHEMICAL REACTIONS I: A CLOCK REACTION

A. Preliminary Experiment

1. What are the colors of the following ions: K^+ __________; I^- __________;
2. The color of the starch · I_2 complex is __________

B. Kinetics Experiment

Solution 1. Initial $[S_2O_8^{2-}] = 0.05\ M$; initial $[I^-] = 0.05\ M$. Time experiment started __________

Aliquot no.	Time (s) between appearances of color	Cumulative time(s)	Total moles of $S_2O_8^{2-}$ consumed
1	______	______	2.0×10^{-4}
2	______	______	4.0×10^{-4}
3	______	______	6.0×10^{-4}
4	______	______	8.0×10^{-4}
5	______	______	10×10^{-4}
6	______	______	12×10^{-4}
7	______	______	14×10^{-4}

Solution 2. Initial $[S_2O_8^{2-}] = 0.10\ M$; initial $[I^-] = 0.05\ M$. Time experiment started __________

Aliquot no.	Time (s) between appearances of color	Cumulative time(s)	Total moles of $S_2O_8^{2-}$ consumed
1	______	______	2.0×10^{-4}
2	______	______	4.0×10^{-4}
3	______	______	6.0×10^{-4}
4	______	______	8.0×10^{-4}
5	______	______	10×10^{-4}
6	______	______	12×10^{-4}
7	______	______	14×10^{-4}

Solution 3. Initial $[S_2O_8^{2-}] = 0.05\ M$; initial $[I^-] = 0.10\ M$. Time experiment started ____________

Aliquot no.	Time (s) between appearances of color	Cumulative time(s)	Total moles of $S_2O_8^{2-}$ consumed
1	____________	____________	2.0×10^{-4}
2	____________	____________	4.0×10^{-4}
3	____________	____________	6.0×10^{-4}
4	____________	____________	8.0×10^{-4}
5	____________	____________	10×10^{-4}
6	____________	____________	12×10^{-4}
7	____________	____________	14×10^{-4}

Solution 4. Initial $[S_2O_8^{2-}] = 0.10\ M$; initial $[I^-] = 0.25\ M$. Time experiment started ____________

Aliquot no.	Time (s) between appearances of color	Cumulative time(s)	Total moles of $S_2O_8^{2-}$ consumed
1	____________	____________	2.0×10^{-4}
2	____________	____________	4.0×10^{-4}
3	____________	____________	6.0×10^{-4}
4	____________	____________	8.0×10^{-4}
5	____________	____________	10×10^{-4}
6	____________	____________	12×10^{-4}
7	____________	____________	14×10^{-4}

Calculations

1. Rate of reaction, $\Delta[S_2O_8^{2-}]/\Delta T$, as calculated from graphs (that is, from slopes of lines):

   Solution 1 ____________ Solution 3 ____________

   Solution 2 ____________ Solution 4 ____________

2. What effect does doubling the concentration of I^- have on the rate of this reaction?

3. What effect does changing the $[S_2O_8^{2-}]$ have on the reaction?

4. Write the rate law for this reaction that is consistent with your data.

5. From your knowledge of x and y in the equation (as well as the rate in a given experiment from your graph), calculate k from your data. Rate $= k[S_2O_8^{2-}]^x[I^-]^y$

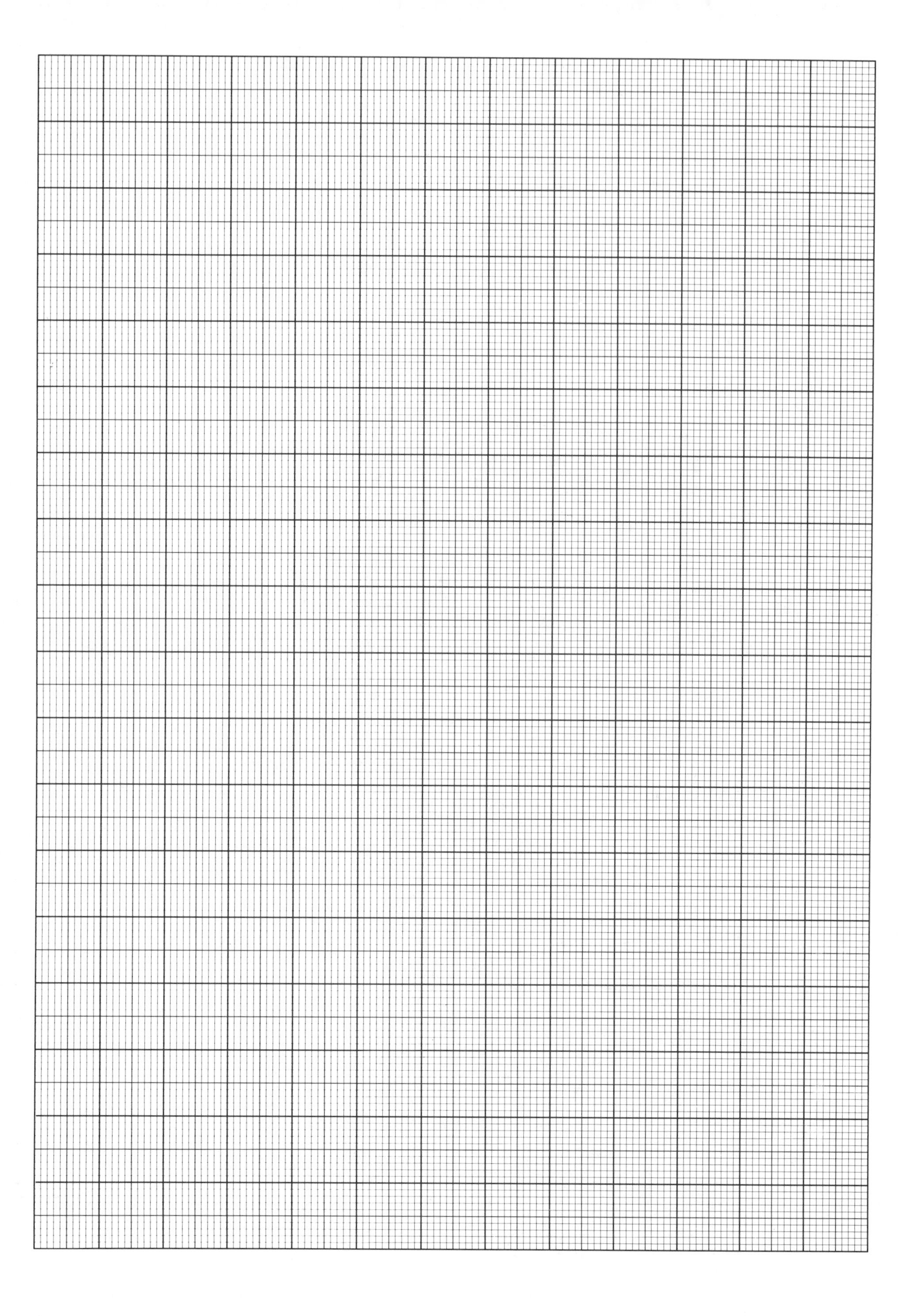

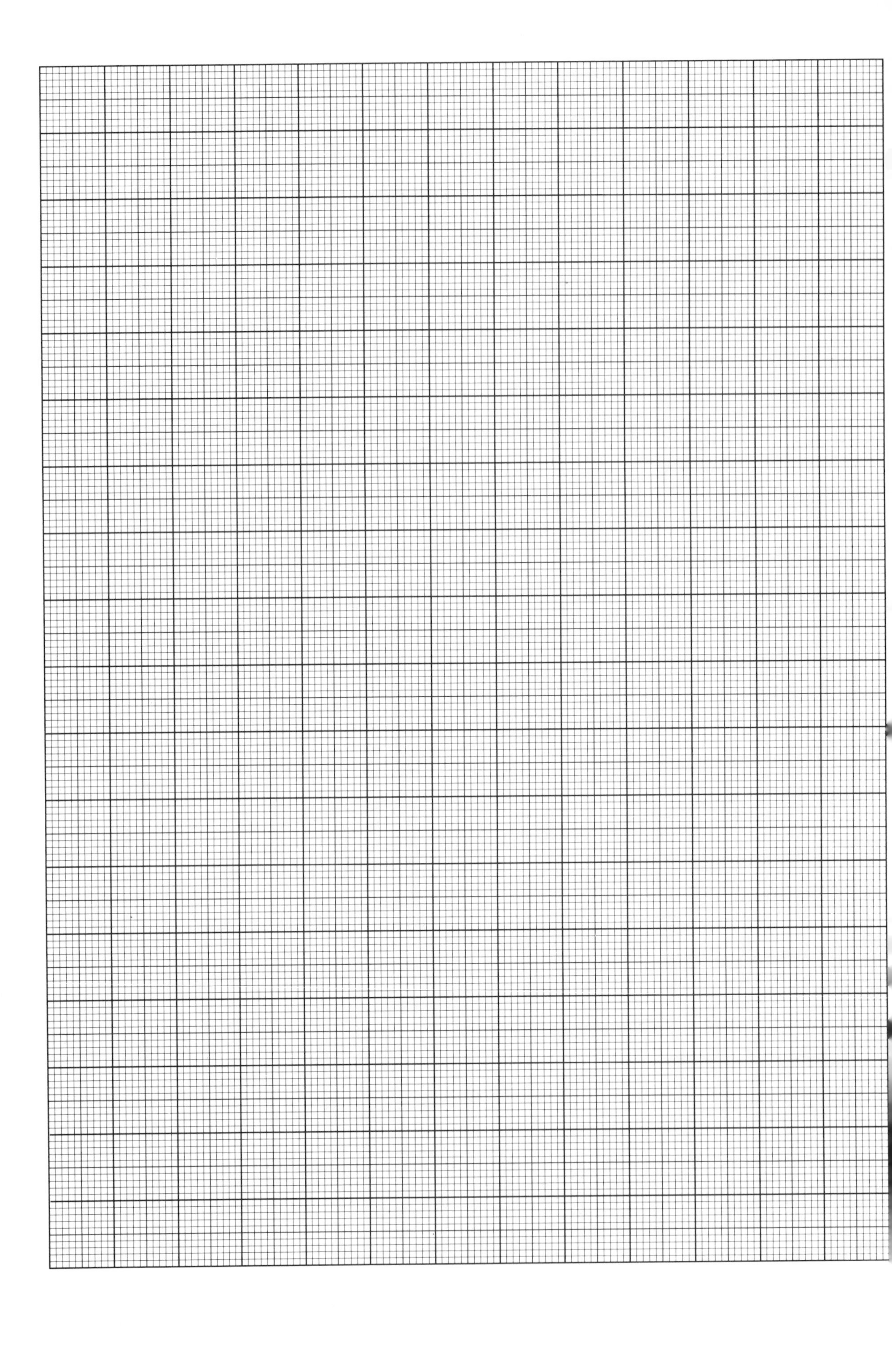

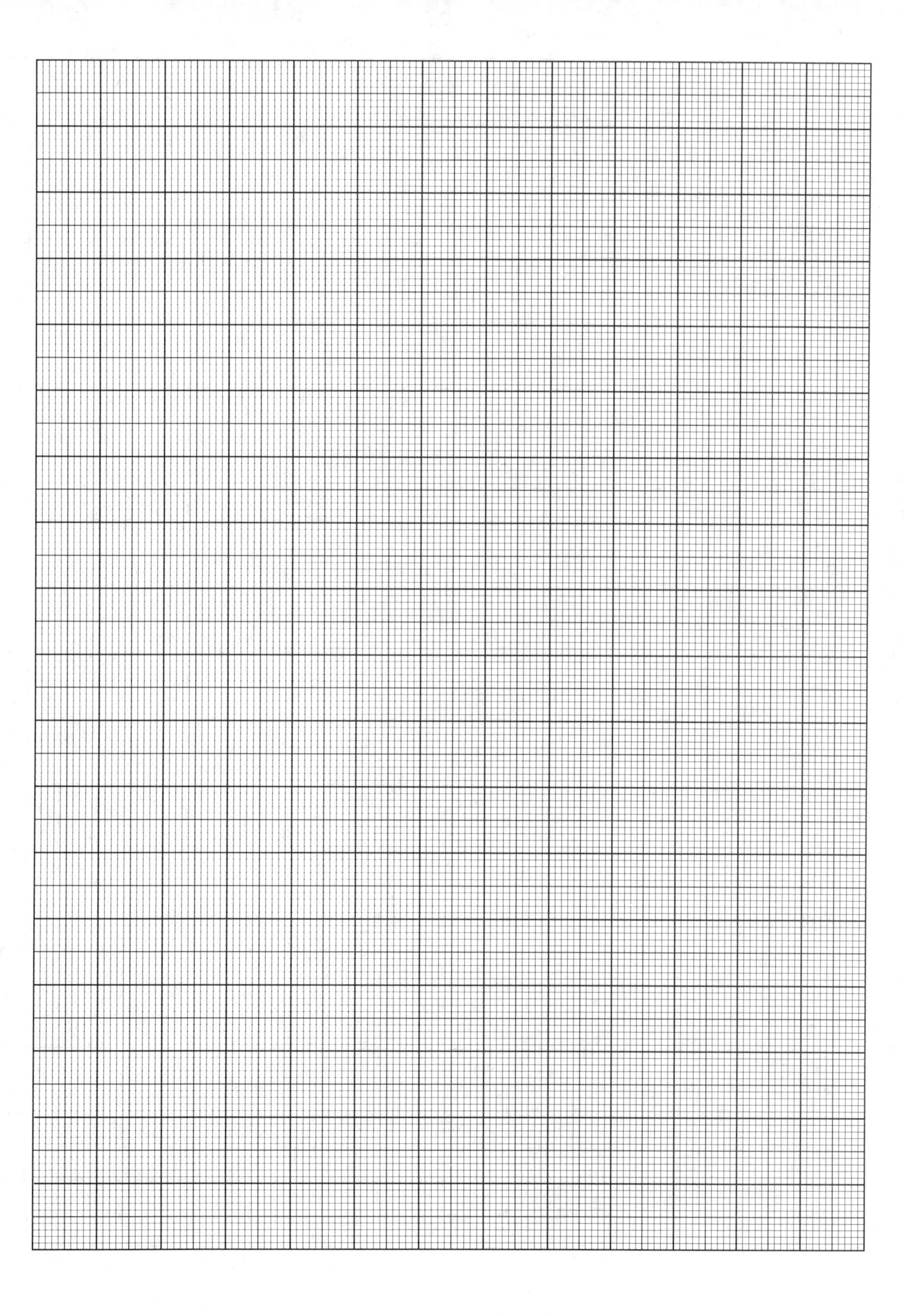

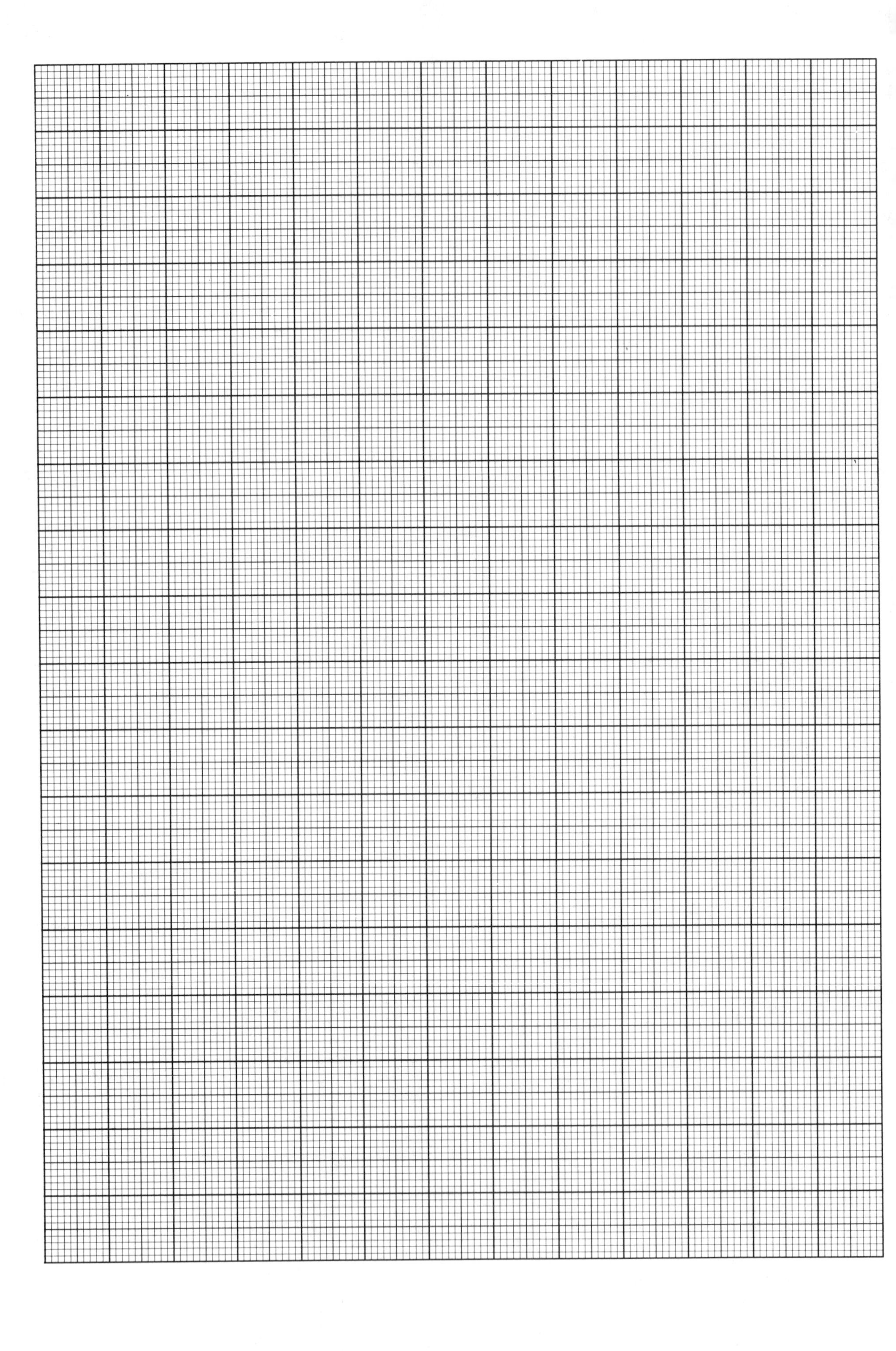

Rates of Chemical Reactions II: Rate and Order of H_2O_2 Decomposition

EXPERIMENT 27

OBJECTIVE

To determine the rate and order of reaction for the decomposition of hydrogen peroxide.

APPARATUS AND CHEMICALS

Mohr buret or standard buret	medicine droppers (2)
leveling bulb or funnel	Bunsen burner and hose
125-mL Erlenmeyer flask	wrist watch or timer
ring stand and iron ring	graduated cylinder
buret clamp	thermometer
rubber tubing	3 percent H_2O_2
pneumatic trough	0.10 *M* KI
no. 00 and no. 6 1-hole stoppers	0.10 *M* $Hg(NO_3)_2$

WORK IN PAIRS BUT EVALUATE YOUR DATA INDEPENDENTLY.

DISCUSSION

We have learned that much information may be obtained from a chemical equation—for example, the identity of the products formed from the specified reactants and the quantitative weight relationships between the substances involved in the reaction. The equation, however, does not tell us anything about the conditions required for the reaction to occur, nor does it give any information about the speed or rate at which the reaction takes place. The area of chemistry concerned with rates of chemical reactions is called *chemical kinetics*. Before proceeding any further you should study the beginning of Experiment 26, where a discussion of factors affecting rates of reaction and an explanation of the term *order of reaction* are given.

The term *rate of reaction* refers to the speed at which reactants are being consumed or products are being formed. This must be determined experimentally by measuring the time rate of change in the concentration of one of the reactants or one of the products or some physical property (such as the volume of a gas or color intensity of a solution) that is directly proportional to one of these concentrations. The rate may be expressed, for example, as moles per liter of product being formed per minute, milliliters of gas being produced per minute, or moles per liter of reactant being consumed per second. In this experiment you will study the rate of the iodide-catalyzed decomposition of hydrogen peroxide to form water and oxygen:

$$2H_2O_2(aq) \xrightarrow{I^-} O_2(g) + 2H_2O(l) \qquad [1]$$

The standard enthalpy change, $\Delta H°_{298}$, and standard free-energy change, $\Delta G°_{298}$, for the reaction in Equation [1] are −189.5 kJ and −211.0 kJ, respectively. Even though $\Delta G°$ is negative and we can conclude that the reaction will proceed spontaneously at room temperature, we cannot predict anything about the *rate* at which H_2O_2 decomposes. Reactions generally involve an energy barrier (see Figure 27.1) that must be overcome in going from reactants to products. This energy barrier is called the *activation energy* and is commonly designated by the term E_a. It corresponds to the minimum energy molecules must have for the reaction to occur. The magnitude of E_a determines the rate of the reaction, whereas the magnitude of the free-energy change determines the extent of the reaction. By observing the volume of oxygen formed as a function of time you will determine how the rate of reaction [1] is affected by different initial concentrations of hydrogen peroxide and iodide ion (which is introduced into the reaction in the form of the strong electrolyte KI). Although iodide does not appear in the chemical equation, it does have a pronounced effect on the rate of the reaction, for it serves as a catalyst.

Your ultimate goal in this experiment is to deduce a rate law for the reaction, showing the dependence of the rate on the concentrations of H_2O_2 and I^-. Your rate law will be of the form

$$\text{Rate of oxygen production} = k[H_2O_2]^x[I^-]^y$$

where k is the specific rate constant and depends only on temperature. Thus your objective is to determine the numerical values of the exponents x and y.

You will also make a qualitative study of the effect that temperature has on the rate of decomposition of H_2O_2. At higher temperatures, the average kinetic energy of molecules is greater. Hence, the fraction of molecules with kinetic energy sufficient for reaction is larger, and the rate of reaction is correspondingly larger. Arrhenius noted a nonlinear relation between rate and temperature and found that most reaction-rate data obeyed the equation

$$\log k = \log A - \frac{E_a}{2.30\,RT}$$

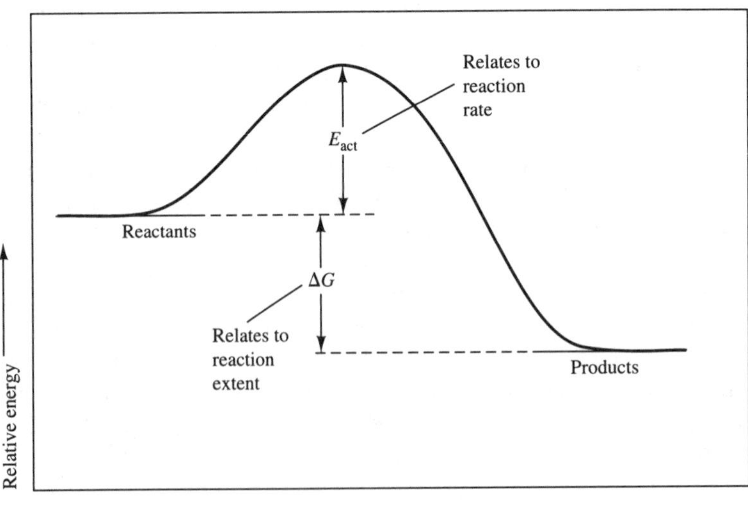

Figure 27.1 Reaction profile for an exothermic reaction illustrating ΔG and E_a.

where k is the specific rate constant, E_a the activation energy, R the gas constant, and T the absolute temperature. A is a constant, or nearly so, as temperature is varied. It is called the frequency factor and is related to the frequency of collisions and the probability that the molecules are suitably oriented for reaction. For many reactions it has been observed that a 10° change in temperature doubles or triples the reaction rate.

PROCEDURE

A. Order of Reaction

Assemble an apparatus for the collection of oxygen similar to that shown in Figure 27.2. Fill the trough with water at room temperature and record the temperature. It may be necessary to add some hot water to achieve room temperature. Add room-temperature water to the assembly until the height of water in the buret is about 10 mL from the top when the water in the leveling bulb is at the same level (see Figure 27.2). Check for leaks in the apparatus by lowering the bulb with the system closed. In the absence of leaks, only small changes in the water level in the buret due to pressure changes will occur as the leveling bulb is lowered.

Solution 1 Add 10 mL of 0.10 *M* KI and 15 mL of distilled water to a clean Erlenmeyer flask. Carefully swirl the flask for a few minutes so that the solution attains the bath temperature in the trough. Add 5 mL of 3 percent H_2O_2 and quickly stopper the flask. One student should keep swirling the flask in the bath as vigorously as possible throughout the remainder of the experiment; the other student should observe the volume of oxygen produced during the reaction at various times. The recording of volume and time should commence after approximately 2 mL of gas has been evolved. Because it is important to measure the volume of oxygen evolved at constant pressure, it is necessary to measure the volume with the level of water in the leveling bulb the same as that in the buret. One student should match the water levels in the buret and

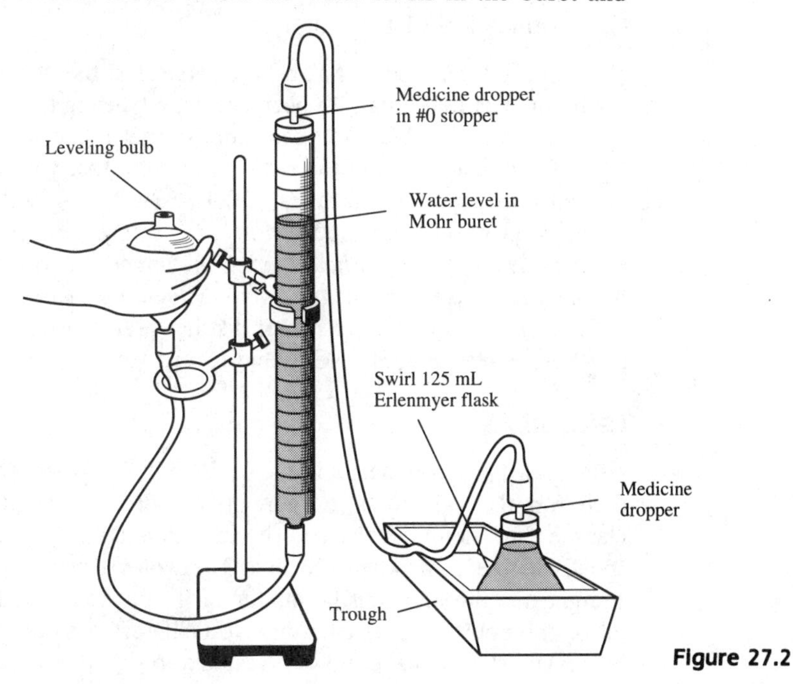

Figure 27.2

leveling bulb (by lowering the leveling bulb) and also read the volume. The other student should continue to swirl the flask and record on the report sheet the total elapsed time at the instant the gas volume is read. Time and volume readings should be taken at approximately 2-mL intervals until a total of about 14 mL of oxygen has been evolved. Although it is not necessary, you will find it convenient to use the stopwatch function of a digital wristwatch for noting the time intervals.

Solution 2 Rinse the Erlenmeyer flask and graduated cylinder thoroughly with distilled water and allow them to drain completely. Be certain that the bath temperature is the same as that used for solution 1. It may be necessary to adjust it by adding a little warm water. Repeat the experiment, this time first adding 10 mL of 0.10 *M* KI and 10 mL of distilled water, swirling, and then adding 10 mL of 3 percent H_2O_2. Quickly stopper the flask and take volume-time readings as before.

Solution 3 Rinse the flask and graduated cylinder. Check bath temperature and adjust if necessary. Repeat as follows: First add 20 mL of 0.10 *M* KI and 5 mL of distilled water, swirl, and then add 5 mL of H_2O_2. Take volume-time readings as before.

B. Effect of Temperature

Adjust the bath temperature so that it is approximately 10° − 12° higher than what it was for Part A above. Then repeat the experiment following the directions given in part A for solution 1. After you have collected 14 mL of O_2, do not take any more volume-time readings, but allow the reaction to continue to completion, that is, until no more oxygen is evolved. Save the solution for Part C. Record the total volume of O_2 collected and the barometric pressure on your report sheet.

C. Identification of the Catalyst

You can show that iodide has been a catalyst in the decomposition of H_2O_2 by noting the fact that it has not been consumed during the reaction even though it appears in the rate law. You can identify the iodide ion by precipitating it as HgI_2. Pour about 1 mL of the reaction mixture that you saved from Part B into a small test tube. Add 10 drops of 0.1 *M* $Hg(NO_3)_2$ to the test tube, mix, and record your observations. (CAUTION: *Mercuric nitrate is toxic and you should not get any on yourself. Should you come in contact with the* $Hg(NO_3)_2$ *solution, immediately wash it off with copious amounts of water*.) Dilute about 0.2 mL of 0.10 *M* KI with about 5 mL of distilled water. Add 2 drops of this solution to 10 drops of 0.1 *M* $Hg(NO_3)_2$ and record your observations.

Calculations

After you have completed Parts A, B, and C, graph your data, showing for each run the volume of oxygen on the ordinate (vertical axis) versus the elapsed time along the abscissa (horizontal axis). Zero time for each run corresponds to the time when the first 2 mL of O_2 were evolved. Draw the best straight line through your points for each run. Your line need not pass through all of the points. Label each line. Calculate the slopes for each of your four lines. The slopes of the lines correspond to the initial rates of the reaction.

Decide how the initial rate of oxygen formation is affected by doubling the concentrations of H_2O_2 and I^-. How do the concentrations of H_2O_2 and I^- change from solution 1 to solution 2 and then solution 3? Write a rate law for the reaction using the values of x and y you have determined by rounding them off to whole numbers.

REVIEW QUESTIONS

Before beginning this experiment in the laboratory, you should be able to answer the following questions:

1. What four factors influence the rate of a reaction?
2. If the rate law for a reaction is rate $= k[A]^2[B]$,
   (a) What is the overall order of the reaction?
   (b) If the concentration of both A and B are doubled, how will this affect the rate of the reaction?
   (c) What name is given to k?
   (d) How will doubling the concentration of A while the concentration of B is kept constant affect the value of k (assume temperature does not change)? How is the rate affected?
3. Define *catalyst*.
4. Write the chemical equation for the iodide-catalyzed decomposition of H_2O_2.
5. It is found for the reaction $2A + B \rightarrow C$ that doubling the concentration of either A or B quadruples the reaction rate. Write a rate law for the reaction.
6. When the concentration of a substance is doubled, what effect does it have on the rate if the order with respect to that reactant is (a) 0; (b) 1; (c) 2; (d) 3; (e) $\frac{1}{2}$?
7. The following data were collected for the volume of O_2 produced in the decomposition of H_2O_2.

Time (s)	*mL O_2*
0.0	0.0
45.0	2.0
88.0	3.9
131.	5.8

Calculate the average rate of reaction for the time intervals between measurements.

NOTES AND CALCULATIONS

Name ______________________________ Desk ____________

Date ____________________ Laboratory Instructor ______________________________

REPORT SHEET FOR EXPERIMENT 27

RATES OF CHEMICAL REACTIONS II: RATE AND ORDER OF H_2O_2 DECOMPOSITION

A. Order of Reaction

Bath temperature

______ °C **Solution 1** **Buret reading (mL)**	**Vol. of O_2 (mL)**	**Time**	______ °C **Solution 2** **Buret reading (mL)**	**Vol. of O_2 (mL)**	**Time**	______ °C **Solution 3** **Buret reading (mL)**	**Vol. of O_2 (mL)**	**Time**
______	______	______	______	______	______	______	______	______
______	______	______	______	______	______	______	______	______
______	______	______	______	______	______	______	______	______
______	______	______	______	______	______	______	______	______
______	______	______	______	______	______	______	______	______
______	______	______	______	______	______	______	______	______
______	______	______	______	______	______	______	______	______
______	______	______	______	______	______	______	______	______
______	______	______	______	______	______	______	______	______

Slope =

Slope =

Slope =

Rate Data

Solution	mL H_2O_2	mL KI	Rate (mL O_2/s)
1	5	10	
2	10	10	
3	5	20	

Rate law:

B. Effect of Temperature

Bath temperature ______________

Buret reading (mL)	Vol. of O_2 (mL)	Time
________	________	________
________	________	________
________	________	________
________	________	________
________	________	________
________	________	________
________	________	________
________	________	________
________	________	________
________	________	________

Total volume of O_2 collected ____________ mL
Barometric pressure ____________ mm Hg
Vapor pressure of water at bath temperature (see Appendix L) ____________ mm Hg

Slope =

Compared with the rate found for solution 1, the rate is ____________ times faster.
Using the ideal-gas law, calculate the moles of O_2 collected ____________ mol O_2
(show calculations)

Based upon the moles of O_2 involved, calculate the molar concentration of the *original* 3 percent H_2O_2 solution (show calculations).

C. Identification of Catalyst

Observation when $Hg(NO_3)_2$ is added to
(a) Solution in which reaction has gone to completion:

(b) original KI solution.

Conclusion regarding nature of KI.

QUESTIONS

1. How does the rate of formation of O_2 compare with (a) the rate of formation of H_2O; (b) the rate of disappearance of H_2O_2 for reaction [1]?

2. Why should the levels of water in the leveling bulb and buret be kept the same?

3. Why were you instructed to keep swirling the Erlenmeyer flask?

4. If you use 0.20 *M* KI instead of 0.10 *M* KI, how would this affect (a) the slopes of your curves, (b) the rate of the reactions, and (c) the numerical value of k for the reaction?

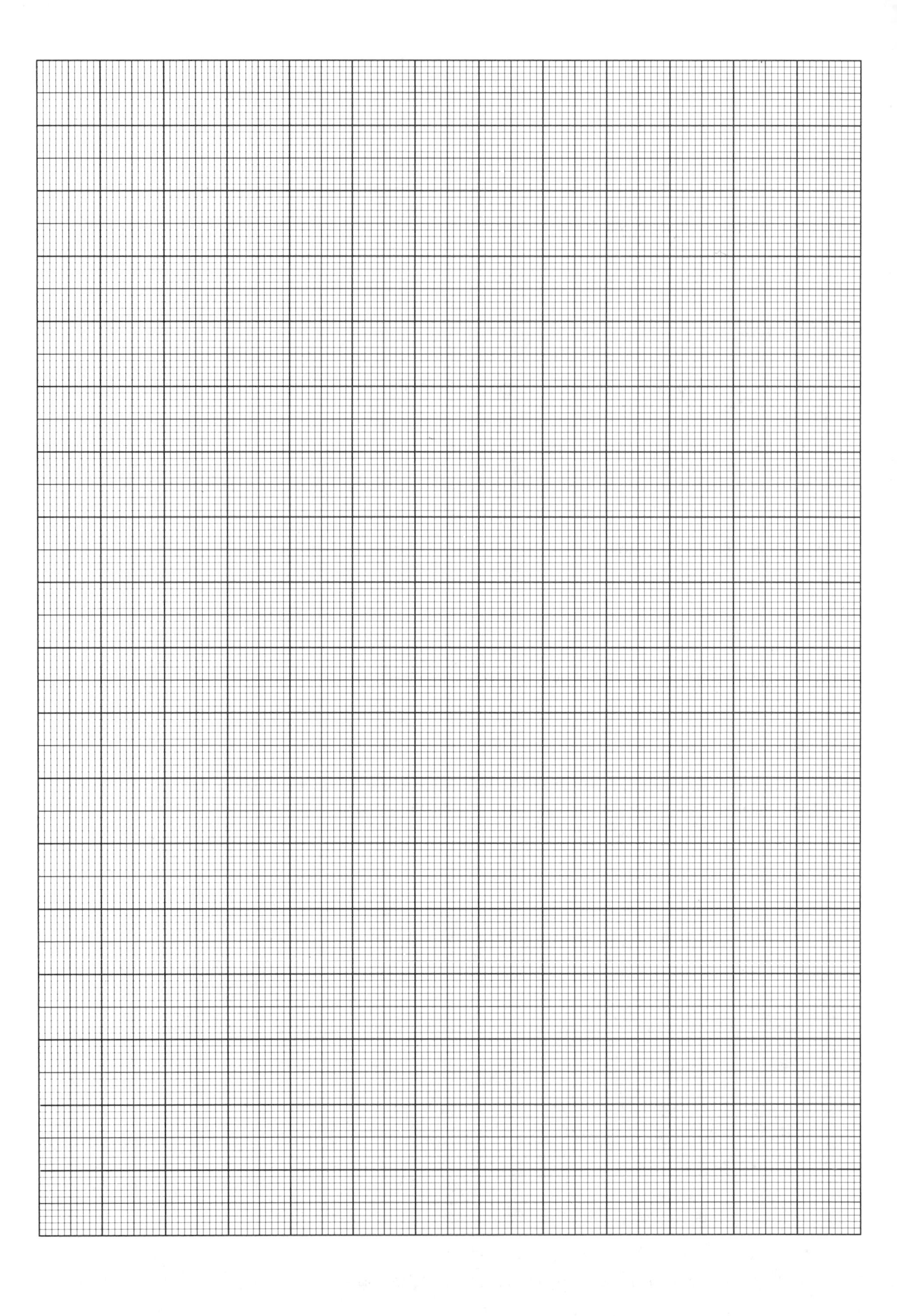

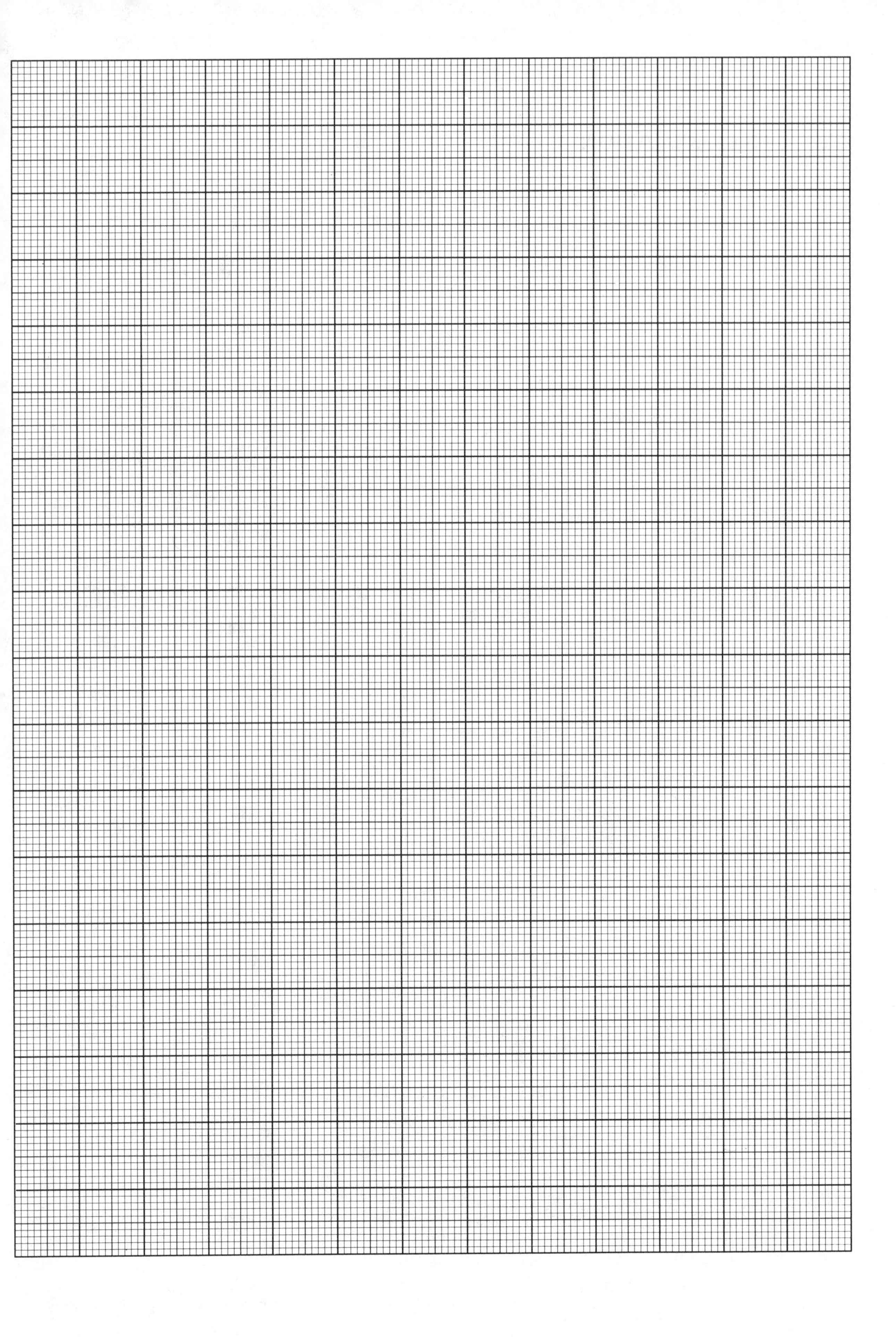

NOTES AND CALCULATIONS

EXPERIMENT

28

Introduction to Qualitative Analysis

CATIONS: Na^+, NH_4^+, Ag^+, Fe^{3+}, Al^{3+}, Cr^{3+}, Ca^{2+}, Mg^{2+}, Ni^{2+}, Zn^{2+}
ANIONS: SO_4^{2-}, NO_3^-, CO_3^{2-}, Cl^-, Br^-, I^-

OBJECTIVE

To become acquainted with the chemistry of several elements and the principles of qualitative analysis.

APPARATUS AND CHEMICALS

small test tubes(12)
centrifuge
unknown cation mixture (solution)
unknown anion salt (as solid)
casserole
litmus
6 droppers
10-cm Nichrome wire
6 *M* H_2SO_4
18 *M* H_2SO_4
3 *M* NaOH
6 *M* HCl
6 *M* NH_3
15 *M* NH_3
6 *M* HNO_3
15 *M* HNO_3
5 *M* NH_4Cl
3% H_2O_2
0.2 *M* $K_4Fe(CN)_6$
0.1% aluminon reagent (1 g of ammonium aurintricarboxylic acid in 1 L of H_2O)
1% dimethylglyoxime in 95% ethanol
Bunsen burner and hose
0.2 *M* $BaCl_2$
0.2 *M* $(NH_4)_2C_2O_4$
0.01% (0.1 g/L) *p*-nitrobenzene-azoresorcinol in 0.025 *M* NaOH(magnesium reagent)
0.2 *M* $FeSO_4$ (stabilized with iron wire and 0.01 *M* H_2SO_4)
0.2 *M* $AgNO_3$
Cl_2 water
mineral oil
0.1 *M* $Ba(OH)_2$
solid: $ZnSO_4$, $NaNO_3$, Na_2CO_3, NaCl, NaBr, NaI
0.1 *M* NH_4NO_3
0.1 *M* $AgNO_3$
0.1 *M* $Fe(NO_3)_3$
0.1 *M* $Cr(NO_3)_3$
0.1 *M* $Al(NO_3)_3$
0.1 *M* $Ca(NO_3)_2$
0.1 *M* $NaNO_3$
0.1 *M* $Zn(NO_3)_2$
0.1 *M* $Ni(NO_3)_2$
0.1 *M* $Mg(NO_3)_2$

ALL SOLUTIONS SHOULD BE PROVIDED IN DROPPER BOTTLES.

PART I : CATIONS

DISCUSSION

Qualitative analysis is concerned with the identification of the constituents contained in a sample of unknown composition. Inorganic qualitative analysis deals with the detection and identification of the elements that are present in a sample of material. Frequently this is accomplished by making an aqueous solution of the sample and then determining which cations and anions are present on the basis of chemical and physical properties. In this experiment you will be exposed to some of the chemistry of ten cations (Ag^+, Fe^{3+}, Cr^{3+}, Al^{3+}, Ca^{2+}, Mg^{2+}, Ni^{2+}, Zn^{2+}, Na^+, NH_4^+) and six anions (SO_4^{2-}, NO_3^-, Cl^-, Br^-, I^-, CO_3^{2-}), and you will learn how to test for their presence or absence. Because there are many other elements and ions than those which we shall consider, we call this experiment an "abbreviated" qualitative-analysis scheme.

If a substance contains only a single cation (or anion), its identification is a fairly simple and straightforward process, as you may have witnessed in Experiments 7 and 10. However, even in this instance additional confirmatory tests are sometimes required to distinguish between two cations (or anions) that have similar chemical properties. The detection of a particular ion in a sample that contains several ions is somewhat more difficult because the presence of the other ions may interfere with the test. For example, if you are testing for Ba^{2+} with K_2CrO_4 and obtain a yellow precipitate, you may draw an erroneous conclusion because, if Pb^{2+} is present, it also will form a yellow precipitate. Thus the presence of lead ions interferes with this test for barium ions. This problem can be circumvented by first precipitating the lead as PbS with H_2S, thereby removing the lead ions from solution prior to testing for Ba^{2+}.

The successful analysis of a mixture containing ten or more cations centers about the systematic separation of the ions into groups containing only a few ions. It is a much simpler task to work with two or three ions than with ten or more. Ultimately, the separation of cations depends upon the differences in their tendencies to form precipitates, to form complex ions, or to exhibit amphoterism.

The chart in Figure 28.1 illustrates how the ten cations you will study are separated into groups. Note that only three of the ten cations are colored: Fe^{3+} (rust to yellow), Cr^{3+} (blue-green), and Ni^{2+} (green). Therefore, a preliminary examination of an unknown that can contain any of the ten cations under consideration yields valuable information. If the solution is colorless, you know immediately that iron, chromium, and nickel are absent. You are to take advantage of all clues that will aid you in identifying ions. However, in your role as a detective in identifying ions, be aware that clues can sometimes be misleading. For example, if Fe^{3+} and Cr^{3+} are present together, what color would you expect this mixture to display? Would the color depend upon the proportions of Fe^{3+} and Cr^{3+} present? Could you assign a *definite* color to such a mixture? The group-separation chart (Figure 28.1) shows that silver can be separated from all the other cations (what are they?) as an insoluble chloride by the addition of hydrochloric acid. Iron, chromium, and aluminum are separated from the decantate as their corresponding hydroxides by precipitating them from a buffered ammonium hydroxide solution. Next, calcium can be isolated by precipitating it as insoluble calcium oxalate. Finally, magnesium and nickel can be separated from the remaining ions (Zn^{2+} and Na^+) as their insoluble hydroxides. Examination of the chart shows that in achieving these separations

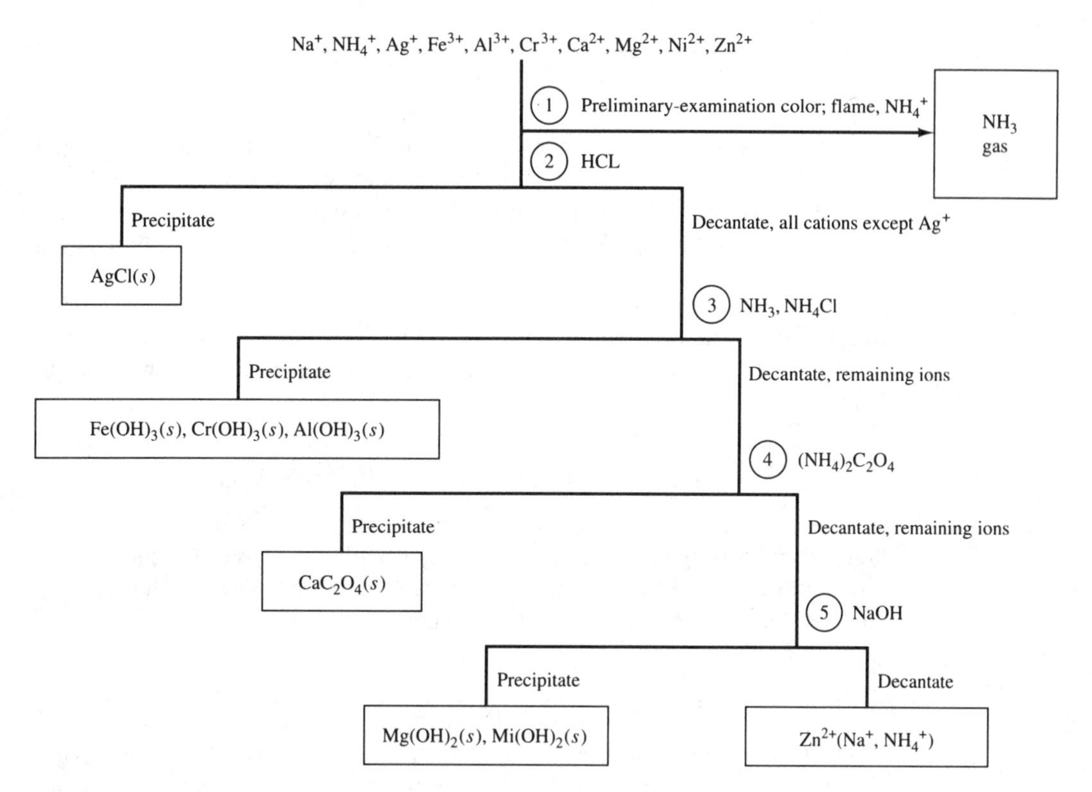

Figure 28.1 Flow chart for groups separations.

reagents containing sodium and ammonium ions are used; therefore, tests for these ions must be made prior to their introduction into the solution.

In order to derive the maximum benefit from this exercise, you should be thoroughly familiar with the group-separation chart in Figure 28.1. You should know not only which ten cations (by formula and charge) you are studying, but also how they are separated into groups. More details regarding the identification of these cations are provided in the flow chart for cations (Figure 28.2) and in the following discussion about the chemistry of the analytical scheme. Make frequent referrals to the flow charts while learning about the chemistry of the qualitative analysis scheme.

CHEMISTRY OF THE QUALITATIVE ANALYSIS SCHEME

(1) Detection of Sodium and Ammonium

Sodium salts and ammonium salts are added as reagents in the analysis of your general unknown. Hence, tests for Na^+ and NH_4^+, must be made on the original sample before performing tests for the other cations. Remember, your unknown may contain up to 10 cations, and you do not want inadvertently to introduce any of them into your unknown.

Sodium Most sodium salts are water soluble. The simplest test for the sodium ion is a flame test. Sodium salts impart a characteristic yellow color to a flame. The test is *very* sensitive, and because of the prevalence of sodium ions, much care must be exercised to keep equipment clean and free from contamination by these ions.

Ammonium The ammonium ion, NH_4^+, is the conjugate acid of the base ammonia, NH_3. The test for NH_4^+ takes advantage of the following equilibrium:

$$NH_4^+(aq) + OH^-(aq) \rightleftharpoons NH_3(g) + H_2O(l)$$

Thus, when a strong base such as sodium hydroxide is added to a solution of an ammonium salt, and this solution is heated, NH_3 gas is evolved. The NH_3 gas can easily be detected by its effect upon moist red litmus paper.

(2) Separation and Detection of Silver

All chloride salts are soluble in water except those of Pb^{2+}, Hg_2^{2+} and Ag^+. Silver can be precipitated and separated from the other nine cations that we are considering by the addition of HCl to the original unknown:

$$Ag^+(aq) + Cl^-(aq) \rightleftharpoons \underset{\text{white}}{\mathbf{AgCl}(s)}$$

A slight excess of HCl is used to ensure the complete precipitation of silver ions and to reduce their solubility by the common ion effect; excess chloride ions drives the above equilibrium to the right. However, a large excess of chloride ions must be avoided because AgCl tends to dissolve by forming a *soluble-complex ion*:

$$\mathbf{AgCl}(s) + Cl^-(aq) \rightleftharpoons AgCl_2^-(aq)$$

To be absolutely certain that the white precipitate is AgCl ($PbCl_2$ and Hg_2Cl_2 are also insoluble, and they are also white), NH_3 is added to the precipitate. If the precipitate is indeed AgCl, it will dissolve and then reprecipitate when the ammonical solution is made acidic:

$$\mathbf{AgCl}(s) + 3NH_3(aq) \rightleftharpoons Ag(NH_3)_2^+(aq) + Cl^-(aq)$$
$$Ag(NH_3)_2^+(aq) + 2H^+(aq) + Cl^-(aq) \rightleftharpoons \mathbf{AgCl}(s) + 2NH_4^+(aq)$$

The other two insoluble chlorides do not behave this way. Thus we can be assured that the white chloride precipitate is silver chloride.

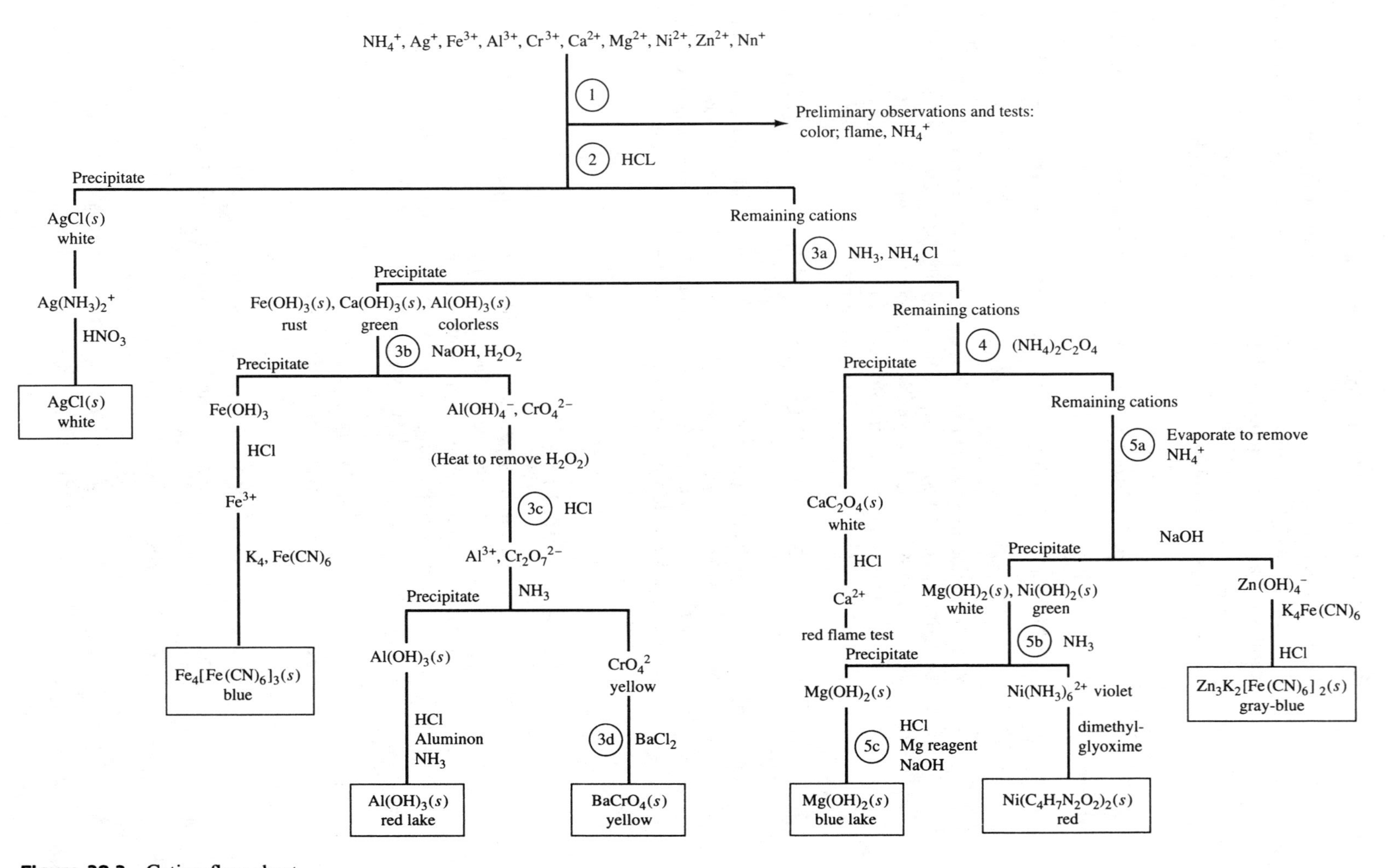

Figure 28.2 Cation flow chart.

(3a) Separation and Detection of Iron, Aluminum and Chromium

Iron, chromium and aluminum can be separated from the other ions (Ca^{2+}, Mg^{2+}, Ni^{2+}, and Zn^{2+}) by making the solution alkaline and precipitating these cations as their corresponding hydoxides:

$$Fe^{3+}(aq) + 3NH_3(aq) + 3H_2O(l) \rightleftharpoons \underset{\text{rust}}{\mathbf{Fe(OH)_3}(s)} + 3NH_4^+(aq)$$

$$Cr^{3+}(aq) + 3NH_3(aq) + 3H_2O(l) \rightleftharpoons \underset{\text{grayish green}}{\mathbf{Cr(OH)_3}(s)} + 3NH_4^+(aq)$$

$$Al^{3+}(aq) + 3NH_3(aq) + 3H_2O(l) \rightleftharpoons \underset{\text{colorless}}{\mathbf{Al(OH)_3}(s)} + 3NH_4^+(aq)$$

The hydroxide ion concentration required to precipitate these three ions must be carefully controlled, because if it is too high, $Mg(OH)_2$ will also precipitate. An alkaline buffer of NH_3 and NH_4Cl provides a hydroxide ion concentration that is high enough to precipitate Fe^{3+}, Cr^{3+}, and Al^{3+} and, yet low enough to prevent precipitation of $Mg(OH)_2$. Aqueous ammonia is a weak base:

$$NH_3(aq) + H_2O(l) \rightleftharpoons NH_4^+(aq) + OH^-(aq)$$

By itself, it would provide too high of a hydroxide ion concentration and $Mg(OH)_2$ would precipitate along with the other cations. However, the NH_4^+ ions derived from the NH_4Cl causes this equilibrium to shift to the left; this reduces the hydroxide ion concentration sufficiently to prevent Mg^{2+} from precipitating.

(3b) Separation of Iron

Iron hydroxide can be separated from the other hydroxides by treating the precipitate with the strong base NaOH and hydrogen peroxide, H_2O_2. These reagents do not react with the insoluble $Fe(OH)_3$; however, $Al(OH)_3$ is amphoteric and dissolves forming the soluble complex ion $Al(OH)_4^-$. The $Cr(OH)_3$ also dissolves, being oxidized by H_2O_2 to form CrO_4^{2-}:

$$\mathbf{Al(OH)_3}(s) + OH^-(aq) \rightleftharpoons Al(OH)_4^-(aq)$$

$$\mathbf{2Cr(OH)_3}(s) + 3H_2O_2(aq) + 4OH^-(aq) \rightleftharpoons \underset{\text{yellow}}{2CrO_4^{2-}(aq)} + 8H_2O(l)$$

The rust colored $Fe(OH)_3$ remains undissolved. That the rust colored precipitate is in fact iron hydroxide can be confirmed by dissolving it in acid and then adding potassium hexacyanoferrate(II), $K_4Fe(CN)_6$, and noting the formation of a dark blue precipitate (Prussian blue):

$$\mathbf{Fe(OH)_3}(s) + 3H^+(aq) \rightleftharpoons Fe^{3+}(aq) + 3H_2O(l)$$

$$4Fe^{3+}(aq) + 3Fe(CN)_6^-(aq) \rightleftharpoons \underset{\text{blue}}{\mathbf{Fe_4[Fe(CN)_6]_3}(s)}$$

(3c) and (3d) Separation of Chromium and Aluminum

While iron was being precipitated as the hydroxide, chromium was oxidized to the yellow chromate ion, CrO_4^{2-} and aluminum was converted to the soluble complex aluminate ion, $Al(OH)_4^-$, which is colorless. These two ions are in the supernatant liquid, which upon acidification converts the chromate ion to the orange dichromate ion and the aluminate to the colorless solvated aluminum ion:

$$2CrO_4^{2-}(aq) + 2H^+(aq) \rightleftharpoons H_2O(l) + Cr_2O_7^{2-}(aq)$$
$$4H^+(aq) + Al(OH)_4^-(aq) \rightleftharpoons 4H_2O(l) + Al^{3+}(aq)$$

When this solution is treated with aqueous ammonia, aluminum precipitates as $Al(OH)_3$ which can be separated from the chromate ion in the supernatant liquid. (Note, the CrO_4^{2-} ion is stable in neutral or alkaline solution but is reversibly converted to the dichromate ion, $Cr_2O_7^{2-}$ in acidic solution.) The formation of a yellow precipitate, $BaCrO_4$ upon the addition of barium chloride, confirms the presence of chromium;

$$Ba^{2+}(aq) + CrO_4^{2-}(aq) \rightleftharpoons \underset{\text{yellow}}{\mathbf{BaCrO_4}(s)}$$

Aluminum hydroxide is a clear-colorless substance and is difficult to see in this analysis. A confirmatory test for aluminum involves dissolving the aluminum hydroxide in acid and then reprecipitating it again with ammonia in the presence of Aluminon reagent. As the aluminum hydroxide precipitates, it absorbs the Aluminon reagent and assumes a red coloration known as a "lake".

$$Al^{3+}(aq) + 3NH_3(aq) + 3H_2O(l) + \underset{\text{reagent}}{\text{Aluminon}} \rightleftharpoons \underset{\text{red lake}}{\mathbf{Al(OH)_3}(s)} + 3NH_4^+(aq)$$

(4) Separation and Detection of Calcium

Calcium Calcium can be separated from the remaining cations (Mg^{2+}, Ni^{2+} and Zn^{2+}) by precipitating it as an insoluble oxalate salt:

$$Ca^{2+}(aq) + C_2O_4^{2-}(aq) \rightleftharpoons \mathbf{CaC_2O_4}(s)$$

If magnesium is present, it is possible that it can precipitate and be mistaken for calcium. To confirm that the precipitate is that of calcium, it is dissolved in acid, and a flame test is performed on the solution.

$$\mathbf{CaC_2O_4}(s) + 2H^+(aq) \rightleftharpoons Ca^{2+}(aq) + H_2C_2O_4(aq)$$

A brick-red flame verifies the presence of calcium ions. Magnesium ions do not impart any color to a flame.

(5a) Separation And Detection of Zinc

Separation of zinc from magnesium and nickel can be accomplished by precipitating $Mg(OH)_2$ and $Ni(OH)_2$ from a strongly alkaline solution. Excess ammonium ions must be removed before the precipitation of these hydroxides because ammonium ions interfere with the confirmatory test for magnesium. Ammonium ions are easily removed by evaporating the solution which has been acidified with nitric acid to dryness:

$$NH_4Cl(s) \xrightarrow{\Delta} NH_3(g) + HCl(g)$$

$$NH_4NO_3(s) \xrightarrow{\Delta} N_2O(g) + 2H_2O(g)$$

While magnesium and nickel form insoluble hydroxides in the presence of a strong base, zinc is amphoteric and forms the soluble complex ion $Zn(OH)_4^{2-}$:

$$Ni^{2+}(aq) + 2OH^-(aq) \rightleftharpoons \mathbf{Ni(OH)_2}(s)$$

$$Mg^{2+}(aq) + 2OH^-(aq) \rightleftharpoons \mathbf{Mg(OH)_2}(s)$$

$$Zn^{2+}(aq) + 4OH^-(aq) \rightleftharpoons Zn(OH)_4^{2-}(aq)$$

Zinc is confirmed by precipitating it from an acidic solution as a blue salt, $Zn_3K_2[Fe(CN)_6]_2$:

$$3Zn(OH)_4^{2-}(aq) + 12H^+(aq) + 2K^+(aq) + 2Fe(CN)_6^{4-}(aq) \rightleftharpoons \underset{\text{blue}}{\mathbf{Zn_3K_2[Fe(CN)_6]_2}(s)} + 12H_2O(l)$$

(5b) Separation and Detection of Nickel

Because nickel forms the soluble complex ion hexaamminenickel (II), $Ni(NH_3)_6^{2+}$, in the presence of aqueous ammonia, it can be separated from $Mg(OH)_2$:

$$\mathbf{Ni(OH)_2}(s) + 6NH_3(aq) \rightleftharpoons Ni(NH_3)_6^{2+}(aq) + 2OH^-(aq)$$

It is confirmed by forming a strawberry red precipitate, $Ni(C_4H_7N_2O_2)_2$, with an organic reagent, dimethylglyoxime (see Experiment 10):

$$Ni(NH_3)_6^{2+}(aq) + 2HC_4H_7N_2O_2(aq) \rightleftharpoons \underset{\text{strawberry red}}{\mathbf{Ni(C_4H_2N_2O_2)_2}(s)} + 2NH_3(aq) + 2NH_4^+(aq)$$

(5c) Detection of Magnesium

Magnesium is confirmed by dissolving the hydroxide with acid and then reprecipitating it in the presence of an organic compound, which is called Magnesium Reagent. The present of Mg^{2+} is indicated by formation of a blue lake.

$$\mathbf{Mg(OH)_2}(s) + 2H^+(aq) \rightleftharpoons Mg^{2+}(aq) + 2H_2O(l)$$

$$Mg^{2+}(aq) + 2OH^-(aq) + \text{Mg reagent} \rightleftharpoons \underset{\text{blue lake}}{\mathbf{Mg(OH)_2}(s)}$$

PROCEDURE

First you will analyze a known that contains all 10 cations. Record on your report sheet the reagents used in each step, your observations, and the equations for each precipitation reaction. After completing this practice analysis, obtain an unknown. Follow the same procedures as with the known, again recording reagents and observations. Also record conclusions regarding the presence or absence of all cations. Before beginning this experiment ***study*** Appendix J on techniques used in qualitative analysis: heating solutions, precipitation, centrifugation, washing precipitates, and testing acidity.

(1) Procedure

Initial Observations and Tests for Sodium and Ammonium Note the color of your sample and record any conclusions about what cations are present or absent on your report sheet.

The flame test for sodium is very sensitive, and traces of sodium ion will impart a characteristic yellow color to the flame. Just about every solution has a trace of sodium and thus will give a positive test. On the basis of the intensity and duration of the yellow color, you can decide whether Na^+ is merely a contaminant or present in substantial quantity. To perform the flame test, obtain a piece of platinum or Nichrome wire that has been sealed in a piece of glass tubing. Clean the wire by dipping it in 12 *M* HCl that is contained in a small test tube and heat the wire in the hottest part of your Bunsen burner flame. Repeat this operation until no color is seen when the wire is placed in the flame. Several cleanings will be required before this is achieved. Then place 10 drops of the solution to be analyzed in a clean test tube and perform a flame test on it. If the sample being tested is your unknown, run a flame test on distilled water and then another on a 0.2 *M* NaCl solution. Compare the tests; this should help you make a decision as to the presence of sodium in your unknown.

Place 10 drops of the original sample to be analyzed in a porcelain casserole or crucible. Moisten a strip of red or neutral litmus paper with distilled water and place the paper on the bottom of a small watch glass. Add 10 drops of 3 *M* NaOH to the unknown, swirl the casserole or crucible, and immediately place the watch glass on it with the litmus paper down. Let stand for a few minutes. The presence of NH_4^+ ions is confirmed if the paper turns blue.

(2) Procedure

Separation and Detection of Silver Place 10 drops of the original solution to be analyzed in a small test tube, add 5 drops of distilled water, and 2 drops of 6 *M* HCl. Stir well, centrifuge (consult appendix J for techniques), and reserve the decantate for Procedure 3. Wash the precipitate with 10 drops of distilled water, centrifuge, and add the washings to the decantate. Dissolve the precipitate in 2 drops of 6 *M* NH_3; then add 6 *M* HNO_3 to the ammoniacal solution until it is acidic to litmus. A curdy white precipitate of AgCl confirms the presence of Ag^+ ions.

(3a) Procedure

Separation of Iron, Aluminum and Chromium To the decantate from Procedure 2 add 2 drops of NH_4Cl solution and 6 *M* NH_3 until the solution is basic to litmus. Centrifuge and reserve the decantate for Procedure 4. Wash the precipitate with 10 drops of distilled water, centrifuge, and add the washings to the decantate.

(3b) Procedure

Separation and Detection of Iron To the precipitate from Procedure 3a, add 5 drops of distilled water, 10 drops of 3 *M* NaOH, and 5 drops of 3-percent-H_2O_2. Stir well, centrifuge, and reserve the decantate for Procedure 3c. Wash the precipitate with 10 drops of distilled water and add the washings to the decantate. Dissolve the reddish-brown precipitate in 2 drops of 6 *M* HCl and add 10 drops of $K_4Fe(CN)_6$ solution. A blue precipitate of $Fe_4[Fe(CN)_6]_3$ confirms the presence of Fe^{3+} ions.

(3c) Procedure

Separation and Detection of Aluminum If the decantate from Procedure 3b is yellow in color, place it in a casserole or crucible and evaporate almost to dryness to remove excess H_2O_2 (see Note 1 below). Add 10 drops of distilled water and 6 *M* HCl until acid to litmus. Then add 6 *M* NH_3 until the solution is basic to litmus. Centrifuge and reserve the decantate for Procedure 3d. Wash the precipitate with 10 drops of distilled water, centrifuge, and add the washings to the decantate. Dissolve the precipitate in 2 drops of 6 *M* HCl, add 2 drops of Aluminon reagent, and 6 *M* NH_3 until basic to litmus. Centrifuge the solution. A red "lake" confirms the presence of Al^{3+} ions.

Note 1 Hydrogen peroxide is a reducing agent in acid solutions, so CrO_4^{2-} ions could be reduced to Cr^{3+} when the solution is acidified. When NH_3 is added, $Cr(OH)_3$ will be precipitated and could be incorrectly reported as $Al(OH)_3$. Also Cr^{3+} would not be confirmed in the proper procedure.

(3d) Procedure

Detection of Chromium Add 2 drops of $BaCl_2$ solution to the decantate from procedure 3c. A yellow precipitate of $BaCrO_4$ confirms the presence of Cr^{3+} ions.

(4) Procedure

Separation and Detection of Calcium Add 3 drops of $(NH_4)_2C_2O_4$ solution to the decantate from procedure 3a and centrifuge. Reserve the decantate for Procedure 5a. Wash the precipitate with 10 drops of distilled water, centrifuge, and add the washings to the decantate. Dissolve the precipitate of CaC_2O_4 in 2 drops of 6 *M* HCl and carry out a flame test with the solution. A brick-red flame confirms Ca^{2+} ions (see Note 2 below).

Note 2 If magnesium ions are present, they may be precipitated as MgC_2O_4. To determine if MgC_2O_4 is precipitated, test the acid solution for Mg^{2+} as described in Procedure 5c.

(5a) Procedure

Separation and Detection of Zinc Place the decantate from Procedure 4 in a porcelain casserole, and carefully evaporate to dryness using a burner flame. Add 5 to 6 drops of concentrated HNO_3 and heat again until no more fumes are observed (you are removing excess NH_4^+ ions that would interfere with the tests for Mg^{2+} ions). Dissolve the residue in 5 drops of 6 *M* HCl. Add 10 drops of distilled water to the HCl solution and transfer it to a small test tube. Wash out the casserole with 10 drops of distilled water, and add the wash to the acid solution. Add 3 *M* NaOH until the solution is basic to litmus; centrifuge, and reserve the precipitate for Procedure 5b. Wash the precipitate with 10 drops of distilled water, and add the washings to the decantate. Add 3 drops of $K_4Fe(CN)_6$ to the decantate, and acidify with 6 *M* HCl; a blue coloration or precipitate of $Zn_3K_2[Fe(CN)_6]_2$ confirms the presence of Zn^{2+} ions.

(5b) Procedure

Separation and Detection of Nickel Add 5 drops of distilled water and 5 drops of 15 *M* NH_3 to the precipitate from Procedure 5a; centrifuge, and reserve the precipitate for Procedure 5c. Wash the precipitate with 10 drops of distilled water, and add the wash to the decantate. Add 1 drop of dimethylglyoxime solution to the decantate; a strawberry-red precipitate of $Ni(C_4H_7N_2O_2)_2$ confirms the presence of Ni^{2+} ions (see Experiment 10).

(5c) Procedure

Detection of Magnesium If Ni^{2+} ions were confirmed in Procedure 5b, add 15 *M* NH_3 to the precipitate saved from Procedure 5b until a negative test for Ni^{2+} is obtained. Dissolve the hydroxide precipitate in 2 drops of 6 *M* HCl; add 2 drops of "Magnesium Reagent" and 3 *M* NaOH until the solution is basic. A blue lake confirms the presence of Mg^{2+} ions.

REVIEW QUESTIONS

Before beginning Part I of this experiment in the laboratory, you should be able to answer the following questions:

1. What are the names and formulas of the ten cations you will identify?
2. Why are confirmatory tests necessary in identifying ions?
3. Which of the ten cations are colored and what are their colors?
4. Which salt is insoluble: $FeCl_3$, $ZnCl_2$, NaCl, or AgCl?
5. How could you separate Fe^{3+} from Ag^+?
6. How could you separate Cr^{3+} from Mg^{2+}?
7. How could you separate Al^{3+} from Ag^+?
8. Complete and balance the following:

$$NH_4^+(aq) + OH^-(aq) \longrightarrow$$
$$AgCl(s) + NH_3(aq) \longrightarrow$$

PART II: ANIONS

DISCUSSION

A systematic scheme based upon the kinds of principles involved in cation analysis can be designed for the analysis of anions. Because we shall limit our consideration to only six anions (sulfate, nitrate, chloride, bromide, iodide, and carbonate) and will not consider mixtures of the ions, our method of analysis is quite simple and straightforward. It is based on specific tests for the individual ions and does not require special precaution to eliminate interferences that may arise in mixtures.

Initially you will make a general test on a solid salt with concentrated sulfuric acid. The results of this test should strongly suggest what the anion is. You will then confirm your suspicions by performing a specific test for the ion you believe to be present.

Table 28.1 summarizes the behavior of anions (as dry salts) with concentrated sulfuric acid.

PROCEDURE

Perform the general sulfuric acid test described below on the individual anions. Then perform the specific tests on each of the ions. Record your observations and equations for the reactions that occur. After completing these tests on the six anions, obtain a solid salt unknown and identify its anion. Record your observations and conclusion. Only one anion is present in the salt.

Sulfuric Acid Test

Place a small amount of the solid (about the size of a pea) in a small test tube. Add 1 or 2 drops of 18 *M* H_2SO_4 and observe everything that occurs, especially the color and odor of gas formed. DO NOT place your nose directly over the mouth of the test tube, but carefully fan gases toward your nose. Then *care-*

Table 28.1 Behavior of Anions with Concentrated Sulfuric Acid

A. Cold H_2SO_4

SO_4^{2-} No reaction.

NO_3^- No reaction.

CO_2^{2-} A colorless odorless gas forms.

$$CO_3^{2-}(s) + 2H^+(q) \longrightarrow H_2O(l) + CO_2(g)$$

Cl^- A colorless gas forms with a sharp-pungent odor which gives an acid test with litmus and fumes in moist air.

$$Cl^-(s) + H^+(aq) \longrightarrow HCl(g)$$

Br^- A brownish-red gas forms with a sharp odor which gives an acidic test with litmus and fumes in moist air. The odor of SO_2 may be detected.

$$2Br^-(s) + 4H^+(aq) + SO_4^{2-}(aq) \longrightarrow Br_2(g) + SO_2(g) + 2H_2O(l) \text{ (HBr is also liberated)}$$

I^- Solid turns dark brown immediately with the slight fromation of violet fumes. The gas has the odor of rotten eggs, gives an acidic test with litmus, and fumes in moist air.

$$2I^-(s) + 4H^+(aq) + SO_4^{2-}(aq) \longrightarrow I_2(g) + SO_2(g) + 2H_2O(l) \text{ (HI is also liberated)}$$

B. Hot concentrated H_2SO_4. There are no additional reactions with any of the anions except NO_3^-, which forms brown fumes of NO_2 gas.

$$4NO_3^-(s) + 4H^+(aq) \longrightarrow 4NO_2(g) + O_2(g) + 2H_2O(l)$$

fully heat the test tube, but not so strongly as to boil the H_2SO_4 (IF YOU HEAT THE ACID TO STRONGLY, IT COULD COME SHOOTING OUT!). Note whether or not brown fumes of NO_2 are produced. CAUTION: Do not look down into the test tube. Do not point the test tube at yourself or at your neighbors. SAFETY GLASSES MUST BE WORN.

Specific Tests for Anions

When an anion is indicated by the preliminary test with concentrated H_2SO_4, it is confirmed using the appropriate specific test. Make an aqueous solution of the solid unknown and perform the following tests on portions of this solution:

1. SO_4^{2-}: Place 10 drops of a solution of the anion salt in a test tube, acidify with 6 *M* HCl, and add a drop of $BaCl_2$ solution. A white precipitate of $BaSO_4$ confirms SO_4^{2-} ions.

$$Ba^{2+}(aq) + SO_4^{2-}(aq) \longrightarrow BaSO_4(s)$$

2. NO_3^-: Place 10 drops of a solution of the anion salt in a small test tube and add 5 drops of $FeSO_4$ solution; mix the solution. Carefully, without agitation, pour concentrated H_2SO_4 down the sides of the test tube. The formation of a brown ring between the two layers confirms NO_3^- ions.

$$(1)\ 3Fe^{2+}(aq) + NO_3^-(aq) + 4H^+(aq) \longrightarrow 3Fe^{3+}(aq) + No(g) + 2H_2O(l)$$

$$(2)\ NO(g) + Fe^{2+}(aq)\ \text{(excess)} \longrightarrow Fe(NO)^{2+}(aq)\ \text{(brown)}$$

3. Cl^-: Place 10 drops of a solution of the anion salt in a test tube and add a drop of $AgNO_3$ solution. A white, curdy precipitate confirms Cl^- ions.

$$Ag^+(aq) + Cl^-(aq) \longrightarrow AgCl(s)$$

4. Br^-: Place 10 drops of a solution of the anion salt in a test tube, add 3 drops of 6 *M* HCl, then add 5 drops of Cl_2 water and 5 drops of mineral oil. Shake well. Br^- ions are confirmed if the mineral oil (top) layer is colored orange to brown. Wait 30 s for the layers to separate.

$$2Br^-(aq) + Cl_2(aq) = Br_2(aq) + 2Cl^-(aq)$$

5. I^-: Repeat the test as described for Br^- ions. If the mineral oil layer is colored violet, I^- ions are confirmed.

$$2I^-(aq) + Cl_2(aq) = I_2(aq) + 2Cl^-(aq)$$

6. CO_3^{2-}: Place a small amount of the solid anion salt in a small test tube and add a few drops of 6 *M* H_2SO_4. If a colorless, odorless gas is evolved, hold a drop of $Ba(OH)_2$ solution over the mouth of the test tube using either an eyedropper or Nichrome wire loop; CO_3^{2-} ions are confirmed if the drop turns milky.

$$(1)\ 2H^+(aq) + CO_2^{2-}(s) \longrightarrow CO_2(g) + H_2O(l)$$

$$(2)\ CO_2(g) + Ba(OH)_2(aq) \longrightarrow BaCO_3(s) + H_2O(l)$$

REVIEW QUESTIONS

Before beginning Part II of this experiment in the laboratory, you should be able to answer the following questions:

1. Give the names and formulas of the anions to be identified.
2. Describe the behavior of each solid containing the anions toward concentrated H_2SO_4.
3. If you had a mixture of NaCl and Na_2CO_3, would the action of concentrated H_2SO_4 allow you to decide that both Cl^- and CO_2^{2-} were present?

Name ______________________________ Desk __________

Date ______________ Laboratory Instructor ______________________________

Unknown number ______________________________

REPORT SHEET FOR EXPERIMENT 28

INTRODUCTION TO QUALITATIVE ANALYSIS

PART I: CATIONS

A. Known

Record the reagent used in each step, your observations, and the equations for each precipitation reaction.

	Reagent	Observations	Equations
1			
2			
3a			
3b			
3c			

	Reagent	Observations	Equations
3d			
4			
5a			
5b			
5c			

B. Unknown Cation

Cations in unknown ______________________________

B. UNKNOWN

Record reagent used, your observations, and the equations for each precipitate formed.

Reagent	Observation	Equation

Cations in unknown ____________________

B. Known

Specific tests.

Ion	**Observations and equations**
SO_4^{2-}	
NO_3^-	
CO_3^{2-}	
Cl^-	
Br^-	
I^-	
NO_3^- (heated)	

C. Unknown

Observations and equations

1 H_2SO_4 test:

2 Specific test(s):

Anion in unknown ______________________________

NOTES AND CALCULATIONS

Abbreviated Qualitative-Analysis Scheme

EXPERIMENT 29

CATIONS:

Ag^+, Pb^{2+}, Hg_2^{2+}, Cu^{2+}, Bi^{3+}, Sn^{4+}, Fe^{3+}, Mn^{2+}, Ni^{2+}, Al^{3+}, Ba^{2+}, NH_4^+, Na^+.

ANIONS:

SO_4^{2-}, NO_3^-, CO_3^{2-}, Cl^-, Br^-, I^-, CrO_4^{2-}, PO_4^{3-}, S^{2-}, SO_3^{2-}.

OBJECTIVE

To become acquainted with the chemistry of several elements and the principles of qualitative analysis.

APPARATUS AND CHEMICALS

small test tubes (12)	6 M HNO_3
centrifuge	3 M HNO_3
unknown cation solution	6 M H_2SO_4
unknown anion salt (as solid mixture)	18 M H_2SO_4
	15 M NH_3
casserole	6 M NH_3
evaporating dish	$Ba(OH)_2$, saturated solution
litmus	6 M NaOH
droppers (6)	aluminon, 0.1 percent[a]
Nichrome wire (10 cm)	aluminum wire, 26 guage
Bunsen burner and hose	1 M $NH_4C_2H_3O_2$
6 M $HC_2H_3O_2$	2 M NH_4Cl
12 M HCl	$(NH_4)_2MoO_4$[b]
6 M HCl	0.2 m $SnCl_2$ (Freshly Prepared)
2 M HCl	$(NH_4)_2S$[c]
0.2 M $BaCl_2$	1 M K_2CrO_4
mineral oil	1 M $K_2C_2O_4$
Cl_2-water	0.2 M KNO_2
diethyl ether	0.2 M KSCN
dimethylglyoxime 1 percent[d]	0.1 M $AgNO_3$

[a] One gram of ammonium aurintricarboxylic acid in 1 L H_2O.

[b] Dissolve 20 g of MoO_3 in a mixture of 60 mL of distilled water and 30 mL of 15 M NH_3. Add this solution slowly, with constant stirring, to a mixture of 230 mL of water and 100 mL of 16 M HNO_3.

[c] Add 1 volume of reagent-grade ammonium sulfide liquid to two volumes of water or saturate 5 M NH_3 with H_2S.

[d] In 95% ethyl alcohol solution.

ethyl alcohol, 95 percent	$NaBiO_3$, solid
0.2 *M* $Fe(NO_3)_2$	Na_2CO_3, saturated solution
0.2 *M* $FeSO_4$	Na_2O_2, solid
H_2O_2, 3 percent	starch solution
0.2 *M* $Pb(C_2H_3O_2)_2$	1 *M* thioacetamide
0.1 *M* $MnCl_2$	Zn, granulated
0.1 *M* $HgCl_2$	Group 1, 2, 3, and 4 known solutions
0.1 *M* $Hg(C_2H_3O_2)_2$	Solid sodium salt anion knowns
16 *M* HNO_3	

ALL SOLUTIONS SHOULD BE PROVIDED IN DROPPER BOTTLES

CATIONS

I Discussion

Qualitative analysis is concerned with the identification of the constituents contained in a sample of unknown composition. Inorganic qualitative analysis deals with the detection and identification of the elements that are present in a sample of material. Frequently this is accomplished by making an aqueous solution of the sample and then determining which cations and anions are present on the basis of chemical and physical properties. In this experiment, you will be exposed to some of the chemistry of 14 cations (Ag^+, Pb^{2+}, Hg_2^{2+}, Cu^{2+}, Bi^{3+}, Sn^{4+}, Fe^{3+}, Mn^{2+}, Ni^{2+}, Al^{3+}, Ba^{2+}, Ca^{2+}, NH_4^+, Na^+) and 10 anions (CrO_4^{2-}, PO_4^{3-}, S^{2-}, SO_3^{2-}, SO_4^{2-}, NO_3^-, Cl^-, Br^-, I^-, CO_3^{2-}), and you will learn how to test for their presence or absence. Because there are many elements and ions other than those which we shall consider, we call this experiment an "abbreviated" qualitative-analysis scheme.

If a sample contains only a single cation (or anion), its identification is a fairly simple and straightforward process, as you may have witnessed in Experiment 28. However, even in this instance additional confirmatory tests are sometimes required to distinguish between two cations (or anions) that have similar chemical properties. The detection of a particular ion in a sample that contains several ions is somewhat more difficult, because the presence of the other ions may interfere with the test. For example, if you are testing for Ba^{2+} with K_2CrO_4 and obtain a yellow precipitate you may draw an erroneous conclusion because if Pb^{2+} is present, it also will form a yellow precipitate. Thus lead ions interfere with this test for barium ions. This problem can be circumvented by precipitating the lead as PbS with H_2S first, thereby removing the lead ions from solution prior to testing for Ba^{2+}.

The successful analysis of a mixture containing 14 or more cations centers upon the systematic separation of the ions into groups containing only a few ions. It is a much simpler task to work with 2 or 3 ions than with 14 or more. Ultimately, the separation of cations depends upon the difference in their tendencies to form precipitates, to form complex ions, or to exhibit amphoterism.

The chart in Figure 29.1 illustrates how the 14 cations you will study are separated into groups. Note that only 4 of the 14 cations are colored: Fe^{3+} (rust to yellow), Cu^{2+} (blue), Mn^{2+} (very faint pink), and Ni^{2+} (green). Therefore a preliminary examination of an unknown that can contain any of the 14 cations under consideration yields valuable information. If the solution is colorless, you

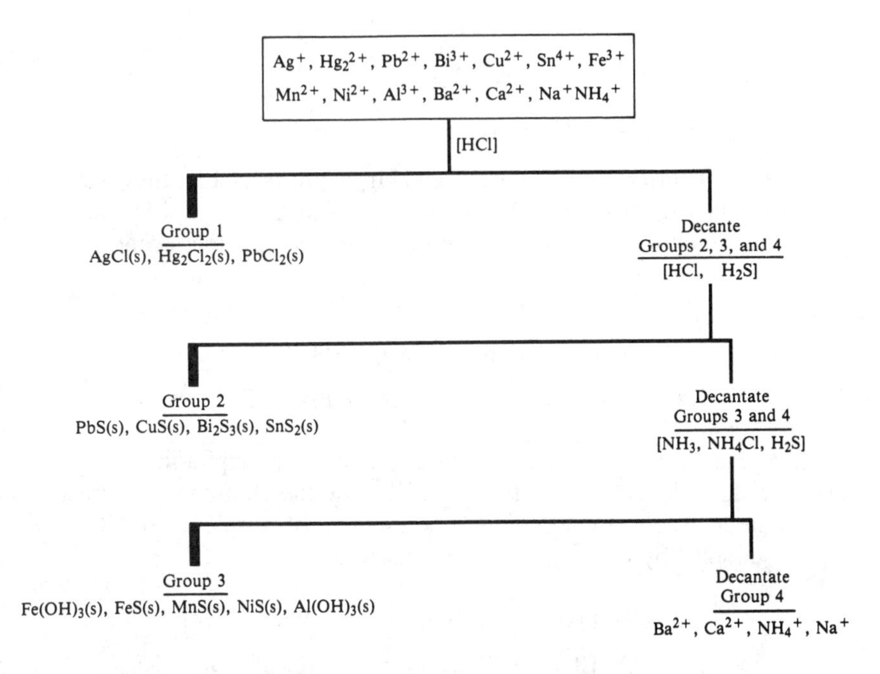

Figure 29.1 Flow chart for group separation.

know immediately that iron, copper, and nickel are absent. You are to take advantage of all clues that will aid you in identifying ions. However, in your role as a detective in identifying ions, be aware that clues can sometimes be misleading. For example, if Fe^{3+} and Ni^{2+} are present together, what color would you expect this mixture to display? Would the color depend upon the proportions of Fe^{3+} and Ni^{2+} present? Could you assign a definite color to such a mixture? The extremely faint pink color of Mn^{2+} may not be discerned unless you carefully examine the solution, and the presence of any of the other colored ions will definitely mask its color.

The group separation chart (Figure 29.1) shows that silver, mercury(I), and lead can be separated from all the other cations as insoluble chlorides by the addition of hydrochloric acid. Addition of hydrogen sulfide and hydrochloric acid to the decantate results in the precipitation of the sulfides of lead, copper, bismuth, and tin(IV). The reason why lead appears in both groups 1 and 2 is explained by the chemistry of these group cations. In order to derive any benefit from this exercise you must become thoroughly familiar with the group-separation chart in Figure 29.1. When you work with the four groups, you will see additional separation charts or flow schemes for the individual groups. Become familiar with these also. It is important to keep in mind what cations you are working with in each group. Know their formulas and charges and become familiar with how they are separated and identified.

You will analyze four known solutions, one for each group. Each of these solutions will contain all of the cations in the group you are studying. Then you will analyze a group unknown for each of the groups. Your instructor may tell you to analyze the known and unknown side by side. If you do, be careful not to mix up your test tubes! As you proceed with your analysis, record on your report sheets the reagents used in each step, your observations, and the equations for each precipitation reaction. After completing the four groups, you will analyze a general cation unknown that may contain any number of the 14 ions studied.

PART A

GROUP 1 CATIONS: Pb^{2+}, Ag^{+}, Hg_2^{2+}

Chemistry of Group 1 Cations

Because the chlorides of Pb^{2+}, Ag^{+}, and Hg_2^{2+} are insoluble, they may be precipitated and separated from the cations of groups 2, 3, and 4 by the addition of HCl. The following equations represent the reactions that occur:

$$Pb^{2+}(aq) + 2Cl^{-}(aq) \longrightarrow \mathbf{PbCl_2}(s) \quad \text{white} \quad [1]$$

$$Ag^{+}(aq) + Cl^{-}(aq) \longrightarrow \mathbf{AgCl}(s) \quad \text{white} \quad [2]$$

$$Hg_2^{2+}(aq) + 2Cl^{-}(aq) \longrightarrow \mathbf{Hg_2Cl_2}(s) \quad \text{white} \quad [3]$$

A slight excess of HCl is used to ensure complete precipitation of the cations and to reduce the solubility of the chlorides by the common-ion effect. However, a large excess of chloride must be avoided, because both AgCl and $PbCl_2$ tend to dissolve by forming soluble complex anions:

$$\mathbf{PbCl_2}(s) + 2Cl^{-}(aq) \longrightarrow PbCl_4^{2-}(aq) \quad [4]$$

$$\mathbf{AgCl}(s) + Cl^{-}(aq) \longrightarrow AgCl_2^{-}(aq) \quad [5]$$

$PbCl_2$ is appreciably more soluble than either AgCl or Hg_2Cl_2. Thus, even when $PbCl_2$ precipitates, a significant amount of Pb^{2+} remains in solution and is subsequently precipitated with the group 2 cations as the sulfide PbS. Because of its solubility, Pb^{2+} sometimes does not precipitate as the chloride, because either its concentration is too small or the solution is too warm.

Lead Lead chloride is much more soluble in hot water than in cold. It is separated from the other two insoluble chlorides by dissolving it in hot water. The presence of Pb^{2+} is confirmed by the formation of a yellow precipitate, $PbCrO_4$, upon the addition of K_2CrO_4:

$$Pb^{2+}(aq) + CrO_4^{2-}(aq) \longrightarrow \mathbf{PbCrO_4}(s) \quad \text{yellow} \quad [6]$$

Mercury(I) Silver chloride is separated from Hg_2Cl_2 by the addition of aqueous NH_3. Silver chloride dissolves because Ag^{+} forms a soluble complex cation with NH_3:

$$\mathbf{AgCl}(s) + 2NH_3(aq) \longrightarrow Ag(NH_3)_2^{+}(aq) + Cl^{-}(aq) \quad [7]$$

Mercury(I) chloride reacts with aqueous ammonia in a disproportionation reaction to form a dark gray precipitate:

$$\mathbf{Hg_2Cl_2}(s) + 2NH_3(aq) \longrightarrow \mathbf{HgNH_2Cl}(s) + \mathbf{Hg}(l) + NH_4^{+}(aq) + Cl^{-}(aq) \quad [8]$$

Although $HgNH_2Cl$ is white, the precipitate appears dark gray because of a colloidal dispersion of Hg(l).

Silver To verify the presence of Ag^{+}, the supernatant liquid from the last reaction is acidified and AgCl reprecipitates if Ag^{+} is present. The acid decom-

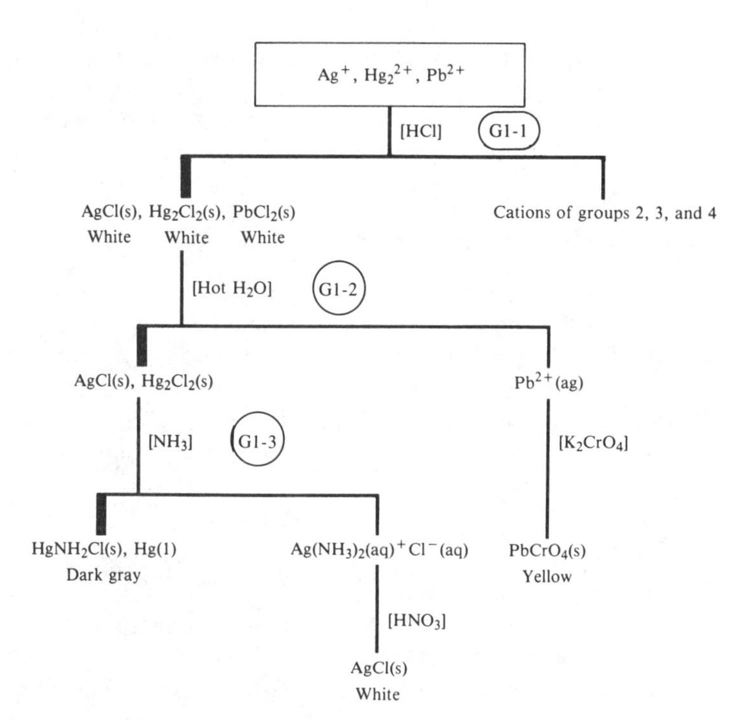

Figure 29.2 Group 1 flow scheme. Procedure numbers are given in the circles.

poses $Ag(NH_3)_2^+$ by neutralizing NH_3 to form NH_4^+. It is necessary that the solution be acidic, or else the AgCl will not precipitate and Ag^+ can be missed.

$$Ag(NH_3)_2^+(aq) + 2H^+(aq) + Cl^-(aq) \longrightarrow \mathbf{AgCl}(s) + 2NH_4^+(aq) \qquad [9]$$

The flow chart in Figure 29.2 shows how the group 1 cations are separated and identified. *You should become very familiar with it and consult it often as you perform your analysis.*

PROCEDURE

First you will analyze a known that contains all three cations of group 1. Record on your report sheet the reagents used in each step, your observations, and the equation for each precipitation reaction. After completing the analysis of a known, obtain an unknown. Follow the same procedures as with the known. Also record conclusions regarding the presence or absence of all cations. *Before beginning this experiment review the techniques used in qualitative analysis found in Appendix J: centrifugation, heating solutions, washing precipitates, and testing acidity.*

(G1-1) Precipitation of Group 1 Cations

Measure out 10 drops (0.5 mL) of the test solution or the unknown into a small (10 mm × 75 mm) test tube. Add 4 drops of 6 *M* HCl, stir thoroughly, and then centrifuge. Test for completeness of precipitation by adding 1 drop of 6 *M* HCl to the clear supernate. If the supernate turns cloudy, this shows that not all of the group 1 cations have precipitated; add another 2 drops of 6 *M* HCl, stir, and centrifuge. Repeat this process until no more precipitate forms. All of the group 1 cations must be precipitated or else they will slip through and interfere with subsequent group analysis. If a general unknown is being analyzed, decant the supernate into a clean test tube and save it for analysis of group 2 cations; otherwise, discard it. Wash the precipitate by adding 5 drops of cold distilled water and stirring. Centrifuge and add the liquid to the supernate. Why is the precipitate washed? See Appendix J.

G1-2 Separation and Identification of Pb^{2+}

Add 15 drops of distilled water to the precipitate and place the test tube in a hot-water bath. Stir using a stirring rod and heat for 1 min or longer. Quickly centrifuge and decant the hot supernate into a clean test tube. Repeat this procedure two more times, combining the supernates, which should contain Pb^{2+} if it is present. Save the precipitate for procedure G1-3. Add 3 drops of 1 *M* K_2CrO_4 to the supernate. The formation of a yellow precipitate, $PbCrO_4$, confirms the presence of Pb^{2+}.

G1-3 Separation and Identification of Ag^+ and Hg_2^{2+}

Add 10 drops of 6 *M* NH_3 to the precipitate from step G1-2. The formation of a dark gray precipitate indicates the presence of mercury. Centrifuge and decant the clear supernate into a clean test tube. Add 20 drops of 6 *M* HNO_3 to the decantate. Stir the solution and test its acidity. Continue to add HNO_3 dropwise until the solution is acidic. A white cloudiness confirms the presence of Ag^+.

REVIEW QUESTIONS

Before beginning Part A of this experiment in the laboratory, you should be able to answer the following questions:

1. What are the symbols and charges of the group 1 cations?
2. Which chloride salt is insoluble in cold water but soluble in hot water?
3. Which chloride salt dissolves in aqueous NH_3?
4. How could you distinguish:
   (a) $BaCl_2$ from $AgCl$?
   (b) HNO_3 from HCl?
5. Complete and balance the following equations:
   (a) $AgCl(s) + NH_3(aq) \rightarrow$
   (b) $Pb^{2+}(aq) + CrO_4^{2-}(aq) \rightarrow$
   (c) $Hg_2Cl_2(s) + NH_3(aq) \rightarrow$
   (d) $Ag(NH_3)_2^+(aq) + H^+(aq) + Cl^-(aq)$.
6. What can you conclude if no precipitate forms when HCl is added to an unknown solution?
7. Why are precipitates washed?
8. How do you decant supernatant liquids from small test tubes?

PART B

GROUP 2 CATIONS: Pb^{2+}, Cu^{2+}, Bi^{3+}, Sn^{4+}

Chemistry of Group 2 Cations

Hydrogen sulfide is the precipitating agent for the group 2 cations, Pb^{2+}, Cu^{2+}, Bi^{3+}, and Sn^{4+}. We will generate H_2S from the hydrolysis of thioacetamide, CH_3CSNH_2:

$$CH_3CSNH_2(aq) + 2H_2O(l) \longrightarrow H_2S(aq) + CH_3CO_2^-(aq) + NH_4^+(aq) \quad [1]$$

By controlling the hydrogen-ion concentration of the solution, we can control the sulfide-ion concentration. We see from Equation [2], which represents the overall ionization of H_2S, that the equilibrium will shift to the left if the hydrogen-ion concentration is increased by adding some strong acid to the solution.

$$H_2S(aq) \rightleftharpoons 2H^+(aq) + S^{2-}(aq) \quad [2]$$

Under these conditions the sulfide-ion concentration is very small. Thus by adjusting the pH of the solution to about 0.5, only the more insoluble sulfides will precipitate. These sulfides are those of group 2, PbS, CuS, Bi_2S_3, and SnS_2:

$Pb^{2+}(aq) + S^{2-}(aq) \longrightarrow$	$\mathbf{PbS}(s)$	black	[3]
$Cu^{2+}(aq) + S^{2-}(aq) \longrightarrow$	$\mathbf{CuS}(s)$	black	[4]
$2Bi^{3+}(aq) + 3S^{2-}(aq) \longrightarrow$	$\mathbf{Bi_2S_3}(s)$	brown	[5]
$Sn^{4+}(aq) + 2S^{2-}(aq) \longrightarrow$	$\mathbf{SnS_2}(s)$	yellow	[6]

The sulfides of group 3 are soluble in such acidic solutions and therefore do not precipitate because the sulfide ion concentration is too small. The flow chart in Figure 29.3 summarizes how the cations of group 2 are separated and identified. Note the colors of the precipitates. *Consult this scheme often to follow the remaining discussion.*

The cations are first treated with hydrogen peroxide (H_2O_2) and HCl to ensure that tin is in the +4 oxidation state:

$$Sn^{2+}(aq) + 2H^+(aq) + H_2O_2(aq) \longrightarrow Sn^{4+}(aq) + 2H_2O(l) \quad [7]$$

Tin must be in the +4 oxidation state to form the soluble SnS_3^{2-} ion and thereby allow it to be separated from the insoluble sulfides PbS, CuS, and Bi_2S_3:

$$\mathbf{SnS_2}(s) + S^{2-}(aq) \longrightarrow SnS_3^{2-} \quad [8]$$

These insoluble sulfides are brought into solution by treating them with hot nitric acid. Elemental sulfur is formed in the oxidation reactions:

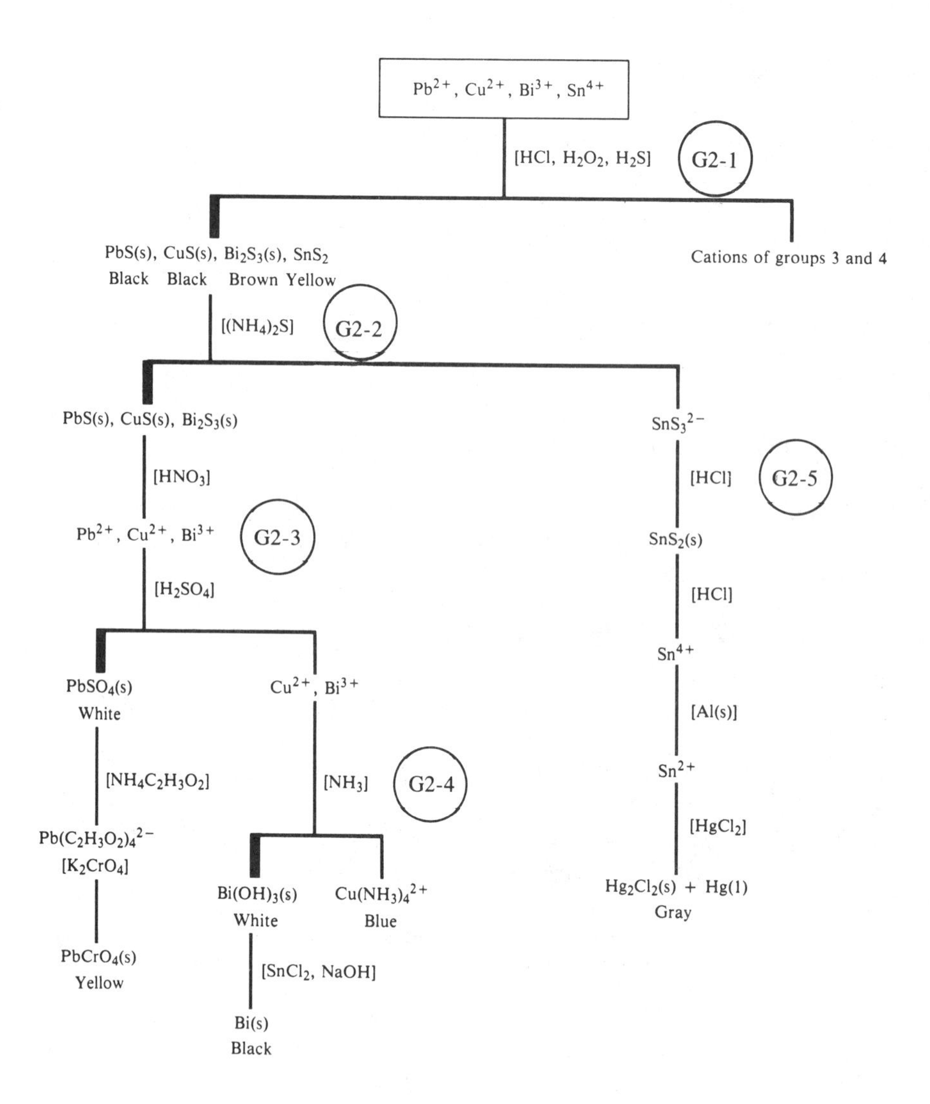

Figure 29.3 Group 2 flow scheme.

$$3\mathbf{PbS}(s) + 8H^+(aq) + 2NO_3^-(aq) \longrightarrow 3Pb^{2+}(aq) + 3\mathbf{S}(s) + 2NO(g) + 4H_2O(l) \quad [9]$$

$$3\mathbf{CuS}(s) + 8H^+(aq) + 2NO_3^-(aq) \longrightarrow 3Cu^{2+}(aq) + 3\mathbf{S}(s) + 2NO(g) + 4H_2O(l) \quad [10]$$

$$\mathbf{Bi_2S}(s) + 8H^+(aq) + 2NO_3^-(aq) \longrightarrow 2Bi^{3+}(aq) + 3\mathbf{S}(s) + 2NO(g) + 4H_2O(l) \quad [11]$$

Lead After the sulfides of lead, copper, and bismuth are brought into solution, lead is precipitated as the white sulfate by the addition of H_2SO_4:

$$Pb^{2+}(aq) + SO_4^{2-}(aq) \longrightarrow \mathbf{PbSO_4}(s) \quad \text{white} \quad [12]$$

The solution is heated strongly to drive off HNO_3, because $PbSO_4$ is soluble in the presence of nitric acid. To confirm the presence of Pb^{2+}, the $PbSO_4$ is dissolved with $NH_4C_2H_3O_2$ and precipitated as yellow $PbCrO_4$:

$$\mathbf{PbSO_4}(s) + 4C_2H_3O_2^-(aq) \longrightarrow Pb(C_2H_3O_2)_4^{2-}(aq) + SO_4^{2-}(aq) \quad [13]$$

$$Pb(C_2H_3O_2)_4^{2-} + CrO_4^{2-}(aq) \longrightarrow 4C_2H_3O_2^-(aq) + \mathbf{PbCrO_4}(s) \quad \text{yellow} \quad [14]$$

Copper The flow chart of Figure 29.3 shows that addition of aqueous NH_3 to the solution containing Cu^{2+} and Bi^{3+} precipitates Bi^{3+} as white $Bi(OH)_3$ and forms a soluble deep blue amine complex of copper, $Cu(NH_3)_4^{2+}$. The deep blue color confirms the presence of copper and can be seen even when the concentration of copper is very low.

$$Bi^{3+}(aq) + 3NH_3(aq) + 3H_2O(l) \longrightarrow 3NH_4^+(aq) + \mathbf{Bi(OH)_3}(s) \quad \text{white} \quad [15]$$

$$Cu^{2+}(aq) + 4NH_3(aq) \longrightarrow Cu(NH_3)_4^{2+}(aq) \quad \text{deep blue} \quad [16]$$

Bismuth The $Bi(OH)_3$ precipitate is often difficult to observe if the solution is blue. Bismuth is confirmed by separating the $Bi(OH)_3$ from the solution and reducing it with $SnCl_2$ in an alkaline solution. A black powder of finely divided bismuth is formed:

$$2Bi(OH)_3(s) + 3Sn(OH)_4^{2-}(aq) \longrightarrow 3Sn(OH)_6^{2-} + 2\mathbf{Bi}(s) \quad \text{black} \quad [17]$$

Tin Addition of concentrated HCl to a solution of SnS_3^{2-} first yields the precipitate SnS_2, which eventually dissolves in the acid solution:

$$SnS_3^{2-}(aq) + 2H^+(aq) \longrightarrow \mathbf{SnS_2}(s) + H_2S(g) \quad [18]$$

$$\mathbf{SnS_2}(s) + 4H^+(aq) + 6Cl^-(aq) \longrightarrow SnCl_6^{2-}(aq) + 2H_2S(g) \quad [19]$$

Aluminum metal reduces tin from the +4 to the +2 oxidation state. The Sn^{2+} (in the form of the complex $SnCl_4^{2-}$) in turn reduces $HgCl_2$ to insoluble Hg_2Cl_2, which is white, and Hg metal, which is black. Thus the precipitate appears white to gray in color.

$$2SnCl_4^{2-}(aq) + 3HgCl_2(aq) \longrightarrow 2SnCl_6^{2-}(aq) + \mathbf{Hg_2Cl_2}(s) + \mathbf{Hg}(l) \quad [20]$$

PROCEDURE

G2-1 Oxidation of Sn^{2+} and Precipitation of Group 2

Place the decantate from Procedure G1-1 or 7 drops of "known" or "unknown" solution in a casserole. Add 4 drops of 3% H_2O_2 and 4 drops of 2 *M* HCl. Carefully boil the solution by passing it back and forth over the flame of your burner. Formation of brown areas on the bottom of the casserole indicates overheating. If brown areas appear, swish the solution around until the brown area disappears. When the volume of the solution is reduced to about 5 drops, stop heating and allow the heat from the casserole to complete the evaporation. About 3 drops of solution should be present after cooling. Add 10 drops of 6 *M* HCl. IN THE HOOD evaporate the contents of the casserole to a pasty mass, being careful again to avoid overheating. Let the casserole cool, and then add 5 drops of 2 *M* HCl and 5 drops of distilled water. Swish the contents to dissolve or suspend the residue and transfer it to a small test tube. The solution is evaporated to dryness or to a pasty mass in order to remove the unknown quantity of acid that is present. A known amount of HCl is then added that is required for the group precipitation as sulfides.

Add 10 drops of 1 *M* thioacetamide to the solution in the test tube. Stir the mixture and heat the test tube in a boiling-water bath for 10 min. (See Appendix J for the use of a hot-water bath.) If excessive frothing occurs, temporarily remove the test tube from the bath. Occasionally stir the solution while it is being heated. After heating for 10 min, add 10 drops of *hot* water and 10 drops of 1 *M* thioacetamide and 2 drops of 1 *M* $NH_4C_2H_3O_2$ (ammonium acetate). Mix and heat in the boiling-water bath for 10 more minutes, stirring occasionally. Cool, centrifuge, and, using a pipet, decant (see Appendix J for the technique) into a test tube. The precipitate contains the group 2 insoluble sulfides, while the supernatant liquid may contain cations from groups 3 and 4.

Test the decantate for completeness of precipitation by adding 3 drops of H_2O, 1 drop of 1 *M* $NH_4C_2H_3O_2$, and 2 drops of 1 *M* thioacetamide. Mix and heat in the boiling-water bath for 1 min. If a colored precipitate forms, continue heating for 3 additional minutes. A faint cloudiness may develop because of the formation of colloidal sulfur. Repeat the precipitation procedure until precipitation is complete. If the decantate is to be analyzed for groups 3 and 4, transfer it to a casserole and boil it to reduce the volume to about 0.5 mL. Otherwise discard it. Transfer the contents to a labeled test tube and save it for Procedure G3-1. Rinse the casserole with 6 drops of water and add the washing to the test tube and stopper it.

All precipitates should be combined into the test tube containing the original precipitate using a few drops of water to aid in the transfer. Wash the precipitate three times, once with 10 drops of *hot* water and twice with 20-drop portions of a hot solution prepared using equal volumes of water and 1 *M* $NH_4C_2H_3O_2$. Be certain to stir the washing liquid and precipitate with a stirring rod before each centrifugation. Should the precipitate form a colloidal suspension, add 10 drops of 1 *M* $NH_4C_2H_3O_2$ and heat the suspension in the boiling water-bath. Discard the washings. Note, $NH_4C_2H_3O_2$ is often used in washing of precipitates. Its purpose is to help prevent the precipitate from becoming colloidal. Finely divided particles are difficult to settle.

G2-2 Separation of Sn^{4+} from PbS, CuS, and Bi_2S_3

To the precipitate from Procedure G2-1 add 10 drops of $(NH_4)_2S$ (ammonium sulfide) solution and stir well. Then heat for 3 to 4 min in the boiling-water bath. Remove the test tube from the water bath as necessary to avoid excessive frothing. Centrifuge, decant, and save the decantate. Repeat the treatment using 7 drops of $(NH_4)_2S$. Centrifuge and combine the decantate with the first. Stopper the combined decantate and save for analysis for tin (Procedure G2-5).

Wash the precipitate twice with 20-drop portions of a hot solution prepared by mixing equal volumes of water and 1 *M* $NH_4C_2H_3O_2$. The precipitate may contain PbS, CuS, and Bi_2S_3 and should be analyzed according to Procedure G2-3.

G2-3 Separation and Identification of Pb^{2+}

Add 1 mL (20 drops) of 3 *M* HNO_3 to the test tube containing the precipitate from Procedure G2-2. Mix thoroughly and transfer the contents to a casserole.

Boil the mixture gently for 1 min. Add more HNO_3, if necessary, to keep the amount of liquid constant. Cool, centrifuge, and discard any free sulfur that forms. Transfer the decantate to a casserole and add 6 drops of 18 *M* H_2SO_4. *Be careful: Concentrated* H_2SO_4 *causes severe burns*. If you get any on yourself, immediately wash the area with copious amounts of water. In a hood, evaporate the contents until the volume is about 1 drop and dense white fumes of SO_3 are formed. The fumes should be so dense that the bottom of the casserole cannot be seen. The appearance of dense white fumes of SO_3 ensures that all HNO_3 has been removed. Cool, add 20 drops of water, and stir. Quickly transfer the contents to a test tube before the suspended material settles. Cool the test tube. A finely divided white precipitate, $PbSO_4$, indicates the presence of lead. Centrifuge and save the decantate for Procedure G2-4. Wash the precipitate twice with 10-drop portions of cold water and discard the washings. Add 6 drops of 1 *M* $NH_4C_2H_3O_2$ to the precipitate and stir for about 15 s. Then add 1 drop of 1 *M* potassium chromate, K_2CrO_4. The formation of a yellow precipitate ($PbCrO_4$) confirms the presence of lead.

G2-4 Separation of Bi^{3+} and Identification of Bi^{3+} and Cu^{2+}

To the decantate from Procedure G2-3 carefully add 15 *M* aqueous NH_3 *(avoid inhalation or skin contact)* dropwise while constantly stirring until the solution is basic to litmus. The appearance of a deep blue color of $Cu(NH_3)_4^{2+}$ confirms the presence of Cu^{2+}. Centrifuge and discard the supernatant liquid, but save the white precipitate of $Bi(OH)_3$. Be careful in observing the white gelatinous $Bi(OH)_3$, for it may be somewhat difficult to see when the solution is colored.

Wash the precipitate once with 10 drops of hot water and discard the washings. Add 6 drops of 6 *M* NaOH and 4 drops of freshly prepared* 0.2 *M* $SnCl_2$ to the precipitate and stir. The formation of a jet-black precipitate confirms the presence of Bi^{3+}.

G2-5 Identification of Sn^{4+}

Transfer the decantate from procedure G2-2 to a casserole and boil for 1 min to expel H_2S; then add 4 drops of cold water. Add a 1-in. piece of 26 gauge aluminum wire and heat gently until the wire has dissolved. Continue to gently heat the solution for about 2 more minutes, replenishing the solution with 6 *M* HCl if necessary. There should be no dark residue at this stage; if there is, continue heating until it dissolves. (You cannot stop at this point.) Transfer the solution to a test tube and cool under running water. Immediately add 3 drops of 0.1 *M* mercuric chloride, $HgCl_2$, and mix. Allow the mixture to stand for 1 min. The formation of a white or gray precipitate confirms the presence of tin.

REVIEW QUESTIONS

Before beginning Part B of this experiment in the laboratory, you should be able to answer the following questions:

---

*Solid tin should be present in the bottle to ensure that tin is in the +2 oxidation state.

1. Give the symbols and charges of the cations of group 2.
2. How could CuS be separated from SnS_2?
3. How are Cu^{2+} and Bi^{3+} separated?
4. How can Ag^+ be separated from Cu^{2+}?
5. Complete and balance the following equations:
   (a) $Bi^{3+}(aq) + S^{2-}(aq) \longrightarrow$
   (b) $SnS_2(s) + S^{2-}(aq) \longrightarrow$
   (c) $PbS(s) + H^+(aq) + NO_3^-(aq) \longrightarrow$
   (d) $Bi^{3+}(aq) + NH_3(aq) + H_2O(l) \longrightarrow$
6. Why is H_2O_2 added in the initial step of the separation of group 2 cations?
7. What is the color of CuS? Of SnS_2? Of $PbSO_4$?

PART C

GROUP 3 CATIONS: Fe^{3+}, Ni^{2+}, Mn^{2+}, Al^{3+}

Chemistry of Group 3 Cations

The group 3 cations that we will consider are Fe^{3+}, Ni^{2+}, Mn^{2+}, and Al^{3+}. These cations do not precipitate as insoluble chlorides (as do those of group 1) or as sulfides in acidic solution (like those of group 2). These ions can be separated from those of group 4 by precipitation as insoluble hydroxides or sulfides under slightly alkaline conditions. The separation is shown in the flow chart of Figure 29.4. As in group 2, the pH of the solution controls the sulfide-ion concentration. The slightly alkaline conditions employed here favor a higher sulfide-ion concentration than that used in the group 2 separation. FeS, NiS, and MnS are more soluble than the sulfides of group 2 and therefore require a higher sulfide-ion concentration for their precipitation. Making the solution slightly alkaline reduces the hydrogen ion concentration. We see that decreasing the hydrogen ion concentration causes the following equilibrium to shift to the right:

$$H_2S(aq) \rightleftharpoons 2H^+(aq) + S^{2-}(aq)$$

This results in an increase in the sulfide ion concentration.

Aqueous ammonia is a weak base:

$$NH_3(aq) + H_2O(l) \rightleftharpoons NH_4^+(aq) + OH^-(aq)$$

A mixture of aqueous NH_3 and NH_4Cl makes a buffer solution whose pH allows the precipitation FeS, MnS, and NiS. Moreover, in this slightly alkaline solution, the insoluble hydroxides of $Fe(OH)_3$ and $Al(OH)_3$ also precipitate. Although $Mg(OH)_2$ is insoluble, it does not precipitate with the group 3 hydroxides because the OH^- ion concentration is too small. The common ion NH_4^+, from NH_4Cl controls the OH^- concentration and keeps it sufficiently low, preventing the $Mg(OH)_2$ from precipitating.

As you analyze the group 3 ions, pay particular attention to the colors of the solutions and the precipitates. Aqueous solutions of Al^{3+} ions are colorless, while those of Fe^{3+} appear yellow to reddish-brown; Ni^{2+} solutions are green, and Mn^{2+}, *very* faint pink.

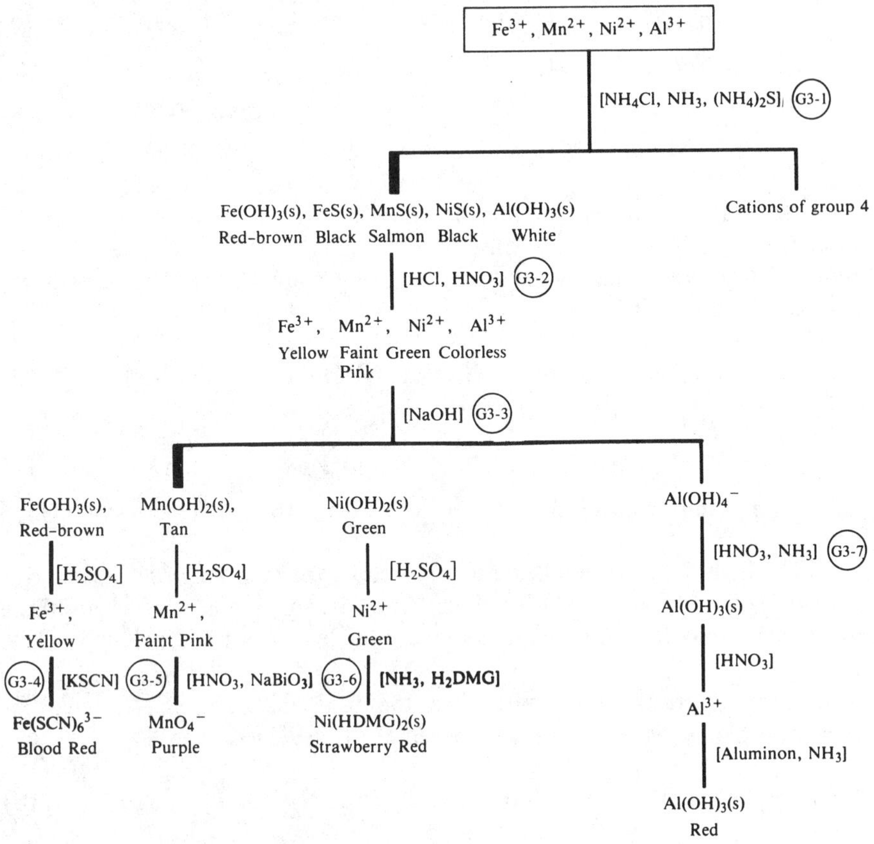

Fe^{3+}, Mn^{2+}, Ni^{2+}, Al^{3+}

[NH_4Cl, NH_3, $(NH_4)_2S$] (G3-1)

$Fe(OH)_3(s)$, $FeS(s)$, $MnS(s)$, $NiS(s)$, $Al(OH)_3(s)$
Red-brown Black Salmon Black White

Cations of group 4

[HCl, HNO_3] (G3-2)

Fe^{3+}, Mn^{2+}, Ni^{2+}, Al^{3+}
Yellow Faint Pink Green Colorless

[NaOH] (G3-3)

$Fe(OH)_3(s)$, Red-brown — [H_2SO_4] — Fe^{3+}, Yellow — (G3-4) [KSCN] — $Fe(SCN)_6^{3-}$ Blood Red

$Mn(OH)_2(s)$, Tan — [H_2SO_4] — Mn^{2+}, Faint Pink — (G3-5) [HNO_3, $NaBiO_3$] — MnO_4^- Purple

$Ni(OH)_2(s)$ Green — [H_2SO_4] — Ni^{2+} Green — (G3-6) [NH_3, H_2DMG] — $Ni(HDMG)_2(s)$ Strawberry Red

$Al(OH)_4^-$ — [HNO_3, NH_3] (G3-7) — $Al(OH)_3(s)$ — [HNO_3] — Al^{3+} — [Aluminon, NH_3] — $Al(OH)_3(s)$ Red

Figure 29.4 Group 3 flow scheme.

The reactions involved in the precipitation of the group 3 cations are:

$$Fe^{3+}(aq) + 3OH^-(aq) \longrightarrow \mathbf{Fe(OH)_3}(s) \quad \text{red-brown} \quad [1]$$

$$Al^{3+}(aq) + 3OH^-(aq) \longrightarrow \mathbf{Al(OH)_3}(s) \quad \text{white} \quad [2]$$

$$Ni(NH_3)_6^{2+}(aq) + S^{2-}(aq) \longrightarrow 6NH_3(aq) + \mathbf{NiS}(s) \quad \text{black} \quad [3]$$

$$Mn^{2+}(aq) + S^{2-}(aq) \longrightarrow \mathbf{MnS}(s) \quad \text{salmon} \quad [4]$$

In aqueous NH_3 solution, Ni^{2+} ions exist as the ammine complex $Ni(NH_3)_6^{2+}$, and addition of $(NH_4)_2S$ results in the precipitation of NiS. Some $Fe(OH)_3$ is reduced by sulfide:

$$2\mathbf{Fe(OH)_3}(s) + 3S^{2-}(aq) \longrightarrow 6OH^-(aq) + \mathbf{FeS}(s) \quad \text{black} \quad [5]$$

The group 3 precipitates are dissolved using HCl. Nitric acid is also added to help dissolve NiS by oxidizing S^{2-} ions to elemental sulfur. At the same time, nitric acid oxidizes Fe^{2+} to Fe^{3+}:

$$\mathbf{Fe(OH)_3}(s) + 3H^+(aq) \longrightarrow Fe^{3+}(aq) + 3H_2O(l) \quad [6]$$

$$\mathbf{FeS}(s) + 2H^+(aq) \longrightarrow Fe^{2+}(aq) + H_2S(g) \quad [7]$$

$$3Fe^{2+}(aq) + NO_3^-(aq) + 4H^+(aq) \longrightarrow 3Fe^{3+}(aq) + NO(g) + 2H_2O(l) \quad [8]$$

$$Al(OH)_3(s) + 3H^+(aq) \longrightarrow Al^{3+}(aq) + 3H_2O(l) \quad [9]$$

$$3NiS(s) + 2NO_3^-(aq) + 8H^+(aq) \longrightarrow 3Ni^{2+}(aq) + 2S(s) + 2NO(g) + 4H_2O(l) \quad [10]$$

$$MnS(s) + 2H^+(aq) \longrightarrow Mn^{2+}(aq) + H_2S(g) \quad [11]$$

After all ions are in solution, addition of excess strong base allows the separation of Fe^{3+}, Ni^{2+}, and Mn^{2+} ions from Al^{3+} ions. $Fe(OH)_3$, $Ni(OH)_2$, and $Mn(OH)_2$ precipitate, while $Al(OH)_3$, being amphoteric, redissolves in excess base forming aluminate ions:

$$Fe^{3+}(aq) + 3OH^-(aq) \longrightarrow Fe(OH)_3(s) \quad \text{red-brown} \quad [12]$$

$$Ni^{2+}(aq) + 2OH^-(aq) \longrightarrow Ni(OH)_2(s) \quad \text{green} \quad [13]$$

$$Mn^{2+}(aq) + 2OH^-(aq) \longrightarrow Mn(OH)_2(s) \quad \text{tan} \quad [14]$$

$$Al^{3+}(aq) + 4OH^-(aq) \longrightarrow Al(OH)_4^-(aq) \quad \text{colorless} \quad [15]$$

The hydroxide precipitates are dissolved by the addition of H_2SO_4 to give a solution of Fe^{3+}, Ni^{3+}, and Mn^{2+} ions. After the solution is divided into three equal parts, tests for the individual ions are made as described below.

Iron A very sensitive test for Fe^{3+} uses the thiocyanate ion, SCN^-. If Fe^{3+} is present, a blood-red solution results when SCN^- is added:

$$Fe^{3+}(aq) + 6SCN^-(aq) \longrightarrow Fe(SCN)_6^{3-}(aq) \quad \text{red} \quad [16]$$

Manganese Because of its intense purple color, the permanganate ion, MnO_4^-, affords a suitable confirmatory test for Mn^{2+}. When a solution of Mn^{2+} is acidified with HNO_3 and then treated with sodium bismuthate, $NaBiO_3$, Mn^{2+} is oxidized to MnO_4^-:

$$2Mn^{2+}(aq) + 5NaBiO_3(s) + 14H^+(aq) \longrightarrow 5Bi^{3+}(aq) + 5Na^+(aq) + 7H_2O(l) + 2MnO_4^-(aq) \quad \text{purple} \quad [17]$$

Nickel The presence of Ni^{2+} is confirmed by the formation of a bright red precipitate when an organic compound called dimethylglyoxime (abbreviated as H_2DMG) is added to an ammoniacal solution:

$$Ni(NH_3)_6^{2+}(aq) + 2H_2DMG(aq) \longrightarrow 4NH_3(aq) + 2NH_4^+(aq) + Ni(HDMG)_2(s) \quad \text{red} \quad [18]$$

Aluminum When a solution that contains $Al(OH)_4^-$ is acidified and then made slightly alkaline with the weak base NH_3, $Al(OH)_3$ precipitates:

$$Al(OH)_4^- + 4H^+(aq) \longrightarrow Al^{3+}(aq) + 4H_2O(l) \quad [19]$$

$$Al^{3+}(aq) + 3NH_3(aq) + 3H_2O(l) \longrightarrow 3NH_4^+(aq) + Al(OH)_3(s) \quad [20]$$

Aluminum hydroxide is not easily seen, for it is a gelatinous, translucent substance. To help see the hydroxide it is precipitated in the presence of a red dye. The dye is absorbed on the $Al(OH)_3$, giving it a cherry-red color.

G3-1 Precipitation of Group 3 Cations

Place the decantate from Procedure G2-1 or 7 drops of known or unknown solution in a small (10 mm × 75 mm) test tube. If the decantate from Procedure G2-1 has a precipitate, centrifuge, decant, and discard the precipitate. Add 5 drops of 2 *M* NH_4Cl and stir; add 15 *M* NH_3 dropwise with stirring, until the solution is just basic. Usually this requires only a few drops of the NH_3. Then add 2 additional drops of the 15 *M* NH_3 and 1 mL (20 drops) of water. Stir thoroughly. Next add 10 drops of $(NH_4)_2S$ and mix thoroughly. Heat the test tube in a boiling-water bath for about 5 min. If excessive frothing occurs, temporarily remove the test tube from the hot-water bath. Centrifuge and test for completeness of precipitation using 1 drop of $(NH_4)_2S$. Note the color of the precipitate. Decant and save the decantate for group 4 analysis, or discard, as appropriate.

Wash the precipitate two times with 20 drops of a solution made by mixing equal portions of water and 1 *M* $NH_4C_2H_3O_2$. For each washing, stir the precipitate with the wash solution and heat the mixture in the water bath before centrifuging. Discard the supernatant wash liquid.

G3-2 Dissolution of Group 3 Precipitate

Treat the precipitate from Procedure G3-1 with 12 drops of 12 *M* HCl, and cautiously add 5 drops of 16 *M* HNO_3 and carefully mix the solution (CAUTION: HNO_3 *can cause severe burns. If you come in contact with the acid, wash the area with copious amounts of water*). Heat the test tube in the hot-water bath until the precipitate dissolves and a clear, but not necessarily colorless, solution is obtained. Add 10 drops of water, centrifuge to remove any sulfur that has precipitated, and decant into a casserole. Note the color of the decantate.

G3-3 Separation of Iron, Nickel, and Manganese from Aluminum

Make the solution in the casserole from Procedure G3-2 strongly basic, using 6 *M* NaOH, mixing thoroughly. If the precipitate is pasty and nonfluid, add 12 drops of water. Note the color of the precipitate. Transfer to a test tube and centrifuge. Decant, saving the decantate, which may contain aluminum, for Procedure G3-7. To the precipitate add 20 drops of water and 10 drops of 6 *M* H_2SO_4. Stir and heat in a water bath for 3 min or until the precipitate dissolves. Add 12 drops of water and divide the solution into three approximately equal volumes.

G3-4 Test for Fe^{3+}

To one of the three samples from Procedure G3-3 add 2 drops of 0.2 *M* KSCN (potassium thiocyanate). A blood-red *solution* confirms the presence of iron as $Fe(SCN)_6^{3-}$. Traces of iron that have been introduced as impurities along the way will give a weak test. If you are in doubt about your results, perform this test on 10 drops of your original sample.

G3-5 Test for Mn^{2+}

To the second portion of a solution from Procedure G3-3 add an equal volume of water and 4 drops of 3 *M* HNO_3. Mix, and then add a few grains of solid sodium bismuthate, $NaBiO_3$. Mix thoroughly with a stirring rod and centrifuge. A pink or purple color is due to MnO_4^- and confirms the presence of manganese.

(G3-6) Test for Ni^{2+}

To a third portion of the sample from Procedure G3-3 add 6 M NH_3 until the solution is basic. If a precipitate forms, remove it by centrifuging and decanting, keeping the decantate. Add about 4 drops of dimethylglyoxime reagent mix, and allow to stand. The formation of a strawberry-red *precipitate* indicates the presence of nickel.

(G3-7) Test for Al^{3+}

Treat only half of the decantate from Procedure G3-3 with 16 M HNO_3 until the solution is slightly acidic. (CAUTION: HNO_3 *can cause severe burns. Wash with water immediately if you come in contact with the acid.*) Then add 15 M NH_3 while stirring until the solution is distinctly alkaline. Allow at least 1 min for the formation of $Al(OH)_3$. Centrifuge and carefully remove the supernatant liquid with a capillary pipet without disturbing the gelatinous $Al(OH)_3$. Discard the supernate. Wash the precipitate two times with 20 drops of hot water, discarding the decantate. Dissolve the precipitate in 7 drops of 3 M HNO_3. Add 3 drops of aluminon reagent, which colors the solution; stir; and add 6 M NH_3 dropwise until the solution is just alkaline (avoid an excess). Stir and centrifuge. The formation of a cherry-red *precipitate,* not solution, confirms the presence of aluminum.

REVIEW QUESTIONS

Before beginning Part C of this experiment in the laboratory, you should be able to answer the following questions:

1. What are the symbols and charges of the group 3 cations?
2. What are the colors of the following ions: Cu^{2+}, Fe^{3+}, Al^{3+}, Ni^{2+}?
3. What are the colors of the following solids: $Fe(OH)_3$, MnS, $Al(OH)_3$, $Ni(OH)_2$?
4. How can Fe^{3+} be separated from Al^{3+}?
5. How can $Ni(OH)_2$ be separated from $Al(OH)_3$?
6. Complete and balance the following equations:
   (a) $Fe^{3+}(aq) + OH^-(aq) \longrightarrow$
   (b) $Al(OH)_3(s) + H^+(aq) \longrightarrow$
   (c) $FeS(s) + H^+(aq) \longrightarrow$
   (d) $NiS(s) + NO_3^-(aq) + H^+(aq) \longrightarrow$
7. Give the formula for a reagent that precipitates
   (a) Pb^{2+} but not Ni^{2+}
   (b) Fe^{3+} but not Al^{3+}
8. What cation forms a
   (a) Blood-red solution with thiocyanate ion?
   (b) Bright red precipitate with dimethylglyoxime?
9. If solid NH_4Cl is added to 3 M NH_3, does the pH increase, decrease or remain the same?

PART D

GROUP 4 CATIONS: Ba^{2+}, Ca^{2+}, NH_4^+, Na^+

Chemistry of Group 4 Cations

In addition to the ammonium ion, the cations of group 4 consist of ions of the alkali and alkaline earth metals. The cations we will consider in this group are Ba^{2+}, Ca^{2+}, NH_4^+, and Na^+. Because their chlorides and sulfides are soluble, these ions do not precipitate with groups 1, 2, or 3.

Sodium ions are a common impurity and were even introduced (as was ammonium ion) in some of the reagents that were used in the analysis of groups 1, 2, and 3. Hence, in the analysis of a general unknown mixture, tests for these ions must be made on the original sample even before performing the group analysis. The flow scheme for group 4 is shown in Figure 29.5.

Barium Because barium chromate, $BaCrO_4$ ($K_{sp} = 1.2 \times 10^{-10}$), is more insoluble than calcium chromate, $CaCrO_4$ ($K_{sp} = 7.1 \times 10^{-4}$), Ba^{2+} can be separated from Ca^{2+} by precipitation as the insoluble yellow chromate salt:

$$Ba^{2+}(aq) + CrO_4^{2-}(aq) \longrightarrow \mathbf{BaCrO_4}(s) \quad \text{yellow} \qquad [1]$$

$BaCrO_4$ is insoluble in the weak acid $HC_2H_3O_2$, but it is soluble in the presence of the strong acid HCl, because it is the salt of a weak acid. After $BaCrO_4$ is dissolved in HCl, a flame test is performed on the resulting solution. A green-yellow flame is indicative of Ba^{2+}. Further confirmation of Ba^{2+} is precipitation of $BaSO_4$, which is white:

$$Ba^{2+}(aq) + SO_4^{2-}(aq) \longrightarrow \mathbf{BaSO_4}(s) \quad \text{white} \qquad [2]$$

Calcium Calcium oxalate, CaC_2O_4, is very insoluble ($K_{sp} = 4.0 \times 10^{-9}$). The formation of a white precipitate when oxalate ion is added to a slightly alkaline solution confirms the presence of Ca^{2+}:

$$Ca^{2+}(aq) + C_2O_4^{2-}(aq) \longrightarrow \mathbf{CaC_2O_4}(s) \quad \text{white} \qquad [3]$$

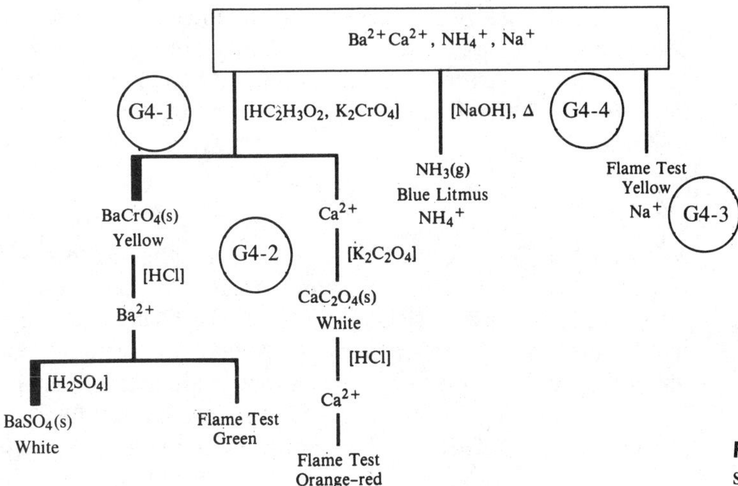

Figure 29.5 Group 4 flow scheme.

Additional evidence for the calcium ion is obtained from a flame test. Dissolution of CaC_2O_4 with HCl, followed by a flame test, produces an orange-red flame that is characteristic of calcium ions.

Sodium Most sodium salts are soluble. The simplest test for sodium ion is a flame test. Sodium salts impart a characteristic yellow color to a flame; the test is very sensitive, and because of the prevalence of sodium ions, much care must be exercised to keep equipment clean and free from contamination by these ions.

Ammonium The ammonium ion, NH_4^+, is the conjugate acid of the base ammonia, NH_3. The test for NH_4^+ takes advantage of the following equilibrium:

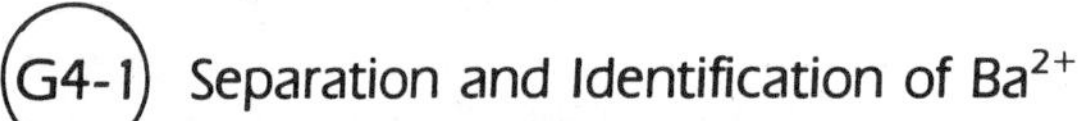

$$NH_4^+(aq) + OH^-(aq) \rightleftharpoons NH_3(aq) + H_2O(l) \qquad [4]$$

Thus, when a strong base is added to a solution of an ammonium salt and this solution is heated, NH_3 gas is evolved. The NH_3 can easily be detected by its effect upon red litmus.

PROCEDURE

G4-1 Separation and Identification of Ba^{2+}

If the solution, "known," or "unknown" contains only cations of group 4, place 7 drops of the solution in a small test tube; otherwise, use the supernate from group 3 analysis, Procedure G3-1. Add 8 drops of 6 *M* acetic acid, $HC_2H_3O_2$, and 1 drop of 1 *M* K_2CrO_4 and mix. The formation of a yellow precipitate indicates the presence of Ba^{2+}. Centrifuge, saving the decantate for Procedure G4-2 to test for calcium. Dissolve the precipitate with 6 *M* HCl and perform a flame test as described below.

To perform the flame test, obtain a piece of platinum or Nichrome wire that has been sealed in a piece of glass tubing. Clean the wire by dipping it in 12 *M* HCl that is contained in a small test tube and heat the wire in the hottest part of your Bunsen burner flame. Repeat this operation until no color is seen when the wire is placed in the flame. Several cleanings will be required before this is achieved. Then dip the wire into the solution to be tested and place the wire in the flame. A pale green flame confirms the presence of Ba^{2+}. If the concentration of Ba^{2+} is very low, you may not detect the green color.

As further confirmation of barium, add 10 drops of 6 *M* H_2SO_4 to the solution on which the flame test was performed. A white precipitate confirms the presence of Ba^{2+}.

G4-2 Test for Ca^{2+}

Make the decantate from Procedure G4-1 alkaline to litmus with 15 *M* NH_3. If a precipitate forms, centrifuge and discard the precipitate. Add 7 drops of 1 *M* $K_2C_2O_4$ (potassium oxalate) and stir. The formation of a white precipitate indicates the presence of calcium ion. Should no precipitate form immediately, warm the test tube briefly in the hot-water bath and then cool.

Additional evidence for Ca^{2+} is obtained from a flame test. Dissolve the precipitate in 6 *M* HCl and then perform a flame test. A transitory red-orange color that appears when the wire is first placed in the flame and later reappears somewhat more red as the wire is heated is characteristic of the calcium ion. If the concentration of Ca^{2+} is very low, you may not observe the red color.

(G4-3) Test for Na^+

The flame test for sodium is sensitive, and traces of sodium ion will impart a characteristic yellow color to the flame. Just about every solution has a trace of sodium and thus will give a positive test. On the basis of the intensity and duration of the yellow color, you can decide whether Na^+ is merely a contaminant or present in substantial quantity. Using a clean wire, perform a flame test on your original (untreated) unknown. To help you make a decision as to the presence of sodium, run a flame test on distilled water and then on a 0.2 *M* NaCl solution. Compare the tests.

(G4-4) Test for NH_4^+

Place 2 mL of the original (untreated) unknown or known in a 100-mL beaker (or casserole) and add 2 mL of 6 *M* NaOH. Moisten a piece of red litmus paper with water and stick it to the convex side of a small watch glass. Cover the beaker with the watch glass convex side down. (The litmus paper must not come into contact with any NaOH.) Gently warm the beaker with a small burner flame; do not boil. Allow the covered beaker to stand 3 min. A change in the color of the litmus paper from red to blue confirms the presence of ammonium ion.

REVIEW QUESTIONS

Before beginning Part D of this experiment in the laboratory, you should be able to answer the following questions:

1. What are the symbols and charges of the group 4 cations?
2. Which is less soluble: $BaCrO_4$ or $CaCrO_4$?
3. What is the color of
   (a) $BaSO_4$?
   (b) $BaCrO_4$?
   (c) CaC_2O_4?
4. What color do the following ions impart to a flame:
   (a) Ba^{2+}?
   (b) Ca^{2+}?
   (c) Na^+
5. What reagent will precipitate
   (a) Cu^{2+} but not Ba^{2+}?
   (b) Ag^+ but not Ca^{2+}?
   (c) Ba^{2+} but not NH_4^+?
6. Complete and balance the following equations:
   (a) $Ba^{2+}(aq) + CrO_4^{2-}(aq)$
   (b) $NH_4^+(aq) + OH^-(aq)$

PART E: ANIONS

DISCUSSION

A systematic scheme based upon the kinds of principles involved in cation analysis can be designed for the analysis of anions. This would involve separation of the anions followed by their identification. However, it is generally easier to take another approach to the identification of anions. An effort is made either to eliminate or verify the presence of certain anions on the basis of the color and solubility of the samples; then the material being analyzed is subjected to a series of preliminary tests. From the results of the preliminary tests and observations, certain of the anions may definitely be shown to be present or absent; then specific tests are performed for those anions not definitely eliminated in the preliminary tests and observations. The preliminary tests include treating the solid with concentrated sulfuric acid and using silver nitrate and $BaCl_2$ as precipitating agents. The behavior of the anions in these tests in conveniently summarized in Tables 29.1 and 29.2.

In this experiment the following 10 anions are considered: sulfate (SO_4^{2-}), nitrate (NO_3^-), carbonate (CO_3^{2-}), chloride (Cl^-), bromide (Br^-), iodide (I^-), chromate (CrO_4^{2-}), phosphate (PO_4^{3-}), sulfide (S^{2-}), and sulfite (SO_3^{2-}).

Table 29.1 Summary of Preliminary Tests

Anion	Conc. H_2SO_4 solid salt	$AgNO_3$ Neutral	$AgNO_3$ Acidic	$BaCl_2$ Neutral	$BaCl_2$ Acidic
NO_3^-	Cold: no observable reaction Hot: brown gas	No reaction		No reaction	
Cl^-	$HCl(g)$, colorless; pungent	$AgCl(s)$ / white	Insol.	No reaction	
Br^-	Br_2, red-brown; pungent	$AgBr(s)$ / cream	Insol.	No reaction	
I^-	Solids turn dark brown, violet vapors; H_2S odor	$AgI(s)$ / yellow	Insol.	No reaction	
S^{2-}	$H_2S(g)$, odor of rotten eggs free S deposited	$Ag_2S(s)$ / black	Insol.	No reaction	
SO_4^{2-}	No evidence of reaction	No reaction		$BaSO_4(s)$ / white	Insol.
SO_3^{2-}	$SO_2(g)$, colorless; choking odor	$Ag_2SO_3(s)$ / white	Sol.	$BaSO_3(s)$ / white	Sol.
CO_3^{2-}	$CO_2(g)$, colorless; odorless	$Ag_2CO_3(s)$ / white → black	Sol.	$BaCO_3(s)$ / white	Sol.
PO_4^{3-}	No evidence of reaction	$Ag_3PO_4(s)$ / yellow	Sol.	$Ba_3(PO_4)_2(s)$ / white	Sol.
CrO_4^{2-}	Solid changes from yellow to orange-red	$Ag_2CrO_4(s)$ / red-brown	Sol.	$BaCrO_4(s)$ / yellow	Sol.

Table 29.2 Behavior of Anions with Concentrated Sulfuric Acid, H_2SO_4

A. Cold H_2SO_4

Anion	Behavior
SO_4^{2-}	No reaction
NO_3^-	No reaction
PO_4^{3-}	No reaction
CO_3^{2-}	A colorless, odorless gas forms.

$$CO_3^{2-} + 2H^+ \longrightarrow H_2O + CO_2$$

Cl^- — A colorless gas forms. It has a sharp, pungent odor, gives an acid test with litmus, and fumes in moist air.

$$Cl^- + H^+ \longrightarrow HCl$$

Br^- — A brownish-red gas forms. It has a sharp odor, gives an acid test with litmus, and fumes in moist air. The odor of SO_2 may be detected.

$$Br^- + H_2SO_4 \longrightarrow HSO_4^- + HBr$$

$$H_2SO_4 + 2HBr \longrightarrow 2H_2O + SO_2 + Br_2$$

I^- — Solid turns dark brown immediately, with the slight formation of violet fumes. The gas has the odor of rotten eggs, gives an acidic test with litmus, and fumes in moist air.

$$I^- + H_2SO_4 \longrightarrow HSO_4^- + HI$$

$$H_2SO_4 + 8HI \longrightarrow H_2S + 4H_2O + 4I_2$$

$$H_2SO_4 + 2HI \longrightarrow 2H_2O + SO_2 + I_2$$

S^{2-} — A colorless gas with the odor of rotten eggs forms, and some free sulfur is deposited.

$$S^{2-} + H_2SO_4 \longrightarrow SO_4^{2-} + H_2S$$

$$H_2SO_4 + H_2S \longrightarrow 2H_2O + SO_2 + S$$

SO_3^{2-} — A colorless gas with a sharp, choking odor forms.

$$SO_3^{2-} + H_2SO_4 \longrightarrow SO_4^{2-} + H_2O + SO_2$$

CrO_4^{2-} — Color changes from yellow to orange-red.

$$2K_2CrO_4 + H_2SO_4 \longrightarrow K_2Cr_2O_7 + H_2O + K_2SO_4$$

B. Hot Concentrated H_2SO_4.

There are no additional reactions with any of the anions except NO_3^-, which forms brown fumes of NO_2 gas.

$$4NO_3^- + 4H^+ \longrightarrow 4NO_2 + O_2 + 2H_2O$$

As in the case of cation identification, the physical and chemical properties of the compounds formed by the anions provide the basis for their identification. We will proceed by using the following general procedures.

PROCEDURE

Examination of the Solid

The color of a substance offers a clue to its constituents. For example, many transition metal salts are colored. Those of nickel, Ni^{2+}, are generally green; of iron(III), Fe^{3+}, reddish-brown to yellow; of iron(II), Fe^{2+}, grayish green; of Cr(III), Cr^{3+}, green to bluish-gray to black; of copper(II), Cu^{2+}, blue to green to black; of cobalt(II), Co^{2+}, wine-red to blue; and of manganese(II), Mn^{2+}, pink to tan. By contrast, only a few anions are colored, and these contain transition metals as well. The colored anions are chromate (CrO_4^{2-}), yellow; dichromate ($Cr_2O_7^{2-}$), orange-red; and permanganate (MnO_4^{2-}), violet-purple. Hence, the color of the solid may be used as a very good indicator of the presence of these ions. Observe the color of your unknown and note it on the report sheet.

Solubility of Salt

The solubility of a substance in water and in acidic or basic solutions, together with a knowledge of the cations present, often permits us to narrow down the choice of anions present. For example, a white substance containing Ag^+ that is insoluble in acid solution (HNO_3) but soluble in aqueous ammonia (NH_3) is most probably AgCl. There is some ambiguity, however, for AgBr behaves similarly. Thus tests for both chloride and bromide would be necessary to confirm which anion is present. Test the solubility of your solid unknown anion sample in water, 6 *M* HNO_3, 6 *M* HCl, and 6 *M* NH_3, and note your results on your report sheet. Compare your results with the solubility properties of ions and solids given in Appendix E.

Reactions with $AgNO_3$ and $BaCl_2$

If your unknown substance is soluble in water, you may obtain additional clues as to the presence of certain anions by treating a solution of your unknown separately with solutions of silver nitrate ($AgNO_3$), and barium chloride ($BaCl_2$). The silver salts of all your possible anions except nitrate and sulfate are insoluble in water and have the following colors (see Table 29.1):

AgCl: white	Ag_2CrO_4: red-brown
AgBr: cream	Ag_2S: black
AgI: yellow	Ag_2CO_3: white
Ag_3PO_4: yellow	Ag_2SO_3: white

All these solids except AgCl, AgBr, AgI, and Ag_2S are soluble in dilute nitric acid. Dissolve some of your solid unknown, about the size of a small pea, in water and add 4 drops of 0.1 *M* $AgNO_3$ solution. Note the color of any precipitate that forms and enter your observation on your report sheet. Dissolve another small-pea-sized portion of your sample in water, add a few drops of 6 *M* HNO_3 to make the solution acidic, and then add 4 drops of 0.1 *M* $AgNO_3$ and stir the mixture. Record your observation on your report sheet.

The barium salts $BaCl_2$, $BaBr_2$, BaI_2, BaS, and $Ba(NO_3)_2$ are soluble in water and in slightly basic solution, but $BaSO_4$, $BaSO_3$, $BaCO_3$, $BaCrO_4$, and $Ba_3(PO_4)_2$ are insoluble (see Table 29.1). Only $BaSO_4$ is insoluble in acidic solution, whereas the other barium salts that are insoluble in water dissolve in acidic solution. All the barium salts are white except $BaCrO_4$, which is yellow. Dissolve a small-pea-sized portion of your unknown in water and to this solution add a few drops of 0.2 *M* barium chloride and stir. Record your observations on your report sheet. Acidify the mixture with a few drops of 6 *M* HNO_3 and stir thoroughly. Record your observations on your report sheet.

Reactions of the Solid Unknown with Concentrated H_2SO_4

Several of the anions form volatile weak acids with, or are oxidized by, concentrated sulfuric acid. Careful observation of the reaction of concentrated sulfuric acid with your solid unknown can give you further clues as to the presence of specific anions. A summary of the pertinent reactions of concentrated sulfuric acid with various anions is given in Table 29.2.

Place a small-pea-sized amount of the solid in a dry, small test tube. Add 1 or 2 drops of 18 *M* H_2SO_4 and observe everything that occurs, especially the color and odor of gas formed. DO NOT place your nose directly over the mouth of the test tube, but carefully fan gases toward your nose. Then *carefully* heat the test tube, but not so strongly as to boil the H_2SO_4 (IF YOU HEAT THE ACID TOO STRONGLY, IT COULD COME SHOOTING OUT!) Note whether or not brown fumes of NO_2 are produced. (CAUTION: *Do not look down into the test tube. Do not point the test tube at yourself or at your neighbors.*) SAFETY GLASSES MUST BE WORN. (Do not get any concentrated H_2SO_4 on yourself or your clothing. If you do, immediately wash with copious amounts of water.)

Specific Tests for Anions

When an anion is indicated by the preliminary tests with $AgNO_3$, $BaCl_2$, or concentrated H_2SO_4, it is confirmed using the appropriate specific test. Make an aqueous solution of the solid unknown and perform the following tests on portions of this solution:

1 SO_4^{2-} Place 10 drops of a solution of the anion unknown in a test tube, acidify with 6 *M* HCl, and add a drop of $BaCl_2$ solution. A white precipitate of $BaSO_4$ confirms SO_4^{2-} ions (see Note 1).

$$Ba^{2+}(aq) + SO_4^{2-}(aq) \longrightarrow \mathbf{BaSO_4}(s)$$

NOTE 1: Sulfites are slowly oxidized to sulfates by atmospheric oxygen. Consequently, sulfites commonly show a positive test for sulfates.

$$2SO_3^{2-}(aq) + O_2(g) \longrightarrow 2SO_4^{2-}(aq)$$

$$SO_4^{2-}(aq) + Ba^{2+}(aq) \longrightarrow \mathbf{BaSO_4}(s)$$

2 SO_3^{2-} Place 10 drops of a solution of the anion unknown in a test tube, acidify with 6*M* HCl, add 2–3 drops of 0.2 *M* $BaCl_2$, and mix thoroughly. If a precipitate ($BaSO_4$) forms, remove it by centrifuging and decanting. To the clear decantate, add a drop of 3-percent H_2O_2 (see Note 2). The formation of a white precipitate of $BaSO_4$ confirms SO_3^{2-} ions.
NOTE 2: H_2O_2 oxidizes sulfite to sulfate.

$$SO_3^{2-}(aq) + H_2O_2(aq) \longrightarrow SO_4^{2-}(aq) + H_2O(l)$$
$$Ba^{2+}(aq) + SO_4^{2-}(aq) \longrightarrow \mathbf{BaSO_4}(s)$$

Any SO_4^{2-} originally present was previously removed as $BaSO_4$ by centrifugation.

3 CrO_4^{2-} Place 2 drops of a solution of the anion unknown in a test tube, add 10 drops of water, and make the solution just acidic with 3 *M* HNO_3. Add 5–6 drops of ether (KEEP THE ETHER AWAY FROM FLAMES) and 1 drop of 3 percent H_2O_2, stir well, and then allow the precipitate to settle. A blue coloration of the top ether layer (see Note 3) confirms CrO_4^{2-} ions (see Note 4).
NOTE 3: The blue coloration in the ether layer is due to the presence of chromium peroxide, CrO_5.

$$2CrO_4^{2-}(aq) + 2H^+(aq) \longrightarrow Cr_2O_7^{2-}(aq) + H_2O(l)$$
$$Cr_2O_7^{2-}(aq) + 4H_2O_2(aq) + 2H^+(aq) \longrightarrow 2CrO_5(aq) + 5H_2O(l)$$

NOTE 4: If your anion unknown in not colored or did not indicate CrO_4^{2-} in the H_2SO_4 reaction, or if both conditions apply, this test may be omitted.

4 I^- Place 5 drops of a solution of the anion unknown in a test tube, add 5 drops of 6 *M* $HC_2H_3O_2$, and then add 2 drops of 0.2 *M* KNO_2. A reddish-brown coloration due to the presence of I_2 confirms I^-. If the brown color is very faint, add a few drops of mineral oil, and shake well. A violet color in the top (mineral oil), layer confirms I^-.

$$NO_2^-(aq) + H^+(aq) \longrightarrow HNO_2(aq)$$
$$2HNO_2(aq) + 2I^-(aq) + 2H^+(aq) \longrightarrow 2NO(g) + I_2(aq) + 2H_2O(l)$$

5 Br^- Iodides will interfere with this test (Note 5). Place 5 drops of a solution of the anion unknown in a test tube and add 5 drops of chlorine water. A brown coloration due to the liberation of Br_2 confirms Br^-. If the solution is shaken with a few drops of mineral oil, the brown color will concentrate in the top layer which is mineral oil. Allow about 20 s for the layers to separate.

$$Cl_2(aq) + 2Br^-(aq) \longrightarrow 2Cl^-(aq) + Br_2(aq)$$

NOTE 5: Iodide ions, if present, must be removed before testing for bromide. To remove iodide, acidify the solution with 3 *M* HNO_3 and add 2 *M* KNO_2, dropwise, with constant stirring, until there is no further increase in the depth of the brown color. Extract once by shaking with 5 drops of mineral oil. Discard the mineral oil layer. Boil the water layer carefully until the iodine has been largely driven off. Test the colorless, or nearly colorless, solution for bromide as directed above.

6 NO_3^- Iodides, bromides, and chromates interfere with this test and must be removed (see Note 6) if they are present. Place 10 drops of a solution of the anion salt in a small test tube, add 5 drops of $FeSO_4$ solution, and mix the solution. Carefully, without agitation, pour concentrated H_2SO_4 down the sides of the test tube. Allow to stand for 1 or 2 min. The formation of a brown ring between the two layers confirms NO_3^- ions. (CAUTION: *Do not get the* H_2SO_4 *on yourself or on your clothing. If you do, wash immediately with copious amounts of water.*)

$$\textbf{(1)}\ 3Fe^{2+}(aq) + NO_3^-(aq) + 4H^+(aq) \longrightarrow 3Fe^{3+}(aq) + NO(aq) + 2H_2O(l)$$

$$\textbf{(2)}\ NO(aq) + Fe^{2+}(aq)\ \text{(excess)} \longrightarrow Fe(NO)^{2+}(aq) \quad \text{(brown)}$$

NOTE 6: Iodides and bromides react with concentrated H_2SO_4 to liberate I_2 and Br_2.

$$SO_4^{2-}(aq) + 8I^-(aq) + 10H^+(aq) \longrightarrow H_2S(g) + 4I_2(aq) + 4H_2O(l)$$

$$SO_4^{2-}(aq) + 2I^-(aq) + 4H^+(aq) \longrightarrow SO_2(g) + I_2(aq) + 2H_2O(l)$$

$$SO_4^{2-}(aq) + 2Br^-(aq) + 4H^+(aq) \longrightarrow SO_2(g) + Br_2 + 2H_2O(l)$$

Chromate ions, if present, will be reduced by Fe^{2+} to green Cr^{3+}.

$$2CrO_4^{2-}(aq) + 2H^+(aq) \longrightarrow Cr_2O_7^{2-}(aq) + H_2O(l)$$

$$Cr_2O_7^{2-}(aq) + 6Fe^{2+}(aq) + 14H^+(aq) \longrightarrow 2Cr^{3+}(aq) + 6Fe^{3+}(aq) + 7H_2O(l)$$

The colors of I_2, Br_2, and Cr^{3+} will interfere with detection of the brown color of $Fe(NO)^{2+}$. Consequently, I^-, Br^-, and CrO_4^{2-} must be removed as follows: Place 4 drops of the unknown anion solution in a test tube and add 0.2 *M* $Pb(C_2H_3O_2)_2$ until precipitation is complete. Centrifuge and decant, discarding the precipitate ($PbCrO_4$). Treat the decantate with 0.1 *M* $Hg(C_2H_3O_2)_2$ until precipitation is complete. Centrifuge the solution to separate the precipitate (HgI_2, $HgBr_2$) and treat the decantate as above to test for nitrate.

7 CO_3^{2-} Sulfites will interfere with this test for carbonates.

A When sulfites are absent. Place a small amount of the solid anion unknown in a small test tube and add a few drops of 6 *M* H_2SO_4. If a colorless, odorless gas is evolved, hold a drop of $Ba(OH)_2$ solution over the mouth of the test tube using either an eyedropper or a Nichrome wire loop (see Figure 29.6); CO_3^{2-} ions are confirmed if the drop turns milky.

$$(1)\ 2H^+(aq) + CO_3^{2-}(aq) \longrightarrow CO_2(g) + H_2O(l)$$

$$(2)\ CO_2(g) + Ba(OH)_2(aq) \longrightarrow \mathbf{BaCO_3}(s) + H_2O(l)$$

B When sulfites are present. Place a small amount of the solid anion unknown in a test tube and add an equal amount of solid Na_2O_2. (CAUTION! *Sodium peroxide is a very strong oxidizing agent and all contact with skin should be avoided. If you do get some on yourself, immediately wash it off with large volumes of water.*) Add 3–4 drops of water, mix thoroughly, then proceed as above to test for carbonate.

8 S^{2-} Place a small quantity of the solid unknown anion in a test tube and (UNDER THE HOOD) add 10 drops of 6 *M* HCl. Hold a piece of filter paper

Figure 29.6 Test for carbonate ion.

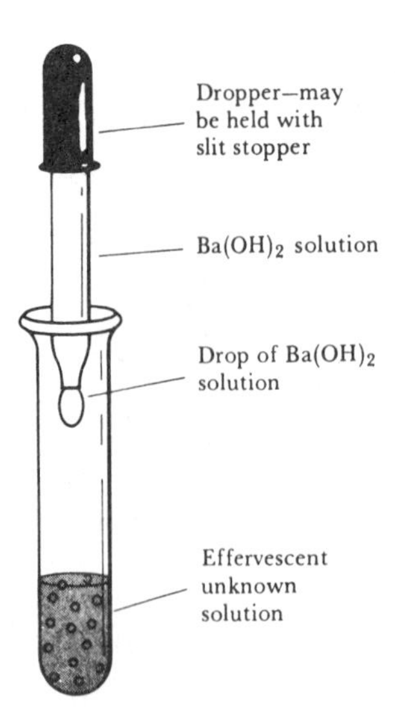

which has been moistened with 0.2 *M* $Pb(C_2H_3O_2)_2$ over the mouth of the test tube so that any gas which escapes comes into contact with the paper. A brownish or silvery black stain (PbS) on the paper confirms the presence of S^{2-}. If no blackening of the lead acetate occurs after 1 min, heat the tube gently; if still no reaction occurs, add a small amount of granulated zinc to the contents of the tube. If the lead acetate is not darkened, S^{2-} is absent.

$$S^{2-}(aq) + 2H^+(aq) \longrightarrow H_2S(g)$$

$$Pb^{2+}(aq) + H_2S(aq) \longrightarrow \mathbf{PbS}(s) + 2H^+(aq)$$

$$\mathbf{Zn}(s) + HgS(aq) + 2H^+(aq) \longrightarrow Zn^{2+}(aq) + \mathbf{Hg}(l) + H_2S(aq)$$

9 Cl^- Sulfides, bromides, and iodides will interfere with this test (see Note 7). Place 10 drops of a solution of the anion unknown in a test tube and add a drop of $AgNO_3$ solution. A white, curdy precipitate confirms Cl^- ions.

$$Ag^+(aq) + Cl^-(aq) \longrightarrow \mathbf{AgCl}(s)$$

NOTE 7: Since Ag_2S, AgBr, and AgI are also insoluble in acid solution, this test is not conclusive unless S^{2-}, Br^-, and I^- are definitely shown to be absent. Chromate ions, if present in high concentrations, may also interfere.

Interference from chromate can be eliminated by dilution with 3 *M* HNO_3. Sulfide ions can be removed by boiling the solution after adding 2 drops of 6 *M* H_2SO_4 until the escaping vapors give no test for H_2S with lead acetate paper. If you suspect the presence of chloride, or bromide, or iodide, or any combination of these, they may all be confirmed as follows:

Place 10 drops of a solution of the anion unknown in a test tube and add 5 drops of $AgNO_3$ solution. Centrifuge and discard the supernate. To the precipitate add 10 drops of concentrated NH_3 solution and 3 drops of yellow ammonium sulfide solution. Stir the mixture with a glass rod and warm gently until the black Ag_2S coagulates. Centrifuge and discard the precipitate. Transfer the solution to a 50-mL beaker and boil to expel NH_3 and to decompose the ammonium sulfide. When the solution becomes cloudy, add 5 or 6 drops of 6 *M* HNO_3 and continue heating until H_2S is completely removed. Then carry out the following tests:

A Place 1 drop of the solution on a piece of filter paper; add 1 drop of iodide-free starch solution and 1 drop of 0.2 *M* KNO_2. A blue color confirms iodide.

$$\mathbf{AgI}(s) + 2NH_3(aq) \longrightarrow Ag(NH_3)_2^+(aq) + I^-(aq)$$

$$2Ag(NH_3)_2^+(aq) + (NH_4)_2S(aq) \longrightarrow \mathbf{Ag_2S}(s) + 2NH_4^+(aq) + 4NH_3(g)$$

$$I^-(aq) + HNO_2(aq) + 2H^+(aq) \longrightarrow 2NO(g) + I_2(aq) + 2H_2O(aq)$$

$$I_2 + \text{starch} \longrightarrow \text{starch} \cdot I_2 \text{ complex (blue)}$$

B If iodide is present, add 3 or 4 drops of 0.2 *M* KNO_2 to the solution in the beaker and boil until no more brown fumes are evolved. If iodide is absent, proceed directly with the test for bromide. Cool the beaker to room temperature by running cold water over its outer surface. To the room-temperature beaker, add 4 or 5 drops of 3 *M* HNO_3, then a small-pea-sized piece of solid Na_2O_2. A brown coloration due to the presence of Br_2 confirms Br^-.

$$4H^+(aq) + O_2^{2-}(aq) + 2Br^-(aq) \longrightarrow 2H_2O(l) + Br_2(aq)$$

C If bromide is present, boil the uncovered contents of the beaker for 30 s to expel the remainder of the bromine. Allow to stand for 30 s and decant the solution into a test tube. Centrifuge if necessary. Add 5 drops of 0.1 *M* $AgNO_3$. A white precipitate of AgCl confirms chloride.

$$Ag^+(aq) + Cl^-(aq) \longrightarrow \mathbf{AgCl}(s)$$

10 PO_4^{3-}

A In the absence of iodides. Place 4–5 drops of a solution of the anion unknown in a test tube, add 2 drops of 6 *M* HNO_3, 3–4 drops of $(NH_4)_2MoO_4$ (ammonium molybdate), mix thoroughly, and heat almost to boiling for 2 min. Formation, sometimes very slow, of a finely divided yellow precipitate confirms phosphate.

$$PO_4^{3-}(aq) + 12MoO_4^{2-}(aq) + 24H^+(aq) + 3NH_4^+(aq) \longrightarrow \mathbf{(NH_4)_3PO_4 \cdot 12MoO_3}(s) + 12H_2O(l)$$

B In the presence of iodides. The phosphate test gives a green solution when iodide is present. If iodide is known to be present, acidify the solution with 6 *M* HNO_3, add 4 drops of 0.1 *M* $AgNO_3$, centrifuge to remove AgI, and treat the decantate as in 10A above.

REVIEW QUESTIONS

Before beginning Part E of this experiment in the laboratory, you should be able to answer the following questions:

1. Give the names and formulas of the anions to be identified.
2. Describe the behavior of each solid containing the anions toward concentrated H_2SO_4.
3. If you had a mixture of NaCl and Na_2CO_3, would the action of concentrated H_2SO_4 allow you to decide that both Cl^- and CO_3^{2-} were present?
4. Identify each of the following anions from the information given (consult Appendix E):
   (a) Its barium salt is insoluble in water but its copper salt is soluble.
   (b) Its copper salt is insoluble in water but its sodium salt is soluble.
   (c) Its mercury(II) salt is soluble in water but its mercury(I) salt is insoluble.
5. A mixture of barium and silver salts is water-soluble. What anions may not be present in this mixture?
6. A white solid unknown is readily soluble in water, and upon treatment of this solution with HCl a colorless, odorless gas is evolved. This gas reacts with $Ba(OH)_2$ to give a white precipitate. What anion is indicated?
7. How would treatment with concentrated H_2SO_4 allow you to distinguish between the following:
   (a) $BaSO_4$ and $Hg(NO_3)_2$?
   (b) $CaBr_2$ and Na_3PO_4?
   (c) ZnS and $BaSO_4$?
   (d) HgI_2 and K_2CrO_4?
8. Five different salts each imparted a yellow color to a flame and reacted with cold H_2SO_4 as follows. Identify each salt.
   (a) Effervescence was observed, and the evolved gas was colorless and odorless and did not fume in moist air.
   (b) Effervescence was observed, and the evolved gas was pale reddish-brown, had a sharp odor, and fumed strongly in moist air.
   (c) No effervescence was observed, and the solid changed color from yellow to orange.
   (d) Effervescence was observed, and the evolved gas was colorless, had a sharp odor, and fumed in moist air.
   (e) Effervescence was observed, and the evolved colorless gas had a very sharp odor but did not fume in moist air and did not discolor a piece of filter paper which had been moistened with a solution of lead acetate.
9. How would you test for NO_3^- in the presence of I^-?
10. How would you test for CO_3^{2-} in the presence of SO_3^{2-}?

Name ____ Desk ____________

Date ________________ Laboratory Instructor ________________________________

Unknown no. ____________________

REPORT SHEET FOR EXPERIMENT 29

ABBREVIATED QUALITATIVE-ANALYSIS SCHEME

Part A Group 1 Cations

Record the reagent used in each step, your observations, and the equations for each precipitation reaction.

Procedure	Reagent	Observations	Equations	Mark (+) if observed in unknown
G1-1				
G1-2				
G1-3				

Cations in group 1 unknown ____

NOTES AND CALCULATIONS

Name ______________________________ Desk __________

Date ______________ Laboratory Instructor ______________________________

Unknown no. ______________

REPORT SHEET FOR EXPERIMENT 29

ABBREVIATED QUALITATIVE-ANALYSIS SCHEME

Part B Group 2 Cations

Record the reagent used in each step, your observations, and the equations for each precipitation reaction.

Procedure	Reagent	Observations	Equations	Mark (+) if observed in unknown
G2-1				
G2-2				
G2-3				
G2-4				
G2-5				

Cations in group 2 unknown ______________________________

NOTES AND CALCULATIONS

Name ______________________________ Desk ____________

Date ______________ Laboratory Instructor ______________________________

Unknown no. ____________________

REPORT SHEET FOR EXPERIMENT 29

ABBREVIATED QUALITATIVE-ANALYSIS SCHEME

Part C Group 3 Cations

Record the reagent used in each step, your observations, and the equations for each precipitation reaction.

Procedure	Reagent	Observations	Equations	Mark (+) if observed in unknown
G3-1				
G3-2				
G3-3				

Procedure	Reagent	Observations	Equations	Mark (+) if observed in unknown
G3-4				
G3-5				
G3-6				
G3-7				

Cations in group 3 unknown ______________________________

Name ______________________________ Desk ____________

Date ______________ Laboratory Instructor ______________________________

Unknown no. ______________

REPORT SHEET FOR EXPERIMENT 29

ABBREVIATED QUALITATIVE-ANALYSIS SCHEME

Part D Group 4 Cations

Record the reagent used in each step, your observations, and the equations for each precipitation reaction.

Procedure	Reagent	Observations	Equations	Mark (+) if observed in unknown
G4-1				
G4-2				
G4-3				
G4-4				

Cations in group 4 unknown ______________________________

NOTES AND CALCULATIONS

Name ____ Desk ____________

Date ____________________ Laboratory Instructor ________________________________

Unknown no. ____________________

REPORT SHEET FOR EXPERIMENT 29

ABBREVIATED QUALITATIVE-ANALYSIS SCHEME

Parts A–D General-Cation Unknown

Record the reagent used in each step, your observations, and the equations for each precipitation reaction.

Procedure	Reagent	Observations	Equations

Cations in general unknown ____

NOTES AND CALCULATIONS

Name ______________________________ Desk ______________

Date ______________ Laboratory Instructor ______________________________

Unknown no. ______________________

REPORT SHEET FOR EXPERIMENT 29

ABBREVIATED QUALITATIVE-ANALYSIS SCHEME

Part E Anions

1. EXAMINATION OF THE SOLID

Color? ______________ Homogeneous? ______________

2. SOLUBILITY

WATER	6 *M* HNO_3	6 *M* HCl	6 *M* NH_3
______	______	______	______

3. REACTIONS WITH $AgNO_3$ AND $BaCl_2$ SOLUTIONS

$AgNO_3$ **Observations**

$AgNO_3$ + 6 *M* HNO_3 ______________________________

Anions indicated ______________________________

Anions eliminated ______________________________

$BaCl_2$ **Observations**

$BaCl_2$ + 6 *M* HNO_3 ______________________________

Anions indicated ______________________________

Anions elimianted ______________________________

4. REACTION OF SOLID WITH CONCENTRATED H_2SO_4

Observations

Anions indicated ______________________________

Anions eliminated ______________________________

5. SPECIFIC TEST RESULTS

Test made	Observation and equation	Anion confirmed

6. ANIONS CONFIRMED IN UNKNOWN SOLID ______________________________

QUESTIONS

1. Write balanced chemical equations for the reactions occurring in each of the following mixtures.

(a) HCl is added to a solid $Mg_3(PO_4)_2$.

(b) HBr is added to solid FeS.

(c) H_2O_2 is added to a solution containing SO_3^{2-}.

(d) $(NH_4)_2S$ solution is added to solid AgBr.

(e) HCl is added to a solution of Na_2CrO_4.

2. What conclusions can be drawn from the following observations?

(a) An acidic solution of unknown anions forms no precipitate when $BaCl_2$ solution is added.

(b) A solution of unknown anions forms a pale yellow precipitate when $AgNO_3$ solution is added.

(c) Addition of 6 *M* NH_3 to a pale blue solution of an unknown gave a deep blue solution and a colorless precipitate.

3. Why is it more difficult to identify all of the components of a mixture than to identify the cation and anion in a simple salt?

4. The observation that a solid is colorless allows you to suggest that a broad category of elements is probably not present in this solid. What is the general name for this category of elements?

5. A colorless compound dissolves in water to give an acidic solution. Addition of either NaOH or NH_3 to this colorless solution produces a flocculent white precipitate which is soluble in acid. If an excess of NaOH is added to a water solutoin of this compound, a colorless precipitate forms which later dissolves as more NaOH is added. Addition of $AgNO_3$ to a water solution of the compound gives no reaction. The solid shows no reaction with concentrated H_2SO_4, and addition of a barium chloride solution to a solution of the compound gives a colorless precipitate which is insoluble in both acid and base. Identify the compound.

6. A white salt dissolves in water to give a neutral solution. No precipitate is formed when the aqueous solution is treated with the buffer NH_4Cl, NH_3. No precipitate forms when $(NH_4)_2S$ is added to the buffer solution. However, a yellow precipitate forms when K_2CrO_4 is added to a solution of the salt that is acidified with $HC_2H_3O_2$ and the precipitate dissolves in concentrated HCl. When $AgNO_3$ is added to a solution of the salt that is acidified with HNO_3, a white precipitate forms. What is the salt?

NOTES AND CALCULATIONS

Colorimetric Determination of Iron

EXPERIMENT 30

OBJECTIVE

To become acquainted with the principles of colorimetric analysis.

APPARATUS AND CHEMICALS

Standard iron solution, $Fe(NO_3)_3$ + HNO_3 (1 mL = 0.50 mg Fe)
1 *M* $NH_4C_2H_3O_2$
10% hydroxylamine hydrochloride
0.30 percent *o*-phenanthroline
50-mL volumetric flask
1-, 2- and 5-mL pipets
6 *M* H_2SO_4
spectrophotometer
cuvettes
unknown iron sample

PREPARE YOUR CALIBRATION CURVE IN TEAMS OF FIVE BUT ANALYZE YOUR UNKNOWN INDIVIDUALLY.

DISCUSSION

The basis for what the chemist calls *colorimetric analysis* is the variation in the intensity of the color of a solution with changes in concentration. The color may be due to an inherent property of the constituent itself—for example, MnO_4^- is purple—or it may be due to the formation of a colored compound as the result of the addition of a suitable reagent. By comparing the intensity of the color of a solution of unknown concentration with the intensities of solutions of known concentrations, the concentration of an unknown solution may be determined.

You will analyze for iron in this experiment by allowing iron(II) to react with an organic compound (*o*-phenanthroline) to form an orange-red complex ion. Note in its structure (shown below) that *o*-phenanthroline has two pairs of unshared electrons that can be used to form coordinate covalent bonds.

The equation for the formation of the complex ion is

$$3C_{12}H_8N_2 + Fe^{2+} \longrightarrow \underset{\text{Orange-red}}{[(C_{12}H_8N_2)_3\,Fe]^{2+}}$$

Unshared electron pairs

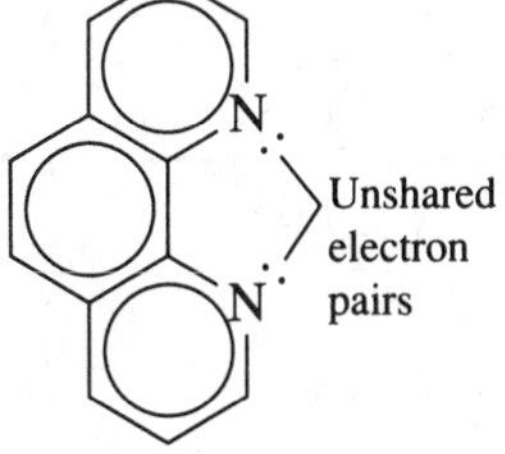

o-phenanthroline

Before the colored iron(II) complex is formed, however, all the Fe^{3+} present must be reduced to Fe^{2+}. This reduction is accomplished by the use of an excess of hydroxylamine hydrochloride:

$$\underset{\text{Hydroxylamine}}{4Fe^{3+} + 2NH_2OH} \longrightarrow 4Fe^{2+} + N_2O + 6H^+ + H_2O$$

Although the eye can discern differences in color intensity with reasonable accuracy, an instrument known as a *spectrophotometer,* which eliminates the "human" error, is commonly used for this purpose. Basically, it is an instrument that measures the fraction I/I_0 of an incident beam of light of a particular wavelength and of intensity I_0 that is transmitted by a sample. (Here, I is the intensity of the light transmitted by the sample.) A schematic representation of a spectrophotometer is shown in Figure 30.1. The instrument has these five fundamental components:

1. A light source that produces light with a wavelength range from about 375 to 650 nm.
2. A monochromator, which *selects* a particular wavelength of light and sends it to the sample cell with an intensity of I_0.
3. The sample cell, which contains the solution being analyzed.
4. A detector that measures the intensity I, of the light transmitted, from the sample cell. If the intensity of the incident light is I_0 and the solution absorbs light, the intensity of the transmitted light, I, is less than I_0.
5. A meter that indicates the intensity of the transmitted light.

For a given substance, the amount of light absorbed depends upon

1. The concentration;
2. The cell or path length;
3. The wavelength of light; and
4. The solvent.

Plots of the amount of light absorbed versus wavelength are called ***absorption*** *spectra.* There are two common ways of expressing the amount of light absorbed. One is in terms of *percent transmittance, T*, which is defined as

Figure 30.1 Schematic representation of a spectrophotometer.

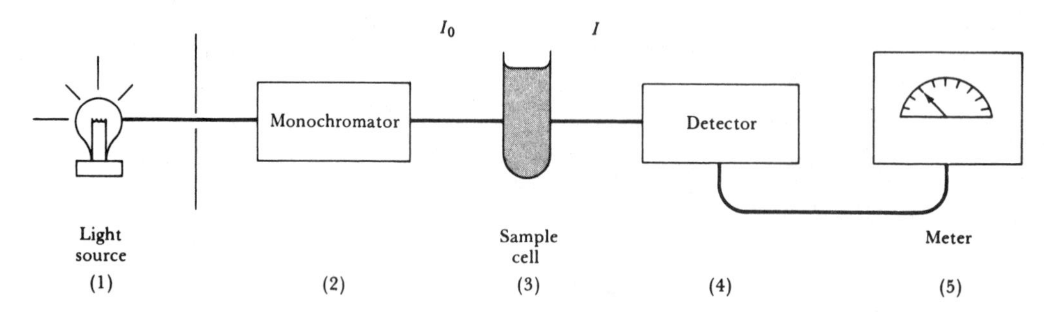

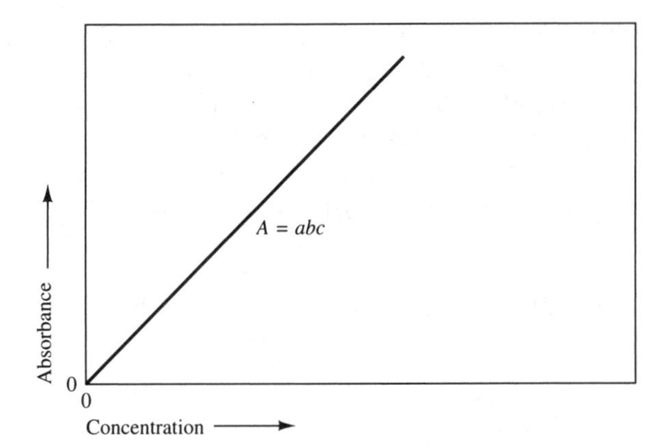

Figure 30.2 Relationship between absorbance and concentration according to the Beer-Lambert law.

$$\%T = \frac{I}{I_0} \times 100 \qquad [1]$$

As the term implies, percent transmittance corresponds to the percentage of light transmitted. When the sample in the cell is a solution, I is the intensity of light transmitted by the solution and I_0 is intensity of light transmitted when the cell only contains solvent. Another method of expressing the amount of light absorbed is in terms of *absorbance*, A, which is defined by

$$A = \log \frac{I_0}{I} \qquad [2]$$

The term *optical density*, O.D., is synonymous with absorbance. If there is no absorption of light by a sample at a given wavelength, the percent transmittance is 100, and the absorbance is 0. On the other hand, if the sample absorbs all of the light, $T = 0$ and $A = \infty$.

Absorbance is related to concentration by the Beer-Lambert law:

$$A = abc$$

where A is absorbance, b is solution path length, c is concentration in moles per liter, and a is molar absorptivity or molar extinction coefficient. There is a linear relationship between absorbance and concentration when the Beer-Lambert law is obeyed, as illustrated in Figure 30.2. However, since deviations from this law sometimes occur, it is wise to construct a calibration curve of absorbance versus concentration.

PROCEDURE

A. Preparation of the Calibration Curve

Accurately pipet 1 mL of standard iron solution (1 mL = 0.050 mg Fe) into a 50-mL volumetric flask. Add 1 mL of 1 *M* ammonium acetate, 1 mL of 10 percent hydroxylamine hydrochloride, and 10 mL of 0.30 percent *o*-phenanthroline solution, and dilute to exactly 50 mL with distilled water. Mix well to develop the characteristic orange-red color of the iron (II) phenanthroline complex. Allow the color to develop for 45 min. Fill halfway a clean and dry cuvette (colorimeter tube) with the colored solution and determine the absorbance at 510 nm using a Spectronic 20 or other colorimeter. Operating instructions for the Spectronic 20 are given below.

Repeat using 2-mL, 3-mL, 4-mL, and 5-mL portions of the standard solution. Plot your results with milligrams of iron along the abscissa and absorbance along the ordinate. This curve should be turned in with your report sheet. (HINT: Time may be saved if these solutions are all made at the same time.)

Operating Instructions for Spectronic 20:

1. Turn the wavelength-control knob (see Figure 30.3) to the desired wavelength.
2. Turn on the instrument by rotating the power control clockwise and allow the instrument to warm up about 5 min. With no sample in the holder but with the cover closed, turn the zero adjust to bring the meter needle to zero on the "percent transmittance" scale.
3. Fill the cuvette about halfway with distilled water (or solvent blank) and insert it in the sample holder, aligning the line on the cuvette with that of the sample holder; close the cover and rotate the light-control knob until the meter reads 100 percent transmittance.
4. Insert the cuvette containing the sample whose absorbance is to be measured in the sample holder in place of the blank. Align lines on the cuvette with the holder and close the cover. Read percent transmittance or optical density from the meter.

B. Determination of Iron

Accurately weigh out about 0.1 g to four significiant figures (0.05 for 12 to 15 percent iron samples) of the iron unknown into a 50-mL volumetric flask, add 5 drops of 6 *M* sulfuric acid, and dilute to exactly 50 mL. Mix thoroughly and then transfer this solution to a clean 125-mL Erlenmeyer flask. Pipet exactly 1 mL of this solution into a thoroughly rinsed 50-mL volumetric flask and repeat the procedure described above in Part A.

Using the observed absorbance and the calibration curve, calculate the milligrams of iron in 1 ml of solution and the percentage of iron in the sample. Repeat this procedure on two additional 1-mL aliquots of unknown solution and calculate the mean and standard deviation of your results.

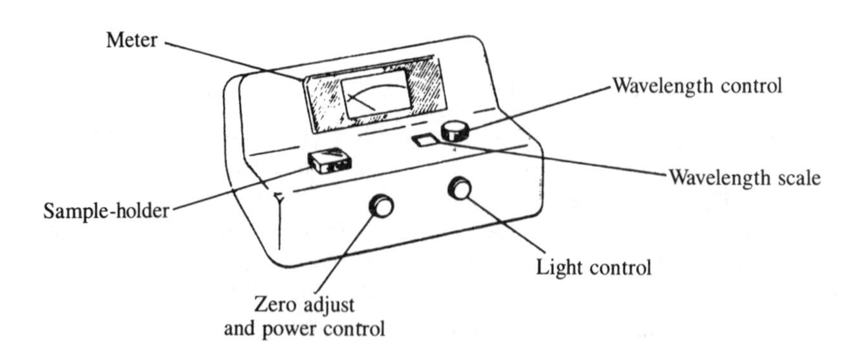

Figure 30.3 Spectrophotometer controls.

REVIEW QUESTIONS

Before beginning this experiment in the laboratory, you should be able to answer the following questions:

1. For a given substance, the amount of light absorbed depends upon what four factors?
2. How are percent transmittance and absorbance related algebraically?
3. What are the five fundamental components of a spectrophotometer?
4. State the Beer-Lambert law and define all terms in it.
5. What is the purpose of preparing a calibration curve?
6. Why is hydroxylamine used in this experiment?
7. If the percent transmittance for a sample is 0 at 350 nm, what is the value of A?
8. Suppose your experimental absorbance is greater than 1. How would you modify your procedure?
9. If 3.0 mL of a standard iron solution (1 mL = 0.060 mg Fe) is diluted to 50 mL, what is the final iron concentration in mg Fe/mL?
10. If aqueous $Co(NO_3)_2$ has an extinction coefficient of 5.1 L/mol-cm at 505 nm, show that a 0.0875 M $Co(NO_3)_2$ solution will give an absorbance of 0.45.

NOTES AND CALCULATIONS

Name ____ Desk ____________

Date ____________________ Laboratory Instructor ______________________________

Unknown no. ____________________

REPORT SHEET FOR EXPERIMENT 30

COLORIMETRIC DETERMINATION OF IRON

A. Calibration Curve

	Concentration of Fe (mg Fe/mL)	Absorbance
1.	____________	____________
2.	____________	____________
3.	____________	____________
4.	____________	____________
5.	____________	____________

Do the above data obey the Beer-Lambert law? ________ Why? ____________________

____

B. Unknown Determination

1. Sample weight ____________ volume of solution ____________

	Absorbance	Concentration of Fe (mg Fe/mL)
Trial 1	____________	____________
Trial 2	____________	____________
Trial 3	____________	____________
Mean	____________	____________

2. Concentration of Fe in mg/mL ____________ ± ____________
3. Standard deviation (show calculations)

4. Percent Fe in original sample (show calculations) ____________ ± ____________

QUESTIONS

1. Why is the line on the cuvette always aligned with that of the sample holder?

2. Why was hydroxylamine hydrochloride added to your sample?

3. Iron (II) reacts with water by a hydrolysis reaction. In order to prevent this hydrolysis, acid has been added to the standard iron solution. How would your final results change if no acid has been added to the standard iron solution?

4. Suppose a solution of $Co(NO_3)_2$ has an extinction coefficient of 5.1 L/mol-cm at 505 nm. Plot, on the graph paper provided, a graph of A versus C (mol/L) for solutions of 0.020, 0.040, 0.060, 0.080, and 0.100 M $Co(NO_3)_2$ in a 1-cm cell. Plot, on the same graph, the percent transmittance, T, of each solution versus concentration.

5. An 8.64 ppm (1 ppm = 1 mg/L) solution of $FeSCN^{2+}$ has a transmittance of 0.295 when measured in a 1.00-cm cell at 580 nm. Calculate the extinction coefficient at this wavelength.

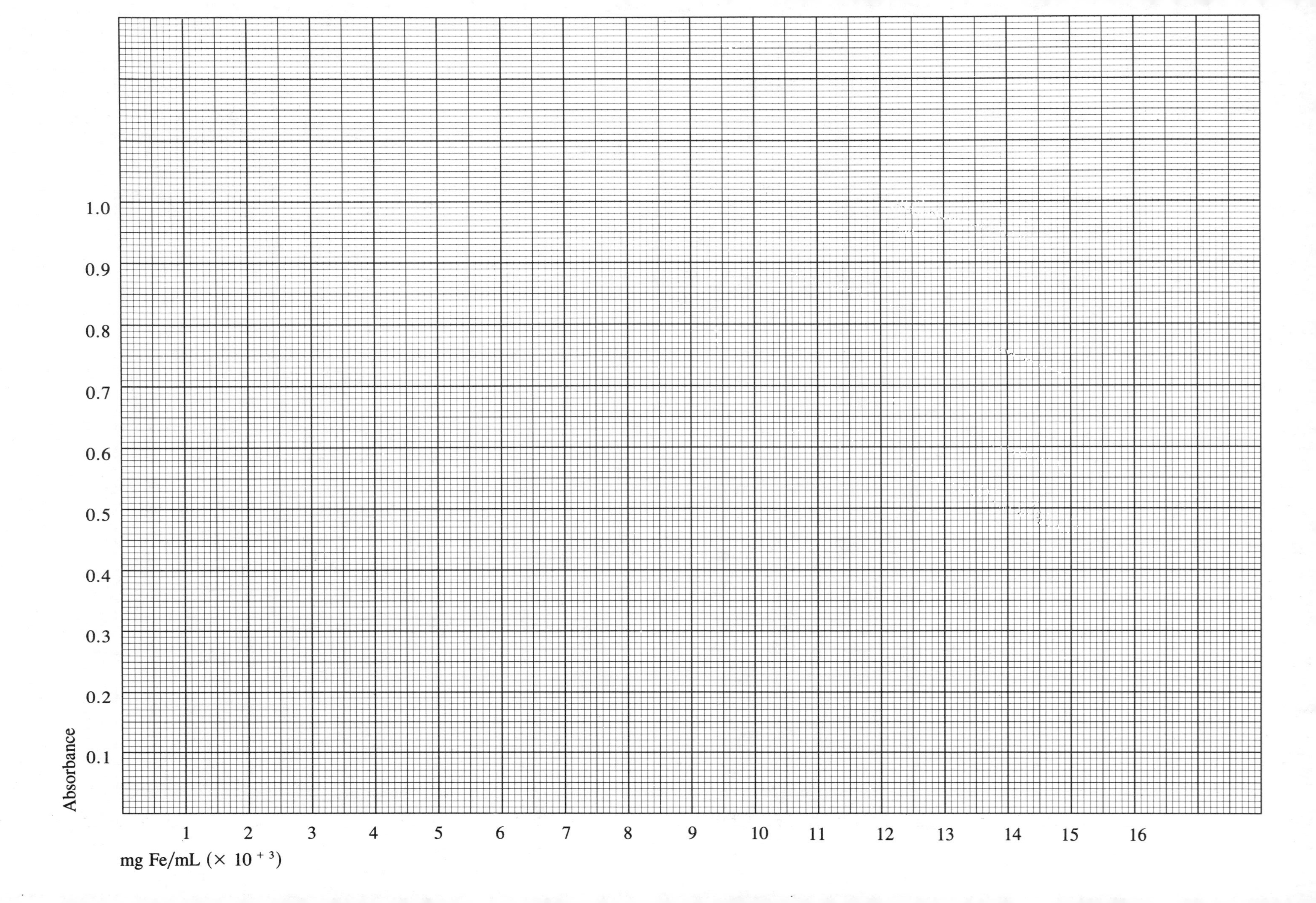

Absorbance
1.0
0.9
0.8
0.7
0.6
0.5
0.4
0.3
0.2
0.1
1
2
3
4
5
6
7
8
9
10
11
12
13
14
15
16
mg Fe/mL (× 10^{+3})

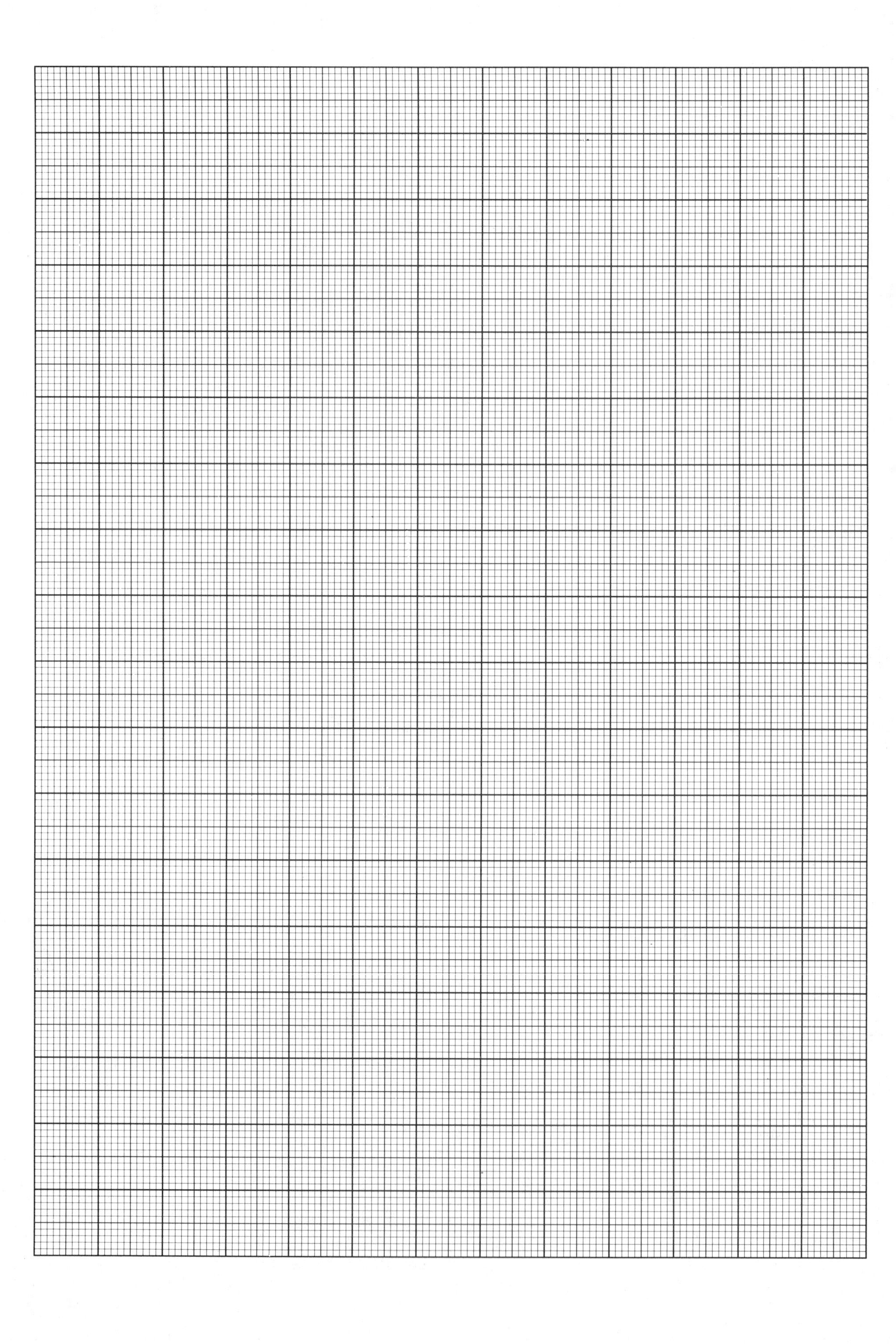

Determination of Orthophosphate in Water

EXPERIMENT 31

OBJECTIVE

To gain some familiarity with the techniques of spectrophotometric analysis by analyzing a water solution for its phosphate content.

APPARATUS AND CHEMICALS

water sample	spectrophotometer
KH_2PO_4 (oven-dried)	1-L volumetric flask
conc. H_2SO_4	1-, 5-, and 10-mL pipets
ammonium vanadomolybdate solution	100-mL volumetric flasks (6)
cuvettes	graduated 5-mL pipet
	$CHCl_3$

PREPARE YOUR CALIBRATION CURVE IN TEAMS OF FOUR BUT EVALUATE YOUR UNKNOWN INDIVIDUALLY.

DISCUSSION

Tripolyphosphates have been found to be extremely effective in enhancing the cleansing ability of detergents; they also are very inexpensive. Their aid in cleaning is probably due, in part, to the stable complexes that they form with Ca^{2+} and Mg^{2+}, thus softening the water. Their extensive use, however, has very serious side effects on nature.

The accelerated eutrophication, or overfertilization, of our lakes has aroused a great deal of ecological concern.* Nutrient enrichment enhances the growth of algae and other microscopic organisms. This produces the green scum of an algal bloom on the water surface, masses of water weeds, and a depletion of dissolved oxygen; it also kills fish and other aquatic organisms and produces malodorous water systems.

When the photosynthetically active algae population near a lake's surface rapidly expands, most of the oxygen produced escapes to the atmosphere. After the algae die, they sink to the lake bottom, where they are biochemically oxidized. This depletes the dissolved oxygen needed to support aquatic life. When oxygen is removed, anaerobic decomposition of the algae continues, producing foul odors.

Although many factors affect algal growth, the only one that is readily subject to preventive control is the supply of nutrients. The many nutrients im-

* Eutrophication is the process whereby a body of water becomes rich in plant nutrient minerals and organisms but often deficient in oxygen.

portant to the growth of algae include phosphorus, carbon, nitrogen, sulfur, potassium, calcium, and magnesium. Many environmentalists have accepted the idea that phosphorus is generally the key nutrient that limits the plant growth that a body of water can support.

There are at least four major sources of phosphorus associated with human activity: human and food wastes, fertilizers, industrial wastes, and detergents. Although detergent products contribute only about one-third of the phosphates entering our water systems, curtailing this particular source is a logical place to begin to combat eutrophication.

The phosphate found in natural waters is present as othophosphate, PO_4^{3-}, as well as the polyphosphates $P_2O_7^{-4}$ and $P_3O_{10}^{5-}$. The species present, PO_4^{3-}, HPO_4^{2-}, $H_2PO_4^{-}$, or H_3PO_4, depend on the pH. Trace amounts are also present as organophosphorus compounds. Detergents usually contain triphosphate, $P_3O_{10}^{5-}$, which slowly hydrolyzes to produce phosphate, PO_4^{3-}, according to the following reaction:

$$P_3O_{10}^{5-} + 2H_2O \longrightarrow 3PO_4^{3-} + 4H^+$$

In this experiment you will determine the amount of orthophosphate present in a sample from a natural body of water, whose source will be given to you by the laboratory instructor.

Analytical Method

In dilute phosphate solutions ammonium metavanadate, (NH_4VO_3), molybdate (MoO_4^{2-}), and phosphate (PO_4^{3-}) condense to form an intensely yellow-colored compound called a *heteropoly acid,* whose formula is thought to be $(NH_4)_3PO_4 \cdot NH_4VO_3 \cdot 16MoO_3$. The intensity of the yellow color is directly proportional to the concentration of phosphate. The relative amount of color developed is measured with a spectrophotometer. The amount of light absorbed by the sample is directly proportional to the concentration of the colored substance. This is stated by the Beer-Lambert law,

$$A = abc$$

where A is absorbance, b is solution path length, c is concentration, and a is absorptivity or extinction coefficient.

The colored solutions that you study in this experiment have been found to obey the Beer-Lambert law in the region of wavelengths ranging from 350 nm to 410 nm. It is convenient to run this experiment at 400 nm. The amount of phosphate in the unknown sample of interest is determined by comparison with a calibration curve constructed by using a distilled-water reference solution and solutions of known phosphate concentration. The minimum detectable concentration of phosphate is about 0.01 mg/L (10 ppb). The usual experimental precision will lie within about ± 1 percent of the result obtained by an experienced analyst.

Comparative Phosphate Levels in Water Systems

Limiting nutrients and their critical concentrations are likely to differ in different bodies of water. Analysis of the waters of 17 Wisconsin lakes has led to the suggestion that an annual average concentration of 0.015 mg/L of inorganic phosphorus (0.05 mg phosphate/L) is the critical level above which algal blooms can be expected if other nutrients, such as nitrogen, are in sufficient supply. During the 1968–69 period, Lake Tahoe in Nevada had an average phosphate level of 0.006 mg/L, while its tributaries averaged 0.018 mg/L. Lake Tahoe is one of the two purest lakes in the world. The other is Lake Baikal in Russia. This figure thus represents the lowest natural-water value of phosphate one is likely to find. In July of 1969 the phosphate level of Lahontan Reservoir (about 55 km east of Reno, Nevada) was 0.52 mg/L. By comparison, Lake Erie's phosphate level increased from 0.014 mg/L in 1942 to 0.40 mg/L in 1967–68. The U.S. Public Health Service has set 0.1 mg/L of phosphorus (0.3 mg phosphate/L) as the maximum value allowable for drinking water. Raw sewage contains an average of about 30 mg/L of phosphate, of which about 25 percent is removed by most secondary sewage-treatment plants.

PROCEDURE

A. Preparation of Calibration Curve

Dissolve about 136 mg of oven-dried KH_2PO_4 (weigh accurately) in about 500 mL of water. Quantitatively transfer this solution to a 1-L volumetric flask, add 0.5 mL of 98 percent H_2SO_4, and dilute to the mark with distilled water. This yields a stock solution that is about 1×10^{-3} *M* in various phosphate species. From this stock solution prepare a series of six solutions with phosphate concentrations 2×10^{-5}, 5×10^{-5}, 1×10^{-4}, 2×10^{-4}, 5×10^{-4}, and 7.5×10^{-4} *M* by appropriate dilution of the stock solution. You must know the precise concentrations of these solutions.

Each point on the calibration curve is obtained by mixing 10 mL of the phosphate solution with 5 mL of the ammonium vanadomolybdate solution (see Note 1, below) and measuring the absorbance on the Spectronic 20 or equivalent spectrophotometer at 400 nm. Your curve is constructed by plotting absorbance as the ordinate versus concentrations of phosphate as the abscissa. A straight line passing through the origin should be obtained. Your calibration curve should be handed in with your report sheet. See Experiment 30 for operating instructions for the Spectronic 20.

B. Analysis of Water Sample

The sample to be analyzed should be taken in a clean bottle, either glass or plastic, which has been rinsed with dilute HCl and distilled water. The bottle should be filled to the top and tightly stoppered. If the sample is particularly turbid, it should be filtered before the analysis is attempted. If the sample is stored for a long period before analysis, it should be preserved by the addition of 5 mL of chloroform per liter of sample. This will arrest the loss of phosphate by microbiological activity. Care must be taken in handling chloroform, since it is very toxic. If you collect and analyze your sample on the same day, chloroform is not needed.

Just prior to the analysis, add 0.5 mL of 98 percent H_2SO_4 to your sample to hydrolyze any polyphosphate present. *CAUTION: Concentrated H_2SO_4 causes severe burns. Do not get any on your skin. If you come in contact with it, wash the area immediately with copious amounts of water.* Transfer 2.0 mL of the hydrolyzed phosphate unknown solution to a clean 100-mL volumetric flask and dilute to the mark with distilled water. Add 5 mL of the ammonium vanadomolybdate solution to 10 mL of this solution and measure the absorbance at 400 nm. If the absorbance is not within your calibration range, prepare either a more dilute or more concentrated sample, whichever is necessary. Measure the absorbance of three separate samples and report the phosphate concentration with its standard deviation. When you make the calculations, *remember* that you have diluted your sample.

Note 1 Ammonium vanadomolybdate solution: Dissolve 40 g of ammonium molybdate (molybdic acid, 85 percent MoO_3) in about 400 mL of distilled water. Dissolve 1 g of ammonium metavanadate, NH_4VO_2, in about 300 mL of distilled water and add 200 mL of concentrated nitric acid. Mix the two solutions and dilute to 1 L. This solution is stable for about 90 days and will be provided for your analysis.

REVIEW QUESTIONS

Before beginning this experiment in the laboratory, you should be able to answer the following questions:

1. What volume of 1×10^{-3} *M* solution is required to make 50 mL of solution with the following concentrations: 2×10^{-5}, 5×10^{-5}, 1×10^{-4}, 2×10^{-4}, 5×10^{-4}, and 7.5×10^{-4} *M*?
2. What species is thought to be the light-absorbing species in this experiment?
3. Write a balanced chemical equation for the formation of the light-absorbing species that results from reaction of phosphate and ammonium vanadomolybdate.
4. State the Beer-Lambert law and define all terms in it.
5. What are the five fundamental components of a spectrophotometer?
6. Why is a calibration curve constructed? How?
7. How do you know whether to measure the absorbance of a more *dilute* or more *concentrated* solution if the absorbance of your unknown solution is not within the limits of your calibration curve?
8. A 0.0750 *M* sample of $Co(NO_3)_2$ gave an absorbance of 0.38 at 505 nm in a 1-cm cell. What is the cobalt concentration of a solution giving an absorbance of 0.26 in the same cell at the same wavelength?
9. An 8.64-ppm (1 ppm = 1 mg/L) solution of $FeSCN^{2+}$ has a transmittance of 0.295 when measured in a 1.00-cm cell at 580 nm. Calculate the extinction coefficient for $FeSCN^{2+}$ at this wavelength.
10. Define eutrophication.
11. If raw sewage contains 30 mg/L phosphate and a secondary sewage-treatment plant removes 25 percent of the phosphate, would a secondary treatment plant provide potable water if 0.3 mg/L is the maximum phosphate concentration allowable in drinking water?
12. Write a balanced chemical equation for the hydrolysis of triphosphate, $P_3O_{10}^{5-}$, to orthophosphate, PO_4^{3-}.

Name ______________________________ Desk ______________

Date ______________ Laboratory Instructor ______________________________

Unknown no. ______________

REPORT SHEET FOR EXPERIMENT 31

DETERMINATION OF ORTHOPHOSPHATE IN WATER

A. Preparation of Calibration Curve

1. Weight KH_2PO_4 ______________________
2. Concentration of phosphate stock solution ______________ mol/L PO_4^{3-}
3. Calibration solutions

	Concentration, mol/L PO_4^{3-}	Absorbance
A	______________	______________
B	______________	______________
C	______________	______________
D	______________	______________
E	______________	______________
F	______________	______________

B. Analysis of Water Sample

1. Water-sample source ______________________

	Absorbance	Concentration of PO_4^{3-} in aliquot	Concentration of PO_4^{3-} in sample
A	______________	______________	______________________
B	______________	______________	______________________
C	______________	______________	______________________

2. Concentration of original solution
(show calculations) ______________ ± ______________

3. Standard deviation (show calculation)

4. Is the sample suitable as drinking water in terms of phosphate concentration? Will algal blooms form in it?

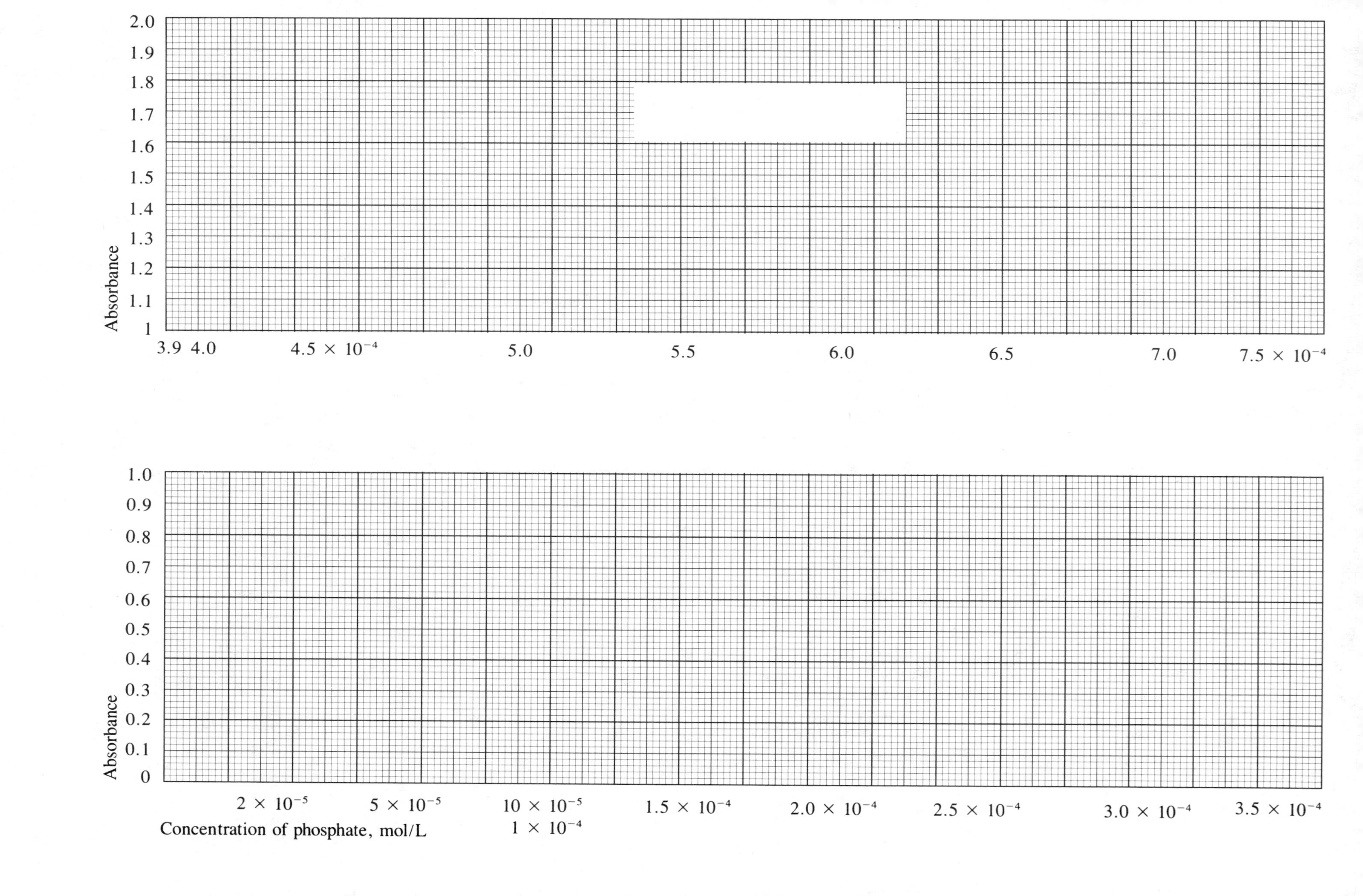

Absorbance
2.0
1.9
1.8
1.7
1.6
1.5
1.4
1.3
1.2
1.1
1
3.9 4.0
4.5 × 10⁻⁴
5.0
5.5
6.0
6.5
7.0
7.5 × 10⁻⁴
Absorbance
1.0
0.9
0.8
0.7
0.6
0.5
0.4
0.3
0.2
0.1
0
2 × 10⁻⁵
5 × 10⁻⁵
10 × 10⁻⁵
1 × 10⁻⁴
1.5 × 10⁻⁴
2.0 × 10⁻⁴
2.5 × 10⁻⁴
3.0 × 10⁻⁴
3.5 × 10⁻⁴
Concentration of phosphate, mol/L

NOTES AND CALCULATIONS

EXPERIMENT 32

Analysis of Water for Dissolved Oxygen

OBJECTIVE

To gain a basic understanding of quantitative techniques of volumetric analysis by determining the dissolved-oxygen content of a water sample.

APPARATUS AND CHEMICALS

- balance
- thermometer
- 50-mL buret
- 250-mL Erlenmeyer flasks (3)
- 500-mL Erlenmeyer flask
- 250-mL narrow-mouth, glass-stoppered bottle or 1-pt bottle
- 2-mL pipets (2)
- 25-mL pipet
- 1-L volumetric flask
- 250-mL volumetric flask

- alkaline iodide-azide reagent*
- conc. H_2SO_4
- water sample (unknown)
- chloroform
- 1 percent boiled starch solution
- $Na_2S_2O_3 \cdot 5H_2O$
- NaN_3 solution (2 g NaN_3/100 mL H_2O)
- 2.15 *M* $MnSO_4$ (freshly prepared)
- 2 *N* H_2SO_4

DISCUSSION

The oxygen normally dissolved in water is indispensable to fish and other water-dwelling organisms. Certain pollutants deplete the dissolved oxygen during the course of their decomposition. This is particularly true of many organic compounds that are present in sewage or dead algae. These are decomposed by the aerobic metabolism of microorganisms, which use these organic compounds for food. The metabolic process is an oxidation of the organic compounds—the dissolved oxygen is the oxidizing agent. Thus while these microorganisms are removing the pollutants, they are also removing the dissolved oxygen that otherwise would be present to support aquatic life. Since the solubility of most gases in solution decreases as the temperature of the solution increases, thermal pollution also decreases the dissolved oxygen content.

As a logical consequence of this, one empirical standard for determining water quality is the dissolved-oxygen content (DO). The survival of aquatic life depends upon the water's ability to maintain certain minimum concentrations of the vital dissolved oxygen. Fish require the highest levels, invertebrates lower levels, and bacteria the least. For a diversified warm-water biota, including game fish, the DO concentration should be at least 5 mg/L (5 ppm). An-

* The alkaline iodide-azide solution is prepared by dissolving 500 g of NaOH and 135 g of NaI in 1 L of distilled water and adding a solution of 10 g NaN_3 in 40 mL of distilled water.

other water quality standard is the biological oxygen demand (BOD). The BOD is the amount of oxygen needed by the microorganisms to remove the pollutant. In order to determine BOD, a sample containing organic pollutants is incubated with its microorganisms for a definite time, usually 5 days, and the amount of oxygen removed is measured. The BOD is taken as the difference in DO before and after incubation. A BOD of 1 ppm is characteristic of nearly pure water. Water is considered fairly pure with a BOD of 3 ppm, and of doubtful purity when the BOD level reaches 5 ppm. Monitoring of water quality therefore logically includes analysis for dissolved oxygen.

This experiment outlines the analysis of water samples for their total dissolved oxygen (DO) content using the azide modification of the iodometric (Winkler) method. This is the procedure most commonly used for analysis of sewage, effluents, and streams. It is based on the use of manganous compounds that are oxidized to manganic compounds by the oxygen in the water sample. The manganic compound in turn reacts with NaI to produce iodine, I_2. The released I_2 is then titrated with standardized sodium thiosulfate, $Na_2S_2O_3$, using starch as an indicator. The chemical reactions involved are as follows:

$$MnSO_4(aq) + 2KOH(aq) \longrightarrow Mn(OH)_2(s) + K_2SO_4(aq)$$
$$2Mn(OH)_2(s) + O_2(aq) \longrightarrow 2MnO(OH)_2(aq)$$
$$MnO(OH)_2(s) + 2H_2SO_4(aq) \longrightarrow Mn(SO_4)_2(aq) + 3H_2O(l)$$
$$Mn(SO_4)_2(aq) + 2NaI(aq) \longrightarrow MnSO_4(aq) + Na_2SO_4(aq) + I_2(aq)$$
$$2Na_2S_2O_3(aq) + I_2(aq) \longrightarrow Na_2S_4O_6(aq) + 2NaI(aq)$$

The net overall chemical equation for this sequence of reactions is:

$$\tfrac{1}{2}O_2(g) + 2S_2O_3^{2-}(aq) + 2H^+(aq) \longrightarrow S_4O_6^{2-}(aq) + H_2O(l)$$

PROCEDURE

A. Standard Sodium Thiosulfate Solution

First prepare and standardize a 0.025 N sodium thiosulfate solution as follows: Begin boiling about 800 mL of distilled water. Weigh out 6.025 g of $Na_2S_2O_3 \cdot 5H_2O$ and dissolve it in 500 mL of water that has been boiled to remove CO_2. Add 5 mL of chloroform and 0.4 g of NaOH to retard bacterial decomposition and dilute to 1 L in a 1-L volumetric flask.

Accurately weigh out about 0.223 g of dry potassium iodate. Dissolve it in 250 mL of previously boiled water in a 250-mL volumetric flask. To each of three 25-mL aliquots of this solution in labeled 250-mL Erlenmeyer flasks add 0.5 g of potassium iodide and about 2 mL of 2 N sulfuric acid. Titrate each of these solutions with your thiosulfate solution, with constant stirring. When the color of the solution has become a pale yellow, dilute to approximately 200 mL with distilled water, add about 2 mL of 1 percent starch solution, and continue the titration until the color changes from blue to colorless for the first time. Ignore any return of color. Record the final buret reading and subtract that value from the initial reading to give the amount of thiosulfate used. Potassium iodate has an equivalent weight of 35.67 g/equiv. The reactions involved in this standardization are

$$IO_3^-(aq) + 5I^-(aq) + 6H^+(aq) \rightleftharpoons 3I_2(aq) + 3H_2O(l)$$
$$2Na_2S_2O_3(aq) + I_2(aq) \rightleftharpoons Na_2S_4O_6(aq) + 2NaI(aq)$$

You are actually titrating with your thiosulfate the iodine formed by the first reaction. Calculate the normality of your thiosulfate from the following equations:

$$\text{Equivalents } Na_2S_2O_3 = \text{equivalents } KIO_3$$
$$\text{Equivalents } Na_2S_2O_3 = V_{Na_2S_2O_3} \times N_{Na_2S_2O_3}$$
$$\text{Equivalents } KIO_3 = \frac{(\text{g } KIO_3)}{(35.67 \text{ g/equiv})} \times \frac{25.00 \text{ mL}}{250.0 \text{ mL}}$$

Thus

$$N_{Na_2S_2O_3} = \frac{(\text{g } KIO_3)(25 \text{ mL})}{(35.67 \text{ g/equiv})(\text{volume } Na_2S_2O_3 \text{ in liters})(250 \text{ mL})}$$

EXAMPLE 32.1

A 0.2264-g sample of KIO_3 was dissolved in 250.0 mL of water. To each 25.00-mL aliquot of this solution were added 0.500 g of KI and 2.00 mL of 2.00 N H_2SO_4. Titration of the liberated iodine required 25.30 mL of $Na_2S_2O_3$ solution. Calculate the normality of the $Na_2S_2O_3$ solution.

Solution: From the analytical reactions

$$IO_3^-(aq) + 5I^-(aq) + 6H^+(aq) \longrightarrow 3I_2(aq) + 3H_2O(aq)$$
$$2Na_2S_2O_3(aq) + I_2(aq) \longrightarrow Na_2S_4O_6(aq) + 2NaI(aq)$$

it is seen that exactly 6 equiv of iodine is liberated for each mole of iodate. Thus KIO_3 (MW = 214.0) has an equivalent weight of

$$\frac{214.0 \text{ g/mol}}{6 \text{ equiv/mol}} = 35.67 \text{ g/equiv}$$

whence, in a 25.00 mL aliquot

$$\text{Equivalents } KIO_3 = \frac{0.02264 \text{ g}}{35.67 \text{ g/equiv}} = 6.347 \times 10^{-4} \text{ equiv}$$

and since

$$\text{Equivalents } Na_2S_2O_3 = \text{equivalents } KIO_3 = 6.347 \times 10^{-4} \text{ equiv}$$

then

$$N_{Na_2S_2O_3} = \frac{\text{equiv } Na_2S_2O_3}{\text{liters of solution}} = \frac{6.347 \times 10^{-4} \text{ equiv}}{0.02530 \text{ L}} = 0.02509\ N$$

B. Water-Sample Analysis

Collection of Sample Collect the sample in a narrow-mouthed, glass-stoppered bottle (250–300 mL capacity). Avoid entrapment or dissolution of atmospheric oxygen. Allow the bottle to overflow its volume and replace the stopper so that no air bubbles are entrained; avoid excessive agitation, which will dissolve atmospheric oxygen. Record the temperature of the water sample in degrees Celsius. The sample should be analyzed as soon as possible. If the method outlined below is used, the sample may be preserved for 4–8 hours by adding 0.7 mL of concentrated H_2SO_4 and 1 mL of sodium azide solution (2 g NaN_3/100 mL H_2O) to the DO bottle. This will arrest biologic activity and maintain the DO if the bottle is stored at the temperature of collection or at 10°–20°C while tightly sealed.

Release of Iodine Open the sample bottle with great care to avoid aeration and add 2 mL of 2.15 *M* manganous sulfate solution from a volumetric pipet. Similarly, add 2 mL of alkaline iodide-azide reagent using another pipet. The neck of the bottle will again have excess liquid, so replace the stopper carefully to avoid splashing. Thoroughly mix the contents of the bottle by inverting the bottle several times. A milky precipitate forms and gradually changes to a yellowish-brown color. Allow the precipitate to settle so that the clear solution occupies the top third of the bottle. Carefully remove the stopper and immediately add 2 mL of concentrated sulfuric acid. This addition should be made by bringing the pipet tip against the neck of the bottle just slightly below the surface of the liquid. Stopper the bottle and then mix the contents by gentle inversion until the precipitate dissolves. At this point the yellowish brown color due to liberated iodine should appear. The sample need not be titrated immediately, but if titration is delayed, the sample should be stored in darkness. The titration should be done within several hours.

Titration Measure accurately 200 mL of the sample into a 500-mL Erlenmeyer flask. Titrate the sample with the standardized thiosulfate solution with constant stirring. When the color of the solution becomes a pale yellow, add about 2 mL of 1 percent starch solution and continue titrating until the color changes from blue to colorless for the first time. Record the volume of titrant necessary. If time permits, analyze another sample.

Calculation of Dissolved Oxygen Content Because $Na_2S_2O_3$ undergoes a one-electron change in its reaction with iodine, a 0.025 *N* solution is also 0.025 *M*:

$$2Na_2S_2O_3 \longrightarrow Na_2S_4O_6 + 2Na^+ + 2e^-$$

According to the above equations, 1 mol of O_2 (32 g) requires 4 mol of $Na_2S_2O_3$ in reaching an end point. The number of moles of $Na_2S_2O_3$ is equal to the volume of $Na_2S_2O_3$ (in liters) times the concentration of the $Na_2S_2O_3$ (in molarity):

$$\text{Moles } Na_2S_2O_3 = V_{Na_2S_2O_3} \times \text{molarity}_{Na_2S_2O_3}$$

From the information given above, calculate the number of grams of O_2

in your 200-mL sample. From this, calculate the number of milligrams of O_2 per liter of solution. Since a liter of solution will weigh approximately 1000 g (the bulk of the solution is water), 1 mg/L is equivalent to 1 mg in 10^6 mg, or 1 million mg, of solution. Therefore, the number of milligrams of O_2 per liter is often referred to as parts per million (ppm).

EXAMPLE 32.2

To a 200-mL water sample were added 0.5000 g of KI, 2.000 mL of 2.000 *N* H_2SO_4, and 2.000 ml of starch solution. The liberated iodine required 7.88 mL of 0.0251 *M* $Na_2S_2O_3$. Calculate the O_2 concentration in the sample in ppm.

Solution: From the analytical reactions

$$\begin{aligned} MnSO_4 + 2KOH &\longrightarrow Mn(OH)_2 + K_2SO_4 \\ 2Mn(OH)_2 + O_2 &\longrightarrow 2MnO(OH)_2 \\ MnO(OH)_2 + 2H_2SO_4 &\longrightarrow Mn(SO_4)_2 + 3H_2O \\ Mn(SO_4)_2 + 2KI &\longrightarrow MnSO_4 + K_2SO_4 + I_2 \\ 2Na_2S_2O_3 + I_2 &\longrightarrow Na_2S_4O_6 + 2NaI \end{aligned}$$

it is seen that each mole of O_2 requires 4 mol of $Na_2S_2O_3$.

$$\begin{aligned} \text{Moles } Na_2S_2O_3 &= (\text{volume } Na_2S_2O_3) \times (\text{molarity } Na_2S_2O_3) \\ &= (0.00788 \text{ L})(0.0251 \text{ mol/L}) \\ &= 1.98 \times 10^{-4} \text{ mol} \\ \text{Moles } O_2 &= \tfrac{1}{4} \text{ moles } Na_2S_2O_3 \\ &= 4.95 \times 10^{-5} \text{ mol} \\ \text{Weight } O_2 &= (4.95 \times 10^{-5} \text{ mol}) \times (32.0 \text{ g/mol}) \\ &= 1.58 \times 10^{-3} \text{ g} \\ \text{Concentration } O_2 &= \frac{1.58 \text{ mg}}{0.200 \text{ L}} = 7.90 \text{ mg/L} \\ &= 7.90 \text{ ppm} \end{aligned}$$

A correction factor may be applied to your answer to correct for solution loss during the addition of manganous sulfate and sulfuric acid. This amounts to multiplication by 204/200 if these reagents were added to a 200-mL bottle.

Comparisons in Dissolved Oxygen Contents The amount of oxygen dissolved in water depends not only upon the amount of chemical pollution but also upon such factors as water temperature and the atmospheric pressure above the water. At temperatures between 0°C and 39°C, the amount of O_2 that will be present in oxygen-saturated distilled water is given by the equation

$$\text{ppm dissolved } O_2 = \frac{(P - p) \times 0.678}{35 + T} = \text{SLDO}$$

where P is the barometric pressure in mm Hg, T is the temperature of the water in °C, and p is the vapor pressure of water at the temperature of the water. Calculate the saturation level (SL) for your water sample. The percent saturation is given by

$$\% \text{ SL} = \frac{100(\text{DO in ppm})}{(\text{SLDO in ppm})}$$

Calculate the percent SL for your sample.

EXAMPLE 32.3

A water sample at 12°C and 652 mm Hg was found to contain 7.90 ppm O_2. Calculate the percent saturation of this sample.

Solution:

$$\text{SLDO} = \frac{(652 \text{ mm} - 10.5 \text{ mm})(0.678 \text{ ppm-°C/mm})}{(35 + 12)\text{°C}}$$

$$= 9.25 \text{ ppm}$$

$$\% \text{ SL} = \frac{(100)(7.90 \text{ ppm})}{9.25 \text{ ppm}} = 85.4\%$$

REVIEW QUESTIONS

Before beginning this experiment in the laboratory, you should be able to answer the following questions:

1. Define molarity and normality in terms of redox reactions. How are the two related?
2. What is the normality of a thiosulfate solution if 23.75 mL of it reacts with 0.1309 g of KIO_3?
3. Suggest how chloroform helps to preserve the water sample.
4. Which species is being oxidized and which reduced in the reaction

$$Mn(SO_4)_2 + 2KI \rightleftharpoons MnSO_4 + K_2SO_4 + I_2$$

5. During the course of the standardization of $S_2O_3^{2-}$ with KIO_3, the text tells you to ignore the return of the blue color with time after the end point has been reached in the reaction. Suggest a reason for the return of the blue color. Write a chemical reaction that is consistent with your suggestion.
6. Calculate the concentration in mg/L (ppm) of a solution that is 1×10^{-4} M in O_2, assuming a density of 1 g/mL for the solution.
7. Some characteristic BOD levels are given below:

Source	**BOD range (ppm)**
Untreated municipal sewage	100–400
Runoff from barnyards and feed lots	100–10,000
Food-processing wastes	100–10,000

Assuming the minimum values above, by what factor must these waters be diluted with pure water to reduce the BOD to a value sufficiently low to support aquatic life (BOD $\leqslant$ 5 ppm)?

8. A 0.130-g sample of KIO_3 required 10.6 mL of $Na_2S_2O_3$ solution for its standardization. What are the normality and molarity of the $Na_2S_2O_3$ solution?

9. In the determination of the BOD of a water sample it was found that the DO level was 7.0 ppm before incubation and 4.9 ppm after 5 days of incubation. What is the BOD of this sample?

10. Would the water sample in question 9 support aquatic life?

NOTES AND CALCULATIONS

Name ______________________________ Desk __________

Date ______________ Laboratory Instructor ______________________

Unknown no. ______________

REPORT SHEET FOR EXPERIMENT 32

ANALYSIS OF WATER FOR DISSOLVED OXYGEN

A. Standard Sodium Thiosulfate Solution

1. Weight in grams of KIO_3 in 250 mL __________ g

	Sample 1	Sample 2	Sample 3
2. Final buret reading	__________ mL	__________ mL	__________ mL
3. Initial buret reading	__________ mL	__________ mL	__________ mL
4. Volume of $Na_2S_2O_3$ solution	__________ mL	__________ mL	__________ mL
	__________ L	__________ L	__________ L
5. Normality of $Na_2S_2O_3$ (show calculations)	__________ *N*	__________ *N*	__________ *N*

Avg. *N* __________ ± __________

6. Standard deviation (show calculations) __________

7. Molarity of $Na_2S_2O_3$ __________ ± __________

B. Water-Sample Analysis

Water temperature __________ °C

1. Volume of $Na_2S_2O_3$ required to titrate water sample

	Sample 1	Sample 2
Final buret reading	________	________
Initial buret reading	________	________
Volume of $Na_2S_2O_3$	________	________

	Sample 1	Sample 2
2. Moles of $Na_2S_2O_3$ required to titrate water sample (show calculations)	________ mol	________ mol

	Sample 1	Sample 2
3. Moles of O_2 present in 200 mL of water sample (show calculations)	________ mol	________ mol

4. Grams of O_2 present in 200 mL of water sample	________ g	________ g
5. Milligrams of O_2 present in 200 mL of water sample	________ mg	________ mg
6. Milligrams of O_2 present in 1000 mL of water sample	________ ppm	________ ppm
7. Saturation level for your water sample (show calculations)	________	

8. Percent saturation level of your water sample ________
(show calculations)

9. Do you think that this water has sufficient DO to sustain aquatic life?

10. Calculate the SLDO for water at 20, 30, 40, 50, and 100°C and 700 mm Hg, and plot the data on the graph paper provided. How would you expect the solubility of a gas in water to change with temperature? Are your calculations in agreement with your expectations?

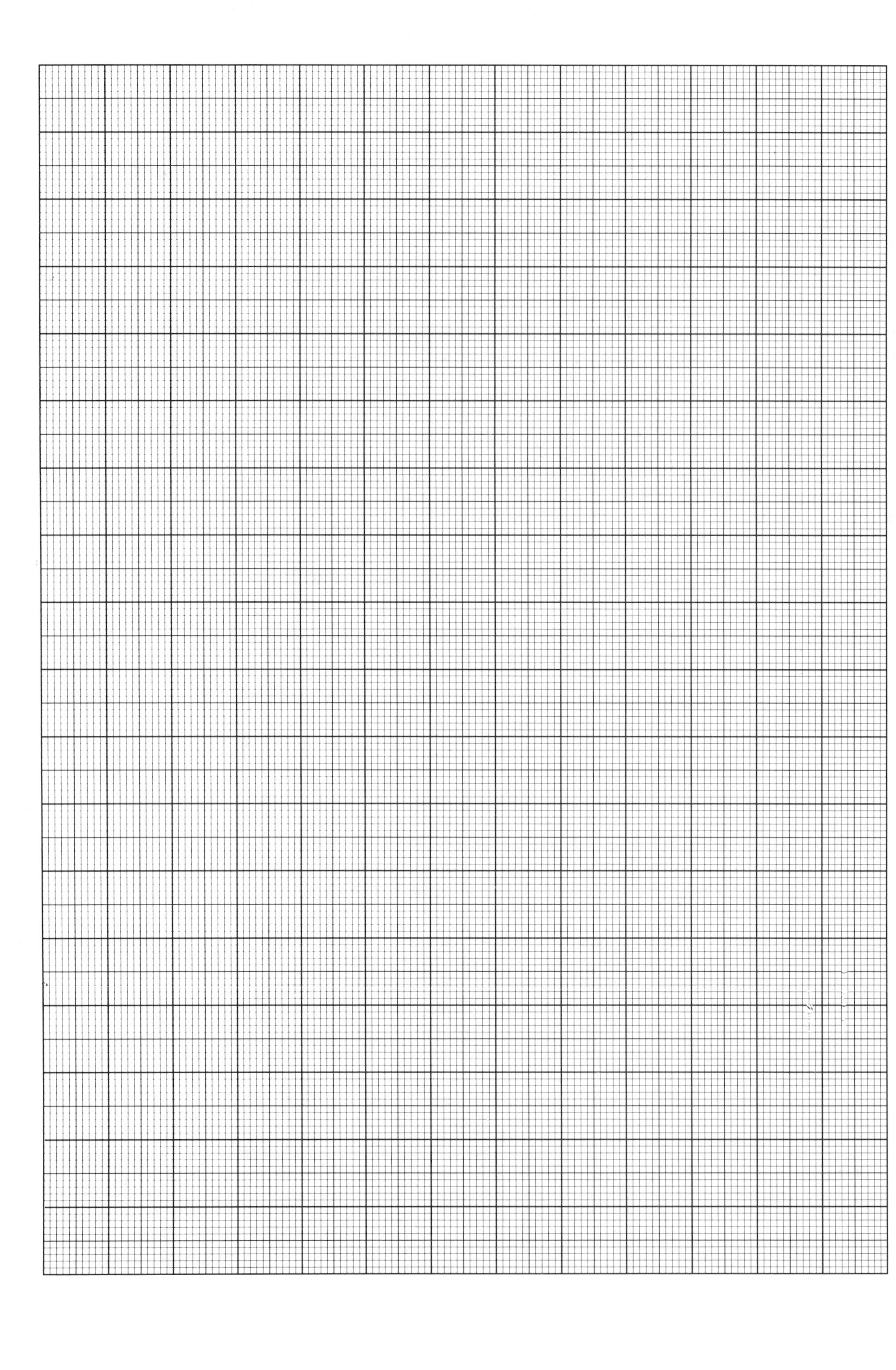

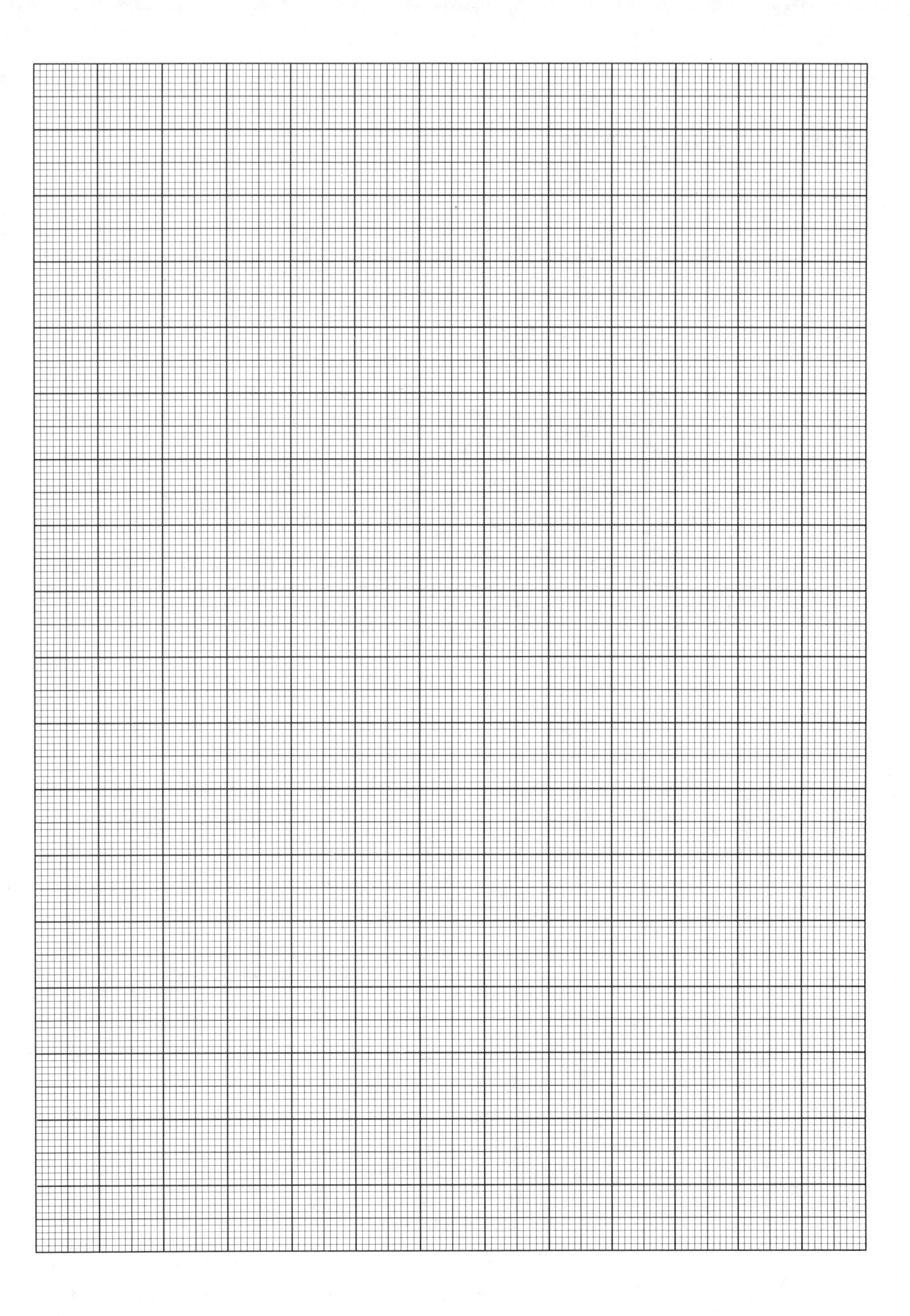

NOTES AND CALCULATIONS

Preparation and Reactions of Coordination Compounds: Oxalate Complexes

EXPERIMENT 33

OBJECTIVE

To gain some familiarity with coordination compounds by preparing a representative compound and witnessing some typical reactions.

APPARATUS AND CHEMICALS

- *cis*- and *trans*-$K[Cr(C_2O_4)_2(H_2O)_2]$ (prep given)
- 6 *M* NH_3 and HCl
- oxalic acid
- potassium oxalate monohydrate
- sodium oxalate
- potasssium dichromate
- 50-percent ethanol
- 95-percent ethanol
- absolute ethanol
- acetone
- copper sulfate pentahydrate
- ferrous ammonium sulfate hexahydrate
- 6-percent hydrogen peroxide
- 6 *M* H_2SO_4
- aluminium turnings
- 6 *M* potassium hydroxide
- 9-cm Büchner funnel
- 250-mL suction flask
- 8-oz wide-mouth bottle
- aspirator
- No. 6 2-hole rubber stopper
- ice
- beakers (100 mL and 250 mL)
- glass stirring rod
- ring stand and iron ring
- wire gauze
- Bunsen burner and hose
- thermometer
- 9-cm filter paper
- glass wool

DISCUSSION

When gaseous boron trifluoride, BF_3, is passed into liquid trimethylamine, $(CH_3)_3N$, a highly exothermic reaction occurs, and a creamy, white solid, $(CH_3)_3N{:}\ BF_3$, separates. This solid, which is an adduct of trimethylamine and boron trifluoride, is a coordination compound. It contains a coordinate covalent, or dative, bond uniting the Lewis acid BF_3 with the Lewis base trimethylamine. Numerous coordination compounds are known, and in fact nearly all compounds of the transition elements are coordination compounds wherein the metal is a Lewis acid and the atoms or molecules joined to the metal are Lewis bases. These Lewis bases are called *ligands,* and the coordination compounds are usually denoted by square brackets when their formulas are written. The metal and the ligands bound to it constitute what is termed the coordination sphere. In writing chemical formulas for coordination compounds, we use square brackets to set off the coordination sphere from other parts of the compound. For example, the salt $NiCl_2 \cdot 6H_2O$ is in reality the coordination compound $[Ni(H_2O)_6]Cl_2$, wherein the hexaaquanickel(II) ion, $[Ni(H_2O)_6]^{2+}$, is a coordination compound possessing an octahedral geometry, as shown in Figure 33.1.

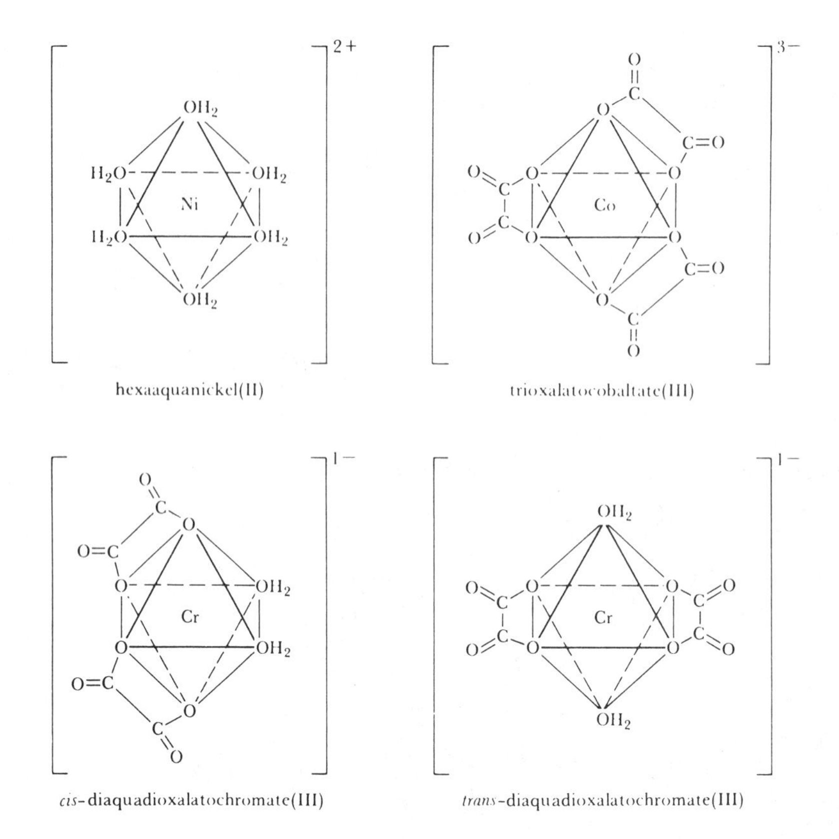

Figure 33.1 Typical octahedral (6-coordinate) coordination compounds.

The apexes of a regular octahedron are all equivalent positions. Thus each of the monodentate (one donor site) H_2O molecules in the $[Ni(H_2O)_6]^{2+}$ ion and the three bidentate (two donor sites) oxalate ions, $C_2O_4^{2-}$, in $[Co(C_2O_4)_3]^{2-}$ are in identical environments. The water molecules in the two isomeric compounds *cis*- and *trans*-$[Cr(C_2O_4)_2(H_2O)_2]^-$ are in equivalent environments within each complex ion (coordination compound), but the two isomeric ions are not equivalent to one another. The two water molecules are adjacent in the *cis* isomer and opposite one another in the *trans* isomer. These two isomers are termed *geometric isomers,* and although they have identical empirical and molecular formulas, their geometrical arrangements in space are different. Consequently, they have different chemical and physical properties, as your laboratory instructor will demonstrate through the reactions shown in Figure 33.2.

Your goal in this experiment is to prepare an oxalate-containing coordination compound. In Experiment 35 you will analyze it by titration for its oxalate content. You will prepare *one* of the following compounds:

1. $K_3\,[Cr(C_2O_4)_3] \cdot 3H_2O$;
2. $K_2\,[Cu(C_2O_4)_2] \cdot 2H_2O$;
3. $K_3\,[Fe(C_2O_4)_3] \cdot 3H_2O$; or
4. $K_3\,[Al(C_2O_4)_3] \cdot 3H_2O$.

Your laboratory instructor will tell you which one to prepare. Each of these compounds will be prepared by someone in your lab section so that you can

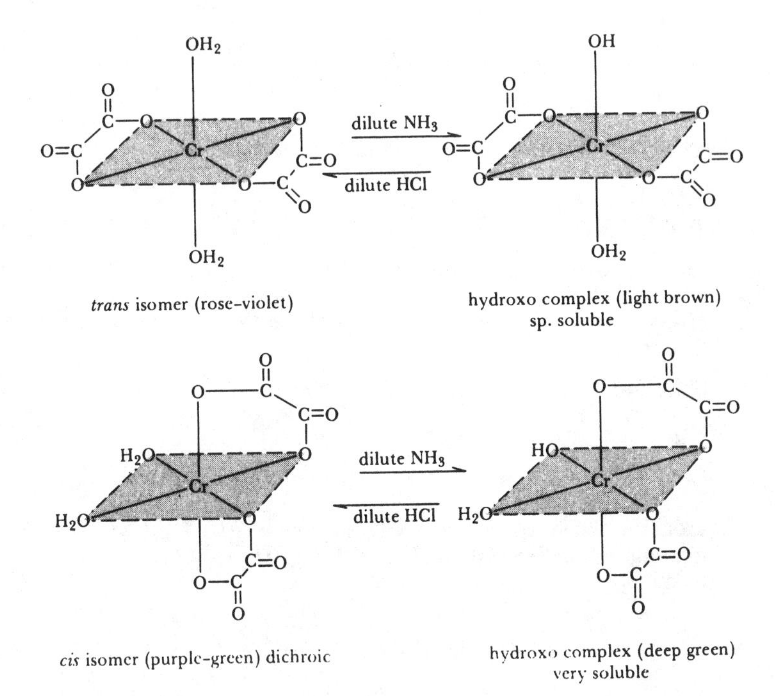

Figure 33.2 Reaction of geometric isomers: *cis*- and *trans*-$[Cr(C_2O_4)_2(H_2O)_2]^-$ ions. The procedure is to place a small amount of an aqueous solution of each complex on separate pieces of filter paper on a watch glass. Let them dry and then add a drop of reagent. Observe the results.

compare their properties. (CAUTION: *Oxalic acid is a toxic compound and is absorbed through the skin. Should any come in contact with your skin, wash it off immediately with copious amounts of water.*)

PROCEDURE

Prepare one of the complexes whose synthesis is given below.

1. Preparation of $K_3[CrC_2O_4)_3] \cdot 3H_2O$

$$K_2Cr_2O_7 + 7H_2C_2O_4 + 2K_2C_2O_4 \longrightarrow 2K_3[Cr(C_2O_4)_3] \cdot 3H_2O + 6CO_2\uparrow + H_2O$$

Slowly add 3.6 g of potassium dichromate to a suspension of 10 g of oxalic acid in 20 mL of H_2O in a 250 mL-beaker. The orange-colored mixture should spontaneously warm up almost to boiling as a vigorous evolution of gas commences. When the reaction has subsided (about 15 min), dissolve 4.2 g of potassium oxalate monohydrate in the hot, green-black liquid and heat to boiling for 10 min. Allow the beaker and its contents to cool to room temperature. Add about 10 mL of 95 percent ethanol, with stirring, into the cooled solution in the beaker. Further cool the beaker and its contents in ice. The cooled liquid should thicken with crystals. After cooling in ice for 15–20 min., the crystals should be collected by filtration with suction using a Buchner funnel and filter flask. Wash the crystals on the funnel with three 10-mL portions of 50-percent aqueous ethanol followed by 25 mL of 95-percent ethanol and dry the product in air. Weigh the air-dried material and store it in a vial. You should obtain about 9 g of product. Calculate the theoretical yield and determine your percentage yield. Reactions of chromium (III) are slow, and your yield will be low if you work too fast.

$$\% \text{ yield} = 100 \times \frac{\text{actual yield in grams}}{\text{theoretical yield in grams}}$$

Save your sample for analysis in Experiment 35.

EXAMPLE 33.1

In the preparation of *cis*-$K[Cr(C_2O_4)_2(H_2O)_2] \cdot 2H_2O$, 12.0 g of oxalic acid was allowed to react with 4.00 g of potassium dichromate, and 8.20 g of *cis*-$K[Cr(C_2O_4)_2(H_2O)_2] \cdot 2H_2O$ was isolated. What is the percent yield in this synthesis?

Solution:

$$K_2Cr_2O_7 + 7H_2C_2O_4 \cdot 2H_2O \longrightarrow 2K[Cr(C_2O_4)_2(H_2O)_2] \cdot 2H_2O + 6CO_2 + 13H_2O$$

From the above reaction we see that 1 mol of $K_2Cr_2O_7$ reacts with 7 mol of $H_2C_2O_4$ to produce 2 mol of $K[Cr(C_2O_4)_2(H_2O)_2] \cdot 2H_2O$. In our synthesis we have used the following:

$$\text{Moles } K_2Cr_2O_7 = \frac{4.00 \text{ g}}{294.19 \text{ g/mol}} = 0.0136 \text{ mol}$$

$$\text{Moles } H_2C_2O_4 \cdot 2H_2O = \frac{12.0 \text{ g}}{126.07 \text{ g/mol}} = 0.0952 \text{ mol}$$

Our reaction requires a 7 : 1 molar ratio of oxalic acid to $K_2Cr_2O_7$, and we have actually used a 6.999 molar ratio or, within experimental error, the stoichiometric amount of each reagent. Hence the number of moles of $K[Cr(C_2O_4)_2(H_2O)_2 \cdot 2H_2O$ formed should be twice the number of moles of $K_2Cr_2O_7$ reacted.

$$\text{Moles } K[Cr(C_2O_4)_2(H_2O)_2] \cdot 2H_2O \text{ (expected)} = (2)(0.0136 \text{ mol}) = 0.0272 \text{ mol}$$

The theoretical yield of $K[Cr(C_2O_4)_2(H_2O)_2] \cdot 2H_2O$ is

$$(0.0272 \text{ mol})(339.0 \text{ g/mol}) = 9.22 \text{ g}$$

Our percentage yield is then

$$\% \text{ yield} = \frac{(100)(8.30 \text{ g})}{(9.22 \text{ g})} = 90.0\%$$

2. Preparation of $K_2[Cu(C_2O_4)_2] \cdot 2H_2O$

$$CuSO_4 \cdot 5H_2O + 2K_2C_2O_4 \cdot H_2O \longrightarrow K_2[Cu(C_2O_4)_2] \cdot 2H_2O + K_2SO_4 + 5H_2O$$

Heat a solution of 6.2 g of copper sulfate pentahydrate in 12 mL of water to about 90°C and add it rapidly, with vigorous stirring, to a hot (~90°C) solution of 10.0 g of potassium oxalate monohydrate in 50 mL of water contained in a 100-mL beaker. Cool the mixture by setting the beaker in an ice bath for 15–30 min. Suction-filter the resultant crystals using a Büchner funnel and filter flask and wash the crystals successively with about 12 mL of cold water, then 10 mL of absolute ethanol, and finally 10 mL of acetone, and air dry. Weigh the air-dried material and store it in a vial. You should obtain about 7 g

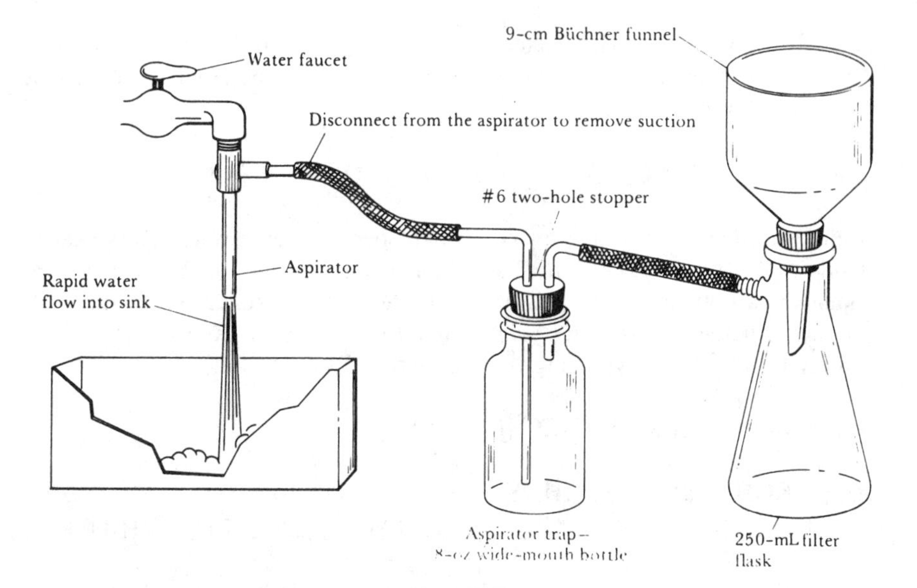

Figure 33.3 Suction filtration assembly.

of product. Calculate the theoretical yield and determine your percentage yield. *Save your sample for analysis in Experiment 35.*

3. Preparation of $K_3[Fe(C_2O_4)_3] \cdot 3H_2O$

$$(NH_4)_2[Fe(H_2O)_2(SO_4)_2] \cdot 4H_2O + H_2C_2O_4 \cdot 2H_2O \longrightarrow FeC_2O_4 + H_2SO_4 + (NH_4)_2SO_4 + 8H_2O$$

$$H_2C_2O_4 \cdot 2H_2O + 2FeC_2O_4 + 3K_2C_2O_4 \cdot H_2O + H_2O_2 \longrightarrow 2K_3[Fe(C_2O_4)_3] \cdot 3H_2O + H_2O$$

This preparation contains two separate parts. Iron(II) oxalate is prepared first and then converted to $K_3[Fe(C_2O_4)_3] \cdot 3H_2O$ by oxidation with hydrogen peroxide in the presence of potassium oxalate.

To a solution of 10 g of ferrous ammonium sulfate hexahydrate in 30 mL of water containing a few drops of 6 *M* H_2SO_4 (to prevent premature oxidation of Fe^{2+} to Fe^{3+} by O_2 in the air), add, with stirring, a solution of 6 g of oxalic acid in 50 mL of H_2O. Yellow iron(II) oxalate forms. Carefully heat the mixture to boiling while stirring constantly to prevent bumping. Decant and discard the supernatant liquid and wash the precipitate several times by adding about 30 mL of hot water, stirring, and decanting the liquid. Filtration is not necessary at this point.

To the wet iron(II) oxalate, add a solution of 6.6 g of potassium oxalate monohydrate in 18 mL of water and heat the mixture to about 40°C. SLOWLY AND CAUTIOUSLY add 17 mL of 6 percent hydrogen peroxide, H_2O_2, while stirring constantly and maintaining the temperature at 40°C. After the addition of peroxide is complete, heat the mixture to boiling and add a solution containing 1.7 g of oxalic acid in 15 mL of water. When adding the oxalic acid solution, add the first 8 mL all at once and the remaining 5 mL dropwise, keeping the temperature near boiling. Filter off any solid by gravity and add 20 mL of 95-percent ethanol to the filtrate. Cover the beaker with a watch glass and store it in your lab desk until the next period. Filter by suction using a Büchner funnel and filter flask and wash the green crystals with a 50-percent aqueous ethanol solution, then with acetone, and air-dry. Weigh the product and store it in a vial in the dark. This complex is photosensitive and reacts with light according to the following:

$$[Fe(C_2O_4)_3]^{3-} \xrightarrow{h\nu} [Fe(C_2O_4)_2]^{2-} + 2CO_2$$

In order to demonstrate this, place a small specimen on a watch glass near the window and observe any changes which occur during the lab period. You should obtain about 8 g of product. Calculate the theoretical yield and determine your percent yield. *Save your sample for analysis in Experiment 35*. Keep it away from light by wrapping the vial with aluminium foil.

4. Preparation of $K_3[Al(C_2O_4)_3] \cdot 3H_2O$

$$Al + 3KOH + 3H_2C_2O_4 \cdot 2H_2O \longrightarrow K_3[Al(C_2O_4)_3] \cdot 3H_2O + 6H_2O + \tfrac{3}{2}H_2$$

Place 1 g of aluminum turnings in a 200-mL beaker and cover with 10 mL of hot water. Add 20 mL of 6 *M* KOH solution in small portions to regulate the vigorous evolution of hydrogen. Finally, heat the liquid almost to boiling in order to dissolve any residual metal. Maintain the heating and add a solution of 13 g of oxalic acid in 100 mL of water in small portions. During the neutralization, hydrated alumina will precipitate, but it will redissolve at the end of the addition after gentle boiling. Cool the solution in an ice bath and add 50 mL of 95 percent ethanol. If oily material separates, stir the solution and scratch the sides of the beaker with your glass rod to induce crystallization. Suction-filter the product using the Büchner funnel and suction flask and wash with a 20-mL portion of ice-cold 50 percent aqueous ethanol and finally with small portions of absolute ethanol. Dry the product in air, weigh it, and store it in a stoppered bottle. You should obtain about 11 g of product. Calculate the theoretical yield and determine your percent yield. *Save your sample for analysis in Experiment 35*.

Preparations of Materials for the Demonstration (These preparations should be done a week before the laboratory period.)

cis-$K[Cr(C_2O_4)_2(H_2O)_2] \cdot 2H_2O$

$$K_2Cr_2O_7 + 7H_2C_2O_4 \cdot 2H_2O \rightleftharpoons 2K[Cr(C_2O_4)_2(H_2O)_2] \cdot 2H_2O + 6CO_2 + 13H_2O$$

Separately powder in a *dry* mortar 12 g of oxalic acid dihydrate and 4 g of potassium dichromate. Mix the powders as intimately as possible by gentle grinding in the mortar. Moisten a large evaporating dish (10 cm) with water and pour off all the water but do not wipe dry. Place the powdered mixture in the evaporating dish as a *compact heap*; it will become moistened by the water that remains in the evaporating dish. Cover the evaporating dish with a large watch glass and warm it gently on a hot plate. A vigorous spontaneous reaction will soon occur and will be accompanied by frothing, as steam and CO_2 escape. The mixture should then liquefy to a deep-colored syrup. Pour about 20 mL of ethanol on the hot liquid and continue to gently warm it on the hot plate. Triturate (grind or crush) the product with a spatula until it solidifies. If complete solidification cannot be effected with one portion of alcohol, decant the liquid, add another 20 mL of alcohol, warm gently, and resume the trituration until the product is entirely crystalline and granular. The yield is essentially quantitative at about 9 g. This compound is intensely dichroic, appearing in the solid state as almost black in diffuse daylight and deep purple in artificial light.

***trans*-$K[Cr(C_2O_4)_2(H_2O)_2] \cdot 3H_2O$** Dissolve 12 g of oxalic acid dihydrate in a minimum of boiling water in a 300-mL (or larger) beaker. Add to this in small portions a solution of 4 g of potassium dichromate in a minimum of hot water and cover the beaker with a watch glass while the violent reaction proceeds. After the addition is complete, cool the contents of the beaker and allow spontaneous evaporation at room temperature to occur so that the solution reduces to about one-third its original volume (this takes 36–48 hours). Collect the deposited crystals by suction filtration and wash several times with cold water and alcohol and air-dry. The yield is about 6.5 g. The complex is rose-colored with a violet tinge and is not dichroic.

REVIEW QUESTIONS

Before beginning this experiment in the laboratory, you should be able to answer the following questions:

1. Define the terms *Lewis acid* and *Lewis base*.
2. Define the terms *ligand* and *coordination sphere*.
3. Define and give an example of a coordination compound.
4. Define the term *geometric isomer*.
5. Draw structures for all possible isomers of the six-coordinate compounds $[Co(NH_3)_4Cl_2]$ and $[Co(NH_3)_3Cl_3]$.
6. Are the chlorine atoms in equivalent environments in each of the compounds $[Co(NH_3)_4Cl_2]$ and $[Co(NH_3)_3Cl_3]$?
7. What is the meaning of the word *dichroism?*
8. What is the meaning of the word *trituration?*
9. Look up the preparation of an oxalate complex of Ni, Mn, or Co. Cite your reference and state whether this preparation would be suitable to add to this experiment. Why or why not?
10. Find an analytical method to determine the amount of Fe, Cu, Cr, or Al in your oxalate complex. Cite the reference to the method. Could you do the determination with the chemicals and equipment available in your laboratory? Why or why not?
11. Oxalic acid is used to remove rust and corrosion from automobile radiators. How do you think it works?

NOTES AND CALCULATIONS

Name ______________________________ Desk __________

Date ______________ Laboratory Instructor ______________________________

REPORT SHEET FOR EXPERIMENT 33

PREPARATION AND REACTIONS OF COORDINATION COMPOUNDS: OXALATE COMPLEXES

1. Complex prepared ______________________

2. Chemical reaction for its preparation ______________________

3. Theoretical yield of oxalate complex (show calculations):

4. Experimental yield of oxalate complex __________

5. Percent yield of oxalate complex (show calculations):

6. Color and general appearance of complex:

7. Describe the reactions of *cis*- and *trans*-$K[Cr(C_2O_4)_2(H_2O)_2]$ with NH_3 and the reverse reactions with HCl using chemical equations. List any observations, such as color changes or apparent solubilities.

8. Is your complex soluble in H_2O? __________ Alcohol? __________ Acetone? __________

QUESTIONS

1. Sodium trioxalatocobaltate(III) trihydrate is prepared by the following reactions:

$$[Co(H_2O)_6]Cl_2 + K_2C_2O_4 \cdot H_2O \longrightarrow CoC_2O_4 + 2KCl + 7H_2O$$

$$2CoC_2O_4 + 4H_2O + H_2O_2 + 4Na_2C_2O_4 \longrightarrow 2Na_3[Co(C_2O_4)_3] \cdot 3H_2O + 2NaOH$$

What is the percent yield of $Na_3[Co(C_2O_4)_3] \cdot 3H_2O$ if 7.6 g is obtained from 12.5 g of $[Co(H_2O)_6]Cl_2$?

2. Why are $K_3[Cr(C_2O_4)_3] \cdot 3H_2O$, $K_2[Cu(C_2O_4)_2] \cdot 2H_2O$, and $K_3[Fe(C_2O_4)_3] \cdot 3H_2O$ colored, whereas $K_3[Al(C_2O_4)_3] \cdot 3H_2O$ is colorless?

3. What are the names of the following compounds?
$K_3[Cr(C_2O_4)_3] \cdot 3H_2O$
$K_2[Cu(C_2O_4)_2] \cdot 2H_2O$
$K_3[Fe(C_2O_4)_3] \cdot 3H_2O$
$K_3[Al(C_2O_4)_3] \cdot 3H_2O$

4. What is the percent oxalate in each of the following compounds?
A. $K_3[Cr(C_2O_4)_3] \cdot 3H_2O$
B. $K_2[Cu(C_2O_4)_2] \cdot 2H_2O$
C. $K_3[Fe(C_2O_4)_3] \cdot 3H_2O$
D. $K_3[Al(C_2O_4)_3] \cdot 3H_2O$

Preparation of Sodium Bicarbonate and Sodium Carbonate

EXPERIMENT 34

OBJECTIVE

To become acquainted with the chemistry involved in the production of $NaHCO_3$ and its conversion to Na_2CO_3.

APPARATUS AND CHEMICALS

marble chips	ring stand
6 M HCl	utility clamp
$Ca(H_2PO_4)_2$ (solid)	thistle tube or long-stem funnel
$(NH_4)_2CO_3$-NaCl saturated solution*	250-mL Erlenmeyer flask
	100-mL beaker
pH paper	rubber stopper
wood splints	test tubes (3)
8-oz wide-mouthed bottle (2)	6 M NH_3
glass tubing	ice
rubber tubing	50 percent ethanol
Bunsen burner	filter flask, filter, and filter paper

DISCUSSION

Sodium bicarbonate ($NaHCO_3$), more properly called sodium hydrogen carbonate, and sodium carbonate (Na_2CO_3) are both important commercial chemicals. Sodium bicarbonate is commonly called baking soda, because it is used in cooking; it is the main ingredient in baking powders, where it serves as a source of carbon dioxide, which causes dough to rise. Most of the sodium bicarbonate manufactured is converted to sodium carbonate by mild heating:

$$2NaHCO_3(s) \xrightarrow{\Delta} Na_2CO_3(s) + H_2O(g) + CO_2(g) \qquad [1]$$

Anhydrous sodium carbonate, commonly known as soda ash, is used in large amounts in the manufacture of glass, enamels, soap, and paper. It is also used in water softening. More than 7 million tons of sodium carbonate is produced annually in the United States alone.

Both sodium bicarbonate and sodium carbonate are prepared industrially by the Solvay process. The Solvay process uses the readily available and relatively inexpensive raw materials salt and limestone; it also uses ammonia, which is not inexpensive. Limestone, a rock composed mainly of $CaCO_3$, is heated to make carbon dioxide:

---

*Prepared by adding 360 mL of water to a mixture of 55 g of $NaHCO_3$, 205 g of NaCl, and 240 mL of 7.4 M NH_3, stirring magnetically overnight, and filtering by gravity.

$$CaCO_3(s) \xrightarrow{\Delta} CaO(s) + CO_2(g) \quad [2]$$

Ammonia and the carbon dioxide from limestone are passed through a saturated sodium chloride solution at about 0°C. Under these conditions, sodium bicarbonate precipitates out as finely divided crystals that can be filtered from the ammonium chloride solution. The overall reaction can be written as follows:

$$CO_2(g) + NH_3(g) + H_2O(l) + Na^+(aq) + Cl^-(aq) \longrightarrow NaHCO_3(s) + NH_4^+(aq) + Cl^-(aq) \quad [3]$$

To a large measure, the Solvay process is economically feasible because the ammonia can be recovered from the ammonium chloride solution and recycled. The calcium oxide (quicklime) produced from the thermal decomposition of limestone according to Equation [1] is used to regenerate the ammonia as follows:

$$2NH_4^+(aq) + CaO(s) \longrightarrow 2NH_3(g) + H_2O(l) + Ca^{2+}(aq) \quad [4]$$

Although it is not practical to carry out the entire process in your laboratory, the process may be illustrated in part by starting with a solution that is saturated with both NaCl and $(NH_4)_2CO_3$ and bubbling in carbon dioxide. The formation of sodium bicarbonate under these conditions may be represented by the following equation:

$$2Na^+(aq) + CO_2(g) + CO_3^{2-}(aq) + H_2O(l) \longrightarrow 2NaHCO_3(s) \quad [5]$$

In this experiment you will make sodium bicarbonate and sodium carbonate and study some of the properties of these two substances.

PROCEDURE

A. Preparation of Sodium Bicarbonate

Set up the apparatus shown in Figure 34.1. CAUTION! *Take care to avoid injury while assembling the apparatus. Use glycerol to lubricate the thistle tube, glass tubing, and holes in the rubber stopper before inserting the glass into the rubber stoppers. Protect your hands with a towel.* Place 55 g of marble chips into the bottle that is fitted with the thistle tube and place 30 mL of water into the bottle that is clamped to the ring stand. BE CERTAIN *that the glass tubing extends down beneath the surface of the water*. Place 50 mL of a saturated solution of $(NH_4)_2CO_3$ that is also saturated with NaCl into the 250-mL Erlenmeyer flask. Ask your instructor to inspect your setup.

Through the thistle tube, add enough 6 *M* HCl to the bottle so that the lower end of the thistle tube extends beneath the surface of the acid. As the carbon dioxide forms and bubbles through the solution, shake the Erlenmeyer flask vigorously, splashing the solution so that it is thoroughly mixed with the gas. Allow the reaction to proceed until a large amount of solid is formed in the flask (about 90 min). Disconnect the flask, add 3 mL of 6 *M* NH_3, and filter the contents. The solid can be transferred completely to the filter funnel by rinsing the flask with the filtrate. Wash the solid with 20 mL of 50 percent ethanol and air dry. Divide your product into four equal portions for later use, as described below.

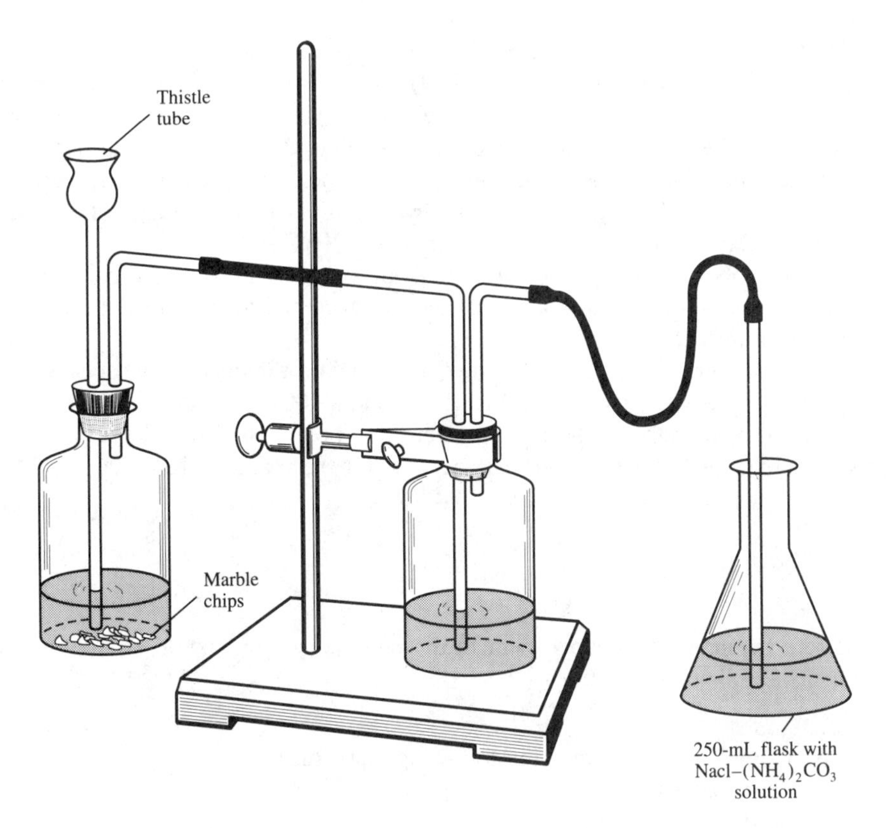

Figure 34.1

B. Properties of Sodium Bicarbonate

On the report sheet, describe the appearance of the sodium bicarbonate that you have prepared (1).

Dissolve one portion of your product in 5 mL of water. Measure the pH of the solution with pH paper and record its value (2). Account for the pH of the solution by means of an ionic equation (3).

Place a second portion of the sodium bicarbonate in the test tube and cautiously add about 2 mL of 6 *M* HCl. Test the gas with a burning splint. What is the gas? (4). Write an equation for the reaction (5). Based upon these observations, do you think that sodium bicarbonate would be suitable to neutralize acid that was spilled on your hand? Justify your answer (6).

C. Preparation of Sodium Carbonate

Place another portion of the sodium bicarbonate in a test tube and heat it. Insert a burning splint in the test tube. Write an equation for the reaction that occurs when sodium bicarbonate is heated (7). After the test tube and contents have cooled, add about 5 mL of water. Measure the pH of the solution and record the value (8). How does the pH of this solution compare with that of the sodium bicarbonate of part B? Which is more alkaline (9)? Write an ionic equation that accounts for the pH of the solution (10).

D. Baking Powders

Baking powders are used because they liberate carbon dioxide, which causes dough to "rise" or to be leavened. Baking powders are mixtures of the following three essential ingredients: (1) baking soda, which furnishes carbon dioxide; (2) an acid substance that furnishes hydrogen ions to react with sodium bicarbonate when water is added; and (3) an inert substance such as starch, to prevent intimate contact between ingredients (1) and (2) and to keep the mixture dry. The various brands of baking powders differ only in the nature of the acid ingredient. The most common acid sources are calcium dihydrogen phosphate, $Ca(H_2PO_4)_2$; potassium hydrogen tartrate (cream of tartar), $KHC_4H_4O_6$; and sodium alum, $NaAl(SO_4)_2 \cdot 12H_2O$.

Dissolve about 0.5 g of $Ca(H_2PO_4)_2$ in water. Measure and record the pH of the solution (11). Write an equation that accounts for the formation of the hydrogen ions from this substance (12).

Mix about 1.5 g of finely powdered calcium dihydrogen phosphate with about 0.5 g of your *dry* sodium bicarbonate in a 100-mL beaker. Add 5 mL of water to the mixture. Is there any evidence of a reaction? Record your observations (13). Write an equation that accounts for the reaction that occured (14).

REVIEW QUESTIONS

Before beginning this experiment in the laboratory, you should be able to answer the following questions:

1. Complete and balance the following equations:
   (a) $NaHCO_3(s) \xrightarrow{\Delta}$
   (b) $CaCO_3(s) \xrightarrow{\Delta}$
   (c) $CaO(s) + H_2O(l) \longrightarrow$
   (d) $NH_3(g) + H_2O(l) \longrightarrow$
   (e) $NH_4^+(aq) + CaO(s) \longrightarrow$
   (f) $H^+(aq) + HCO_3^-(aq) \longrightarrow$
2. How is the solubility of a gas related to its partial pressure above the solution?
3. How could you simply test for the presence of CO_3^{2-} in an unknown?
4. How could you test for the ammonium ion?
5. HCl and NaCl are both very soluble in water. The solubility of NaCl in aqueous HCl is much less than in pure water. Why?
6. How much $NaHCO_3$ could be prepared by the addition of excess CO_2 to a solution containing 15 g of NaCl and 10 g of $(NH_4)_2CO_3$ according to the reaction

$$2NaCl(aq) + CO_2(g) + (NH_4)_2CO_3(aq) + H_2O(l) \longrightarrow 2NaHCO_3(s) + 2NH_4Cl(aq)$$

7. Why should the bottom of the thistle tube extend beneath the surface of the hydrochloric acid?

Name ____ Desk __________

Date ____________________ Laboratory Instructor ______________________________

REPORT SHEET FOR EXPERIMENT 34

PREPARATION OF SODIUM BICARBONATE AND SODIUM CARBONATE

B. Properties of Sodium Bicarbonate

1. Appearance of $NaHCO_3$ ______________________________

2. pH of $NaHCO_3$ solution ______________________________

3. Ionic equation ______________________________

4. Identity of gas ______________________________

5. ______________________________

6. ______________________________

C. Preparation and Properties of Sodium Carbonate

7. ______________________________

8. pH of Na_2CO_3 ______________________________

9. ______________________________

10. ______________________________

D. Baking Powders

11. pH of $Ca(H_2PO_4)_2$ solution ______________________________

12. ______________________________

13. Evidence of reaction ______________________________

14. ______________________________

QUESTIONS

1. Dry-chemical fire extinguishers contain $NaHCO_3$. How might powered $NaHCO_3$ help extinguish a fire?

2. Complete and balance the following equations:

$KHC_4H_4O_6 + NaHCO_3 \rightarrow$

$HC_2H_3O_2 + NaHCO_3 \rightarrow$

3. Why is a sodium carbonate solution more alkaline than a sodium bicarbonate solution?

4. Consult a handbook and determine the solubility of the following in mol/L at 20°C:

(a) NH_4Cl ____________ mol/L

(b) Na_2CO_3 ____________ mol/L

Oxidation-Reduction Titrations I: Determination of Oxalate

EXPERIMENT 35

OBJECTIVE

To gain some familiarity with redox chemistry through analysis of an oxalate sample.

APPARATUS AND CHEMICALS

An oxalate sample (either an unknown or one of the complexes from Experiment 33)
sodium oxalate (primary standard)
1.0 *M* sulfuric acid
~0.1 *N* potassium permanganate solution
glass stirring rods
400-mL beakers (3)
buret clamp
50-mL buret
weighing bottle
thermometer
balance
Bunsen burner and hose

DISCUSSION

Potassium permanganate reacts with oxalate ions to produce carbon dioxide and water in an acidic solution, and the permanganate ion is reduced to manganese(II) as follows:

$$5C_2O_4^{2-}(aq) + 2MnO_4^-(aq) + 16H^+(aq) \longrightarrow 10CO_2(g) + 8H_2O(l) + 2Mn^{2+}(aq) \qquad [1]$$

Because this reaction proceeds slowly at room temperature, it is necessary to heat the solution gently to obtain satisfactory reaction rates. No indicators are necessary in permanganate titrations, since the end points are easily observed. The permanganate ion is intensely purple, whereas the manganese(II) ion is nearly colorless. The first slight excess of permanganate imparts a pink color to the solution, signaling that all of the oxalate has been consumed.

In solving volumetric analysis problems it is convenient to use the concepts of *equivalents* and normality. An *equivalent* is an amount of a reactant. Its definition depends upon the type of reaction being considered. However, the definition is always stated so that *one equivalent of a given reactant will react with exactly one equivalent of another*. The mass of one equivalent of a compound is called an *equivalent weight* (eqw). In the case of neutralization reactions, we saw in Experiment 19 that one equivalent weight of an acid or base is that quantity of acid or base that will release or react with 1 mole of protons.

For redox reactions, an equivalent weight of an oxidizing or reducing agent is that amount of a substance that will either gain or lose one mole of electrons. The balanced half reactions associated with Equation [1] are:

Oxidation:	$C_2O_4^{2-}(aq)$	$\longrightarrow$	$2CO_2(g) + 2e$	[2]
Reduction:	$MnO_4^-(aq) + 8H^+(aq) + 5e$	$\longrightarrow$	$Mn^{2+}(aq) + 4H_2O(l)$	[3]

We see that 1 mol of $C_2O_4^{2-}$ releases 2 moles of electrons. Hence, 1 mole $C_2O_4^{2-}$ is 2 equivalents of $C_2O_4^{2-}$, and thus 1 equivalent of $C_2O_4^{2-}$ is $\frac{1}{2}$ mol $C_2O_4^{2-}$. For MnO_4^-, 1 mol corresponds to 5 equivalents of MnO_4^-; 1 equivalent of MnO_4^- is $\frac{1}{5}$ mol MnO_4^-.

Normality is defined as the number of equivalents of solute in a liter of solution (which is also the number of milliequivalents of solute in a millimeter of solution):

$$N = \frac{\text{equivalents}}{V_L} \qquad [4]$$

where V_L is the volume of solution in liters. To find the number of equivalents in a given mass of a substance we divide the mass of the substance by the equivalent weight:

$$\text{equivalents} = \frac{m}{\text{eqw}} \qquad [5]$$

where m is the mass in grams and eqw is the equivalent weight in grams/equivalent. Substituting equation [5] into equation [4] yields:

$$N = \frac{m}{\text{eqw} \times V_L} \qquad [6]$$

Because equivalent weight and formula weight are related, normality and molarity are related. In general,

$$\text{Equivalent weight} = \frac{\text{formula weight}}{a} \qquad [7]$$

where a is the number of electrons gained or lost per mole of reactant in the balanced half-reaction. Thus, normality of a solution and molarity are related:

$$N = M \times a \qquad [8]$$

For a $KMnO_4$ solution that reacts according to equation [3], a is 5. Hence, a 1*M* $KMnO_4$ solution for this reaction is 5 *N*.

In this experiment you will standardize a $KMnO_4$ solution—that is, you will determine its normality—by titrating it against a very pure sample of sodium oxalate, $Na_2C_2O_4$. You will then use your standardized $KMnO_4$ to determine the percentage of oxalate ion, $C_2O_4^{2-}$, either in an unknown sample or in the complex you prepared in Experiment 33.

The basis of the determination is that equivalents react in a ratio of 1 : 1. An equal number of equivalents of MnO_4^- and $C_2O_4^{2-}$ have reacted when the *equivalence point* is reached:

$$\text{Equivalents } MnO_4^- = \text{equivalents } C_2O_4^{2-} \qquad [9]$$

Rearranging Equation [4] yields

$$N \times V_L = \text{equivalents} = \frac{m}{\text{eqw}} \qquad [10]$$

Because volumes are measured in mL, a more convenient form of Equation [10] is

$$N \times V_{mL} = \text{meq} = \frac{m(10^3)}{\text{eqw}} \qquad [11]$$

where, V_{mL} is the volume in mL and meq is milliequivalents (that is, 10^{-3} equivalents). Example 35.1 illustrates the calculation of normality for a standardization; Example 35.2 illustrates analysis for oxalate.

EXAMPLE 35.1

What is the normality of a $KMnO_4$ solution if 40.41 mL is required to titrate 0.2538 g of $Na_2C_2O_4$?

Solution: The reaction proceeds according to Equation [1]. At the equivalence point:

$$\text{meq } KMnO_4 = \text{meq } Na_2C_2O_4$$

From the half-reaction [2] Equation [7] we see that the equivalent weight of $Na_2C_2O_4$ is (134.00 g)/2 = 67.00 g.

The number of meq $Na_2C_2O_4$ is

$$\frac{0.2538 \text{ g}}{67.00 \text{ g/equiv}} \times \frac{10^3 \text{ meq}}{1 \text{ equiv}} = 3.778 \text{ meq}$$

Now $N \times V_{mL}$ = the number of meq for $KMnO_4$. Thus,

$$N \times 40.41 \text{ mL} = 3.778 \text{ meq}$$

and

$$N = \frac{3.778 \text{ meq}}{40.41 \text{ mL}} = 0.9350\ N$$

(Remember that normality is given in equivalents per liter or milliequivalents per milliliter.)

EXAMPLE 35.2

What is the percent oxalate, $C_2O_4^{2-}$, in a 1.429-g sample if 34.21 mL of 0.1002 N $KMnO_4$ solution is required for titration?

Solution: We want to find:

$$\%\ C_2O_4^{2-} = \frac{\text{weight } C_2O_4^{2-}}{\text{weight of sample}} \times 100$$

Thus, we need to know the mass of $C_2O_4^{2-}$ in the sample that weighs 1.429 g. At the equivalence point,

$$\text{meq KMnO}_4 = \text{meq C}_2\text{O}_4^{2-}$$
$$= N \times V_{\text{mL}}$$
$$= 34.21 \text{ mL} \times 0.1002 \text{ meq/mL}$$
$$= 3.428 \text{ meq}$$

From equations [2] and [11] we see that the equivalent weight of $C_2O_4^{2-}$ is (88.02 g)/2 = 44.01 g. (NOTE: We do not want the percent of sodium oxalate and hence do not use the equivalent weight of it. Rather we use the equivalent weight of the ion $C_2O_4^{2-}$.)

Thus,

$$3.428 \text{ meq C}_2\text{O}_4^{2-} = \frac{m \times (10^3 \text{ meq/equiv})}{44.01 \text{ g/equiv}}$$

and

$$m = 3.428 \text{ meq} \times \frac{1 \text{ equiv}}{10^3 \text{ meq}} \times \frac{44.01 \text{ g}}{1 \text{ equiv}}$$
$$= 0.1509 \text{ g C}_2\text{O}_4^{2-}$$

Thus,

$$\%\text{C}_2\text{O}_4^{2-} = \frac{0.1509 \text{ g}}{1.429 \text{ g}} \times 100 = 10.56\%$$

PROCEDURE

A. Preparation of $KMnO_4$ Solution

Potassium permanganate solutions are not stable over long periods of time, because they react slowly with the organic matter present even in most distilled water. The solution should be protected from heat and light as far as possible, because both induce decomposition, producing manganese dioxide, which seems to act as a catalyst for further decomposition. Because commercial samples of the permanganate usually contain some manganese dioxide, it is advisable to boil and filter the permanganate solutions before standardizing. Almost any form of organic matter will reduce permanganate solutions, so care must be taken to keep the solutions out of contact with rubber, filter paper, dust particles, and so on. For this reason permanganate solutions are filtered through sintered glass.

For this determination you will be provided with a potassium permanganate solution which is approximately 0.1 *N* and which will have to be standardized.

B. Standardization of Permanganate Solution—Sodium Oxalate Method

Several reducing agents can be used as primary standards for permanganate solutions, but sodium oxalate is very commonly used.

The oxidation of the oxalate ion is carried out in an acid solution that is maintained at a temperature between 80 and 90°C. This oxidation is catalyzed by the Mn^{2+} ion, which is a product of the oxidation-reduction reaction. The intense pink color of the permanganate ion may persist until several milliliters of the permanganate reagent have been added; however, when sufficient Mn^{2+}

ions have been formed to catalyze the reaction, the pink color will suddenly disappear and will continue to do so with each drop of reagent until the equivalence point is reached, at which point one drop in excess will turn the solution pink.

The titration can be accomplished more quickly if we know the approximate volume of $KMnO_4$ required. Assume a 0.2-g sample of $Na_2C_2O_4$ is weighed out. This corresponds to:

$$\text{meq } Na_2CO_4 = 0.2 \text{ g} \times \frac{1 \text{ equiv}}{67 \text{ g}} \times \frac{10^3 \text{ meq}}{1 \text{ equiv}} = 3 \text{ meq}$$

The reaction requires the same number of meq of $KMnO_4$.

If a 0.1 N $KMnO_4$ solution is used, the number of mL required in the titration is (see Equation [9]):

$$3 \text{ meq} = V_{mL} \times 0.1 \text{ meq/mL}$$

$$V_{mL} = 30 \text{ mL}$$

A volume of $KMnO_4$ solution up to within about 5.0 mL of this amount, that is, up to 25 mL, can be rapidly added to the hot, acidified oxalate solution without much danger of overrunning the end point. The permanganate solution is then added dropwise until one drop in excess turns the solution pink.

Procedure Weigh out from a weighing bottle, to the nearest 0.1 mg, triplicate portons of about 0.20 g each of pure sodium oxalate into 400-mL beakers. Run each trial as follows: Add about 250 mL of 1.0 M sulfuric acid; then, using a thermometer as a stirring rod, warm the mixture until the oxalate has dissolved and the temperature has been brought to 80–90°C. After taking the initial reading of the permanganate in the buret (see Note 1), titrate with the permanganate, stirring constantly, while keeping the solution above 70°C at all times. Add the permanganate dropwise when near the end point, allowing each drop to decolorize before adding the next. The end point is reached when the faintest visible shade of pink remains even after the solution has been allowed to stand for 15 s (see Note 2). Again read the buret and record the volume of permanganate solution used.

From the data obtained, calculate the normality of the potassium permanganate solution as an oxidizing agent in acid solution. The three results should agree wihin 3 ppt. If they do not agree, titrate another sample.

Notes

1. A 0.1 N permanganate solution is so deeply colored that the bottom of the meniscus is difficult to read. Thus it is necessary to read the top of the surface of the solution, taking the usual precautions to have the eyes level with the solution surface when the reading is made. Remember, the buret reading can be estimated to ±0.01 mL.
2. If there is any doubt whether or not an end point has been reached, it is good practice to take the buret reading and then add another drop of the permanganate. The development of an intense color indicates that the reading did correspond to the true end point.

Ordinarily, a permanganate end point is not permanent. Due to the action of dissolved reducing species in the water, the permanganate is slowly reduced and the color fades. A pink color that remains after stirring the solution for 15 s should be taken as the true end point.

C. The Determination of Oxalate in Your Oxalate Complex or Unknown

The unknown for this determination may be either one that your instructor provides, or the oxalate complex you prepared in Experiment 33, in which case read Note 1 before starting the procedure.

Procedure Into three separate 400-mL beakers, weigh out from a weighing bottle, to the nearest 0.1 mg, triplicate portions of about 0.2–0.3 g of the oxalate complex or about 0.5 g of unknown. Add about 250 mL of 1.0 *M* sulfuric acid. Titrate with the standardized permanganate solution as described above in Procedure B. Calculate the weight percent of $C_2O_4^{2-}$. A standard deviation of 0.3 or better represents acceptable results.

Note 1 Chromium oxalate complexes are difficult to analyze by titration with MnO_4^- because of their dark colors and slow rates of decomposition. In order to analyze them, first decompose the complex by boiling about 0.6–1 g of the complex in 10 mL of water and 10 mL of 3 *M* KOH for 15 min. Filter off the green $Cr(OH)_3$ and wash with distilled water. Combine the filtrate and washings, dilute to 25 mL in a volumetric flask, pipet 5 mL of this solution into a 400 mL beaker and titrate with MnO_4^- as above.

REVIEW QUESTIONS

Before beginning this experiment in the laboratory, you should be able to answer the following questions:

1. For a redox reaction, define the term equivalent weight for an oxidizing agent.
2. If 0.5468 g of sodium oxalate, $Na_2C_2O_4$, requires 35.43 mL of a $KMnO_4$ solution to reach the end point, what are the normality and molarity of the $KMnO_4$ solution?
3. Titration of an oxalate sample gave the following percentages: 15.75%, 15.55%, and 15.70%. Calculate the average and the standard deviation.
4. Why does the solution decolorize on standing after the equivalence point has been reached?
5. What is the normality of 1 *M* $KMnO_4$ solution that is used in an oxalate titration?
6. Why is the $KMnO_4$ solution filtered and why should it not be stored in a rubber-stoppered bottle?
7. What volume of 0.1 *M* $KMnO_4$ would be required to titrate 0.56 g of $K_2[Cu(C_2O_4)_2] \cdot 2H_2O$?

8. Calculate the percent $C_2O_4^{2-}$ in each of the following: $H_2C_2O_4$, $Na_2C_2O_4$, $K_2C_2O_4$, and $K_3[Al(C_2O_4)_3] \cdot 3H_2O$.
9. If 29.00 mL of 0.100 *N* $KMnO_4$ was required to titrate a 0.250-g sample of $K_3[Fe(C_2O_4)_3] \cdot 3H_2O$, what is the percent $C_2O_4^{2-}$ in the complex?
10. What is the percent purity of the complex in question 8?
11. What is the normality of a 0.30 *M* solution of $KMnO_4$ for the reaction

$$2KMnO_4 + 2KOH + H_3AsO_3 \longrightarrow H_3AsO_4 + H_2O + 2K_2MnO_4$$

NOTES AND CALCULATIONS

Name ______________________________ Desk __________

Date ______________ Laboratory Instructor ______________________________

Unknown no. or complex ______________________________

REPORT SHEET FOR EXPERIMENT 35

OXIDATION REDUCTION TITRATIONS I: DETERMINATION OF OXALATE

B. Standardization of $KMnO_4$

Weight of $Na_2C_2O_4$	**Trial 1**	**Trial 2**	**Trial 3**
Weighing bottle initial weight	________	________	________
Weighing bottle final weight	________	________	________
Weight of $Na_2C_2O_4$	________	________	________
Titration volume			
Final reading	________	________	________
Initial reading	________	________	________
Volume of $KMnO_4$	________	________	________
Calculations			
Moles of $Na_2C_2O_4$	________	________	________
Equivalents of $Na_2C_2O_4$	________	________	________
Volume of $KMnO_4$	________	________	________
Normality of $KMnO_4$	________	________	________

Average normality ________

Standard deviation (show calculations) ________

C. Analysis of Oxalate Complex or Unknown

Weight sample	**Trial 1**	**Trial 2**	**Trial 3**
Weighing bottle initial weight	________	________	________
Weighing bottle final weight	________	________	________
Weight of sample	________	________	________

Titration			
Final reading	______	______	______
Initial reading	______	______	______
Volume of $KMnO_4$	______	______	______
Calculations			
Equivalents of oxalate, $C_2O_4^{2-}$	______	______	______
Weight of oxalate, $C_2O_4^{2-}$	______	______	______
Percent oxalate, $C_2O_4^{2-}$	______	______	______

Average percent oxalate ______ Standard deviation ______

If you analyzed your oxalate complex from Experiment 33, complete the following.

Theoretical percent oxalate in your complex (show calculations) ______

Determine the purity of your complex as $\frac{\text{Experimental \% oxalate}}{\text{Theoretical \% oxalate}} \times 100 = \%$ purity.

QUESTIONS

1. What would be the normality of 1 *M* solutions of MnO_4^- according to these reactions (balance them first):
   (a) $MnO_4^- + 5e^- + 8H^+ \longrightarrow Mn^{2+} + 4H_2O$

   (b) $MnO_4^- + 3e^- + 4H^+ \longrightarrow MnO_2 + 2H_2O$

   (c) $MnO_4^- + 4e^- + 8H^+ \longrightarrow Mn^{3+} + 4H_2O$

2. How many grams of $KMnO_4$ are required to prepare 2.000 L of 0.2000 *N* $KMnO_4$ solution that is used in titrating $Na_2C_2O_4$?

3. MnO_4^- reacts with Fe^{2+} in acid solution to produce Fe^{3+} and Mn^{2+}. Write a balanced equation for this reaction.

Oxidation-Reduction Titrations II: Determination of Iron

EXPERIMENT 36

OBJECTIVE

To become familiar with the principles of redox titrations.

APPARATUS AND CHEMICALS

balance
50-mL buret
150-mL beakers (3)
500-mL Erlenmeyer flasks (2)
0.100 *N* standard $K_2Cr_2O_7$
18 *M* H_2SO_4
15 *M* H_3PO_4 (85 percent)
saturated $HgCl_2$ solution (0.2 *M*)
0.50 *M* $SnCl_2$ solution (2.27 g $SnCl_2 \cdot 2H_2O$ in 10 mL conc. HCl, diluted to 20 mL)
0.2 percent sodium diphenylamine sulfonate indicator
soluble unknown-iron sample

DISCUSSION

Because of the metallurgical importance of iron, various analytical procedures have been developed for its determinaion. Experiment 30 illustrates how iron may be determined colorimetrically. In this experiment you will analyze for iron by a volumetric procedure that involves the oxidation of iron(II) to iron(III) by dichromate ions according to Equation [1]:

$$6Fe^{2+}(aq) + Cr_2O_7^{2-}(aq) + 14H^+(aq) \longrightarrow 2Cr^{3+}(aq) + 6Fe^{3+}(aq) + 7H_2O(l) \quad [1]$$

Since so many substances are easily oxidized or reduced, redox titrations are used for more types of analysis than are all other volumetric methods combined.

Potasssium dichromate is an oxidizing agent that can be obtained in a high state of purity. Because of this and because of its relatively low cost, it is used as a primary standard for redox analyses. Its standard electrode potential indicates that it is capable of oxidizing iron(II):

$$Cr_2O_7^{2-}(aq) + 14H^+(aq) + 6e = 2Cr^{3+}(aq) + 7H_2O(l) \qquad E^\circ = 1.33 \text{ V} \quad [2]$$

$$Fe^{3+}(aq) + e = Fe^{2+}(aq) \qquad E^\circ = 0.77 \text{ V} \quad [3]$$

Thus E° for Equation [1] is 1.33 V − 0.77 V = 0.56 V. Because $E^\circ > 0$, this reaction will occur spontaneously.

The potentials of electrochemical reactions are governed by the Nernst equation:

$$E = E^\circ - \frac{0.30\ RT}{n\mathcal{F}} \log Q \qquad [4]$$

where E° is an electrochemical potential characteristic of the reaction, R is the gas-law constant (8.314J/K-mol), T is the absolute temperature, n is the number of electrons transferred in the reaction, $\mathcal{F}$ is the faraday (96,500 J/V-mol of electron), and Q is the reaction quotient. Substituting these values, and $T = 298\ K$, into Equation [4] yields Equation [5]:

$$E = E^\circ - \frac{0.059\ V}{n} \log Q \qquad [5]$$

At equilibrium, $Q = K$ and $E = 0$, and so at equilibrium Equation [5] becomes

$$E^\circ = \frac{0.059\ V}{n} \log K \qquad [6]$$

Thus for Equation [1], whose $E^\circ = 0.56$ V,

$$\log K = \frac{nE^\circ}{0.059V} = \frac{(6)(0.56\ V)}{0.059V} = 56.95$$

$$K = 8.91 \times 10^{56}$$

Because of the very large magnitude of the equilibrium constant, Equation [1] will proceed essentially to completion.

The above qualities make potassium dichromate a suitable reagent for the analysis of iron. Because of its high purity, a standard solution of $K_2Cr_2O_7$ can be readily prepared simply by dissolving a given amount of this substance in a known volume of water. The equivalent weight of $K_2Cr_2O_7$, based upon the half-cell reaction, is the formula weight (294.2) divided by 6, or 49.03 g/equiv. The exact normality of such a solution can be calculated from the following equation:

$$N = \frac{\text{grams of } K_2Cr_2O_7}{49.03\ \text{g/equiv} \times \text{liters}} \qquad [7]$$

A standard solution of $K_2Cr_2O_7$ for use in this experiment has been prepared for you, and the exact normality is recorded on the bottle. One liter of a 0.100 N solution would require 4.903 g of $K_2Cr_2O_7$. See Experiment 35 for a discussion about the use of equivalents and normality in redox volumetric calculations.

Unlike potassium permanganate, potassium dichromate is not a self-indicator. Therefore, a redox indicator must be used to detect the endpoint of the titration. Redox indicators usually are soluble organic compounds that change colors when oxidized or reduced. As is true of any other oxidizing agent, an oxidation potential is associated with each indicator. The oxidation potential of the indicator must lie between those of the oxidizing agent and the reducing agent being titrated, which in this case are $Cr_2O_7^{2-}$ and Fe^{2+}, respectively. This requirement must be met in order that the indicator not be prematurely oxidized and thereby undergo the characteristic color change before all of the reducing agent has been oxidized. The organic salt sodium diphenylamine sulfonate meets this requirement and is therefore a suitable indicator for the $Cr_2O_7^{2-}$ titration of Fe^{2+}. This indicator is colorless in the reduced form and purple to violet in the oxidized form. The purple color is so intense that it can be detected readily even in the presence of green Cr^{3+} formed in the reduction of the dichromate.

Prior to the titration with dichromate, all iron(III) must be quantitatively reduced to iron(II). Stannous chloride is used for this process; the reaction is

$$Sn^{2+}(aq) + 2Fe^{3+}(aq) \longrightarrow Sn^{4+}(aq) + 2Fe^{2+}(aq) \qquad [8]$$

To ensure complete reduction, a slight excess of $SnCl_2$ is used. However, since dichromate can oxidize Sn^{2+}, these ions must be removed before the addition of the standard dichromate solution. Mercuric chloride is used to remove the excess $SnCl_2$. Mercuric chloride is a strong enough oxidizing agent to oxidize $SnCl_2$ (but not Fe^{2+}), according to the following equation:

$$2HgCl_2(aq) + SnCl_2(aq) \longrightarrow Hg_2Cl_2(s) + SnCl_4(aq) \qquad [9]$$

Because iron undergoes a change of oxidation state of one unit (Fe^{2+} to Fe^{3+}), its equivalent weight is the same as its atomic weight, 55.85. At the equivalence point in the titration, the milliequivalents of dichromate equal the milliequivalents of iron. From a knowledge of the milliliters and normality of the dichromate solution used in the titration, the weight of iron contained in a sample can be calculated:

$$\text{Milliequivalents } Cr_2O_7^{2-} = \text{milliequivalents Fe} \qquad [10]$$

$$\text{Milliliters } K_2Cr_2O_7 \times NK_2Cr_2O_7 = \frac{\text{grams Fe} \times 10^3 \text{ meq/equiv}}{55.85 \text{ g/equiv}} \qquad [11]$$

Solving Equation [11] for grams of iron yields

$$\text{Grams Fe} = \frac{\text{milliliters } K_2Cr_2O_7 \times NK_2Cr_2O_7 \times 55.85 \text{ g/equiv}}{10^3 \text{ meq/equiv}} \qquad [12]$$

The percentage of iron in the sample can then be calculated from a knowledge of the weight of sample analyzed:

$$\% \text{ Fe} = \frac{\text{grams Fe}}{\text{grams sample}} \times 10^2$$

$$= \frac{\text{milliliters } K_2Cr_2O_7 \times NK_2Cr_2O_7 \times 55.85 \text{ g/equiv} \times 10^2}{\text{grams sample} \times 10^3 \text{ meq/equiv}} \qquad [13]$$

EXAMPLE 36.1

Find the percentage of iron contained in a 0.5215-g sample that, when dissolved and reduced with $SnCl_2$, required 26.31 mL of 0.1012 $NK_2Cr_2O_7$ for titration.

Solution: At the equivalence point,

$$\text{meq } K_2Cr_2O_7 = \text{meq Fe}$$
$$= 26.31 \text{ mL} \times 0.1012\ N$$
$$= 2.663 \text{ meq}$$

and

$$\text{g Fe} = 2.663 \text{ meq} \times \frac{1 \text{ equiv}}{10^3 \text{ meq}} \times \frac{55.85 \text{ g}}{1 \text{ equiv}}$$
$$= 0.1487 \text{ g Fe}$$

Thus

$$\% \text{ Fe} = \frac{0.1487 \text{ g}}{0.5215 \text{ g}} \times 100$$
$$= 28.51\ \%$$

PROCEDURE

Weigh out accurately to 0.1 mg triplicate samples of about 1.0 g of the unknown iron samples into 150-mL beakers and add to each beaker 10 mL of dilute HCl and 10 mL of distilled water.

Follow this procedure for each sample. Heat the beaker in the hood with a small flame, being careful not to spatter the solution, and evaporate the solution to a volume of about 10 mL. The solution should now show a pronounced yellow color due to $FeCl_4^-$ ions (see Note 1 below).

Reduce the Fe^{3+} to Fe^{2+} by adding stannous chloride solution dropwise to the *hot* solution until the color changes from yellow to a light green (practically colorless). Add only two drops of stannous chloride in excess after the color changes. Allow the solution to cool to room temperature and transfer it quantitatively by washing with distilled water into a 500-mL Erlenmeyer flask. Dilute to about 150 mL; rapidly pour in 10 mL of 0.2 *M* mercuric chloride while swirling the flask and its contents (see Note 2). A white, silky precipitate should form (see Note 3). Allow to stand for 5 min.

Next add 50 mL of sulfuric acid–phosphoric acid reagent (see Note 4) and enough water to make about 200 mL. Cool the Erlenmeyer flask and its contents to room temperature under a water tap. Add 8 drops of sodium diphenylamine sulfonate indicator solution. Remember, the buret reading can be estimated to ± 0.01 mL. Titrate with the standard dichromate solution, adding the reagent very slowly with constant swirling. When the color begins to change, add the dichromate a drop at a time. The endpoint is indicated by the first appearance of a purple or violet tinge that persists after the solution is swirled for 30 s. Repeat this titration procedure with the other two samples.

Calculate the percentage of iron in the three samples and the standard deviation. A standard deviation of 0.1 or better represents acceptable results.

Notes

1. After the solutions have been concentrated, work with only one solution at a time, carrying it completely through the titration before reducing the next sample.

2. The $HgCl_2$ is added all at once, so that it is always in excess. If it is added slowly or if too much $SnCl_2$ is present, the $HgCl_2$ may be reduced to mercury, rather than Hg_2Cl_2, according to this reaction:

$$HgCl_2(aq) + SnCl_2(aq) \longrightarrow Hg(l) + SnCl_4(aq)$$

3. The precipitate should be pure white in color and small in quantity. A grayish precipitate indicates reduction to mercury, which can be slowly oxidized by the dichromate. If this occurs, the sample should be discarded. If no precipitate forms, probably not enough $SnCl_2$ was added, and the sample also should be discarded.
4. The addition of $H_2SO_4 \cdot H_3PO_4$ reagent serves a dual purpose: (1) the oxidation potential of the ferric-ferrous system is lowered enough so that the indicator is not oxidized before the equivalence point is reached; and (2) the yellow color of $FeCl_4^-$ is removed by the formation of the more stable, colorless phosphate complex, and this facilitates the detection of the end point. The reagent is prepared by diluting a mixture containing 120 mL of 6 *M* H_2SO_4 and 90 mL of 85% H_3PO_4 to 1 L with distilled water.

REVIEW QUESTIONS

Before beginning this experiment in the laboratory, you should be able to answer the following questions:

1. Define the term of equivalent weight for a reducing agent.
2. Complete and balance the following redox equations:
   $Fe^{2+} + Cr_2O_7^{2-} + H^+ \longrightarrow$
   $Sn^{2+} + Fe^{3+} \longrightarrow$
3. What is the normality of a $K_2Cr_2O_7$ solution used in the determination of iron that contains 3.42 g of $K_2Cr_2O_7$ in 250 mL of solution?
4. What are two requirements that a suitable redox indicator should possess for a given redox reaction?
5. What are the roles of $SnCl_2$ and $HgCl_2$ in the dichromate determination of iron?
6. Why should you discard your solutions if you notice that they contain elemental mercury after oxidation with $HgCl_2$?
7. A 1.0155-g unknown iron sample required 21.12 mL of 0.1000 *N* $K_2Cr_2O_7$ solution to reach a sodium diphenylamine sulfonate end point. What is the percent iron in this sample?
8. How many milliliters of 0.0982 *N* $K_2Cr_2O_7$ would be required to titrate a 0.865-g iron sample that is 48.2 percent iron?
9. What is the normality of a 0.53 *M* $K_2Cr_2O_7$ solution? (The $K_2Cr_2O_7$ solution is used to titrate Fe^{2+}.)
10. What are the colors of $K_2Cr_2O_7$, K_2CrO_4, and Cr^{3+}?

NOTES AND CALCULATIONS

Name ______________________ Desk ________

Date ____________ Laboratory Instructor ______________________

Unknown no. ____________

REPORT SHEET FOR EXPERIMENT 36

OXIDATION-REDUCTION TITRATIONS II: DETERMINATION OF IRON

Weight of Samples

	Sample 1	**Sample 2**	**Sample 3**
Weight of vial + unknown	________	________	________
Weight of vial	________	________	________
Weight of unknown used	________	________	________
Volume of $K_2Cr_2O_7$			
Final buret reading	________	________	________
Initial buret reading	________	________	________
Milliliters of $K_2Cr_2O_7$ used	________	________	________
Weight of iron in sample	________	________	________
Percentage Fe	________	________	________

Average percent Fe ________ Standard deviation ________

Calculations for percent Fe and standard deviation:

QUESTIONS

1. Calculate $E°$ for the reaction $Cr_2O_7^{2-} + 6Fe^{2+} + 14H^+ \rightleftharpoons 2Cr^{3+} + 6Fe^{3+} + 7H_2O$.

2. Calculate the equilibrium constant for the above reaction.

3. What weight of $K_2Cr_2O_7$ is required to prepare 500 mL of 0.200 N solution?

4. How many millititers of 0.100 N $K_2Cr_20_7$ are required to titrate a 0.912-g iron sample that is 46.3 percent iron?

Molecular Geometry: Experience with Models

EXPERIMENT 37

OBJECTIVE

To become familiar with the three-dimensional aspects of organic molecules.

MATERIALS

Molecular models (ball and stick); carbon is represented by a black sphere with four holes, hydrogen by a yellow sphere with one hole, and chlorine by a green sphere with one hole.

DISCUSSION

Organic compounds are extremely numerous—in fact, there are approximately 2×10^6 known organic compounds. The chemical and physical properties of these compounds depend upon what elements are present, how many atoms of each element are present, and how these atoms are arranged in the molecule. Molecular formulas often, but not always, permit one to distinguish between two compounds. For example, even though there are eight atoms in both C_2H_6 and C_2H_5Cl, we know immediately that these are different substances on the basis of their molecular formulas. Similarly, inspection of the molecular formulas C_2H_6 and C_3H_8 reveals that these are different compounds. However, there are many substances that have identical molecular formulas but are completely different compounds. Consider the molecular formula C_2H_6O. There are two compounds that correspond to this formula: ethyl alcohol and dimethyl ether. While the molecular formula gives no clue as to which compound one may be referring, examination of the *structural formula* immediately reveals a different arrangement of atoms for these substances:

```
    H   H                 H         H
    |   |                 |         |
H — C — C — O — H     H — C — O — C — H
    |   |                 |         |
    H   H                 H         H
  Ethyl alcohol          Dimethyl ether
```

In addition, when molecular models (ball-and-stick type) are used, trial and error will show that there are just two ways that two carbons, six hydrogens, and one oxygen can be arranged without violating the usual valences of these elements. (Remember, carbon has a valence of 4, oxygen 2, and hydrogen 1.) compounds that have the same molecular formula but different structural formulas are termed *isomers*. This difference in molecular structure results in differences in chemical and physical properties of isomers. In the case of ethyl

Table 37.1 Properties of Ethyl Alcohol and Dimethyl Ether

Property	Ethyl Alcohol	Dimethyl ether
Boiling point	78.5°C	−24°C
Melting point	−117°C	−139°C
Solubility in H_2O	Infinite	Slight
Behavior toward sodium	Reacts vigorously, liberating hydrogen	No reaction

alcohol and dimethyl ether, whose molecular formula is C_2H_6O, these differences are very pronounced (see Table 37.1). In other cases, the differences may be more subtle.

The importance of the use of structural formulas in organic chemistry becomes evident when we consider the fact that there are 35 known isomers corresponding to the formula C_9H_{20}! For the sake of convenience, *condensed structural formulas* are often used. The structural and condensed structural formulas for ethyl alcohol and dimethyl ether are

Structural formulas:

$$\begin{array}{ccccccc} & & H & & H & & \\ & & | & & | & & \\ H & — & C & — & C & — & O — H \\ & & | & & | & & \\ & & H & & H & & \end{array} \qquad \begin{array}{ccccccc} & & H & & & & H & & \\ & & | & & & & | & & \\ H & — & C & — & O & — & C & — & H \\ & & | & & & & | & & \\ & & H & & & & H & & \end{array}$$

Condensed structural formulas: CH_3CH_2OH (Ethyl alcohol) $\qquad CH_3OCH_3$ (Dimethyl ether)

A short glance at these formulas readily reveals their difference. The compounds differ in their **connectivity**. The atoms in ethyl alcohol are connected or bonded in a different sequence from those in dimethyl ether.

You must learn to translate these condensed formulas into three-dimensional mental structures and to translate structures represented by molecular models to condensed structural formulas.

PROCEDURE

Construct models of the molecules as directed and determine the number of isomers by trial and error as directed below. Use a black ball with four holes for carbon, a yellow ball with one hole for hydrogen, and a green ball with one hole for chlorine. Answer all questions on the report sheet at the end of this experiment.

A. Methane

Construct a model of methane, CH_4. Place the model on the desk top and note the symmetry of the molecule. Note that the molecule looks the same regardless of which three hydrogens are resting on the desk. All four hydrogens are said to be *equivalent*. Now, grasp the top hydrogen and tilt the molecule so that only two hydrogens rest on the desk and the other two are in a plane parallel to the desk top (see Figure 37.1). Now imagine pressing this methane model flat onto the desk top. The resulting imaginary projection in the plane of the desk is the conventional representation of the structural formula of methane:

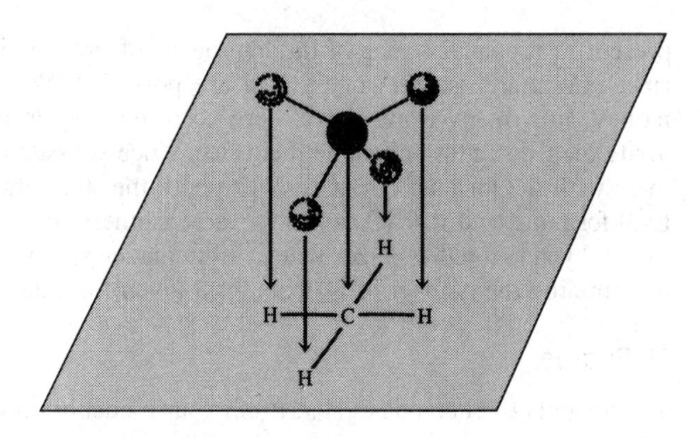

Figure 37.1 Model of methane.

$$
\begin{array}{ccc}
 & H & \\
 & | & \\
H- & C & -H \\
 & | & \\
 & H &
\end{array}
$$

Replace one of the hydrogen atoms with a chlorine atom to construct a model of chloromethane (or methyl chloride), CH_3Cl. Replace a second hydrogen atom by a chlorine atom to make dichloromethane, CH_2Cl_2. Convince yourself that the two formulas

$$
\begin{array}{ccc}
 & H & \\
 & | & \\
Cl- & C & -Cl \\
 & | & \\
 & H &
\end{array}
\quad \text{and} \quad
\begin{array}{ccc}
 & Cl & \\
 & | & \\
Cl- & C & -H \\
 & | & \\
 & H &
\end{array}
$$

represent the same three-dimensional structure and are not isomers. Replace another hydrogen to make $CHCl_3$, chloroform (or trichloromethane). Finally, make CCl_4, carbon tetrachloride.

B. Ethane

Make a model of ethane, C_2H_6, from your model of CH_4 by replacing one of the hydrogens by a CH_3 unit; the $—CH_3$ unit is called the *methyl group*. Note that the hydrogens of ethane are all equivalent. Replace one of the hydrogens in your ethane model by a chlorine. Does it matter which hydrogen you replace? How many compounds result? Now examine your model C_2H_5Cl and note how many different hydrogen atoms are present. If another hydrogen of C_2H_5Cl is replaced by a chlorine atom to yield $C_2H_4Cl_2$, how many isomers would result? Does the formula $C_2H_4Cl_2$ distinguish the possible molecules corresponding to this formula? Write the structural and condensed structural formulas for all isomers of $C_2H_4Cl_2$ and assign the IUPAC names to each.

C. Propane

From your model of ethane, construct a molecular model of propane, C_3H_8, by replacing one of the hydrogen atoms by a methyl group, $—CH_3$. Examine your model of propane and determine how many different hydrogen atoms are present in propane. If one of the hydrogens of propane is replaced with chlorine, how many isomers of C_3H_7Cl are possible? Write them and give their names. How many isomers are there corresponding to the formula $C_3H_6Cl_2$? Write their formulas and name them. Convince yourself that there are five isomers with the formula $C_3H_5Cl_3$. Write both the structural and condensed stuctural formulas and IUPAC names for these isomers. By this stage in this experiment, you should realize that a systematic approach is most useful in determining the number of isomers for a given formula.

D. Butane

The formula of butane is C_4H_{10}. From your model of propane, C_3H_8, construct all of the possible isomers of butane by replacing a hydrogen atom by the methyl group, $—CH_3$. How many isomers of butane are there? Give their structural formulas and IUPAC names. There are four isomers corresponding to the formula C_4H_9Cl. Write their structural formulas and name them. How many isomers of $C_4H_8Cl_2$ are there? Use your models to help answer this question. Write and name all of the isomers of $C_4H_8Cl_2$.

E. Pentane

Use your models in a systematic manner to determine how many isomers there are for the formula C_5H_{12}. Write their structural formulas and name them. Write and name all isomers for the formula $C_5H_{11}Cl$.

F. Cycloalkanes

Cycloalkanes corresponding to the formula C_nH_{2n} exist. Try to construct (but *do not* force too much in your attempt) models of cyclopropane, C_3H_6; cyclobutane, C_4H_8; cyclopentane, C_5H_{10}; and cyclohexane. Although cyclopropane and cyclobutane exist, would you anticipate these to be highly stable molecules? How many isomers of 1,2-dichlorocyclopentane are there? The answer is not obvious. There are three. If you cannot convince yourself that there *are* three, check with your laboratory instructor. This is another aspect of isomerization that is most significant in biochemical systems. Although you may think this is trivial, such differences are of utmost importance in nature!

G. Alkenes

Using two springs for bonds, construct a model of ethene (ethylene), C_2H_4. Note the rigidity of the molecule; note that there is no rotation about the carbon-carbon double bond as there is in the case of carbon-carbon single bonds such as in ethane or propane. How many isomers are there corresponding to the formulas C_2H_3Cl? $C_2H_2Cl_2$? Draw the formulas and name them.

REVIEW QUESTIONS

Before beginning this experiment in the laboratory, you should be able to answer the following questions:

1. Distinguish between molecular and structural formulas.

2. What is a condensed structural formula? Give an example.
3. What is the meaning of the word *isomer*?
4. Why should the properties of structural isomers differ?
5. Draw the structural formulas for ethane and propane.
6. Distinguish between molecular and empirical formulas.
7. The molecular formula of benzene is C_6H_6. What is the empirical formula of benzene?
8. Distinguish between geometric and structural isomers.
9. Carbon has a valence of 4, oxygen 2, and hydrogen 1. How many compounds of C, H, and O containing only one carbon and one oxygen can you make? Draw their structures.
10. Draw Lewis electron-dot formulas for the compounds in question 9 and name them.

NOTES AND CALCULATIONS

Name ______________________________ Desk ____________

Date ______________ Laboratory Instructor ______________________________

REPORT SHEET FOR EXPERIMENT 37

MOLECULAR GEOMETRY: EXPERIENCE WITH MODELS

A. Methane

Write the structure for and name each of the chloromethanes.

B. Ethane

Write the structures for C_2H_5Cl and $C_2H_4Cl_2$ and name each compound.

C. Propane

1. Write the structural and condensed formulas as well as the names for all isomers of C_3H_7Cl and $C_3H_6Cl_2$.

2. Write the condensed and structural formulas as well as the names for all the isomers of $C_3H_5Cl_3$.

D. Butane

1. Write the structural formulas and names for all the butanes.

2. Give the structural formulas and names for all the isomers of:
(a) C_4H_9Cl

(b) $C_4H_8Cl_2$

E. Pentane

1. Write the structural formulas and names for all the isomers of C_5H_{12}.

2. Give the structural formula and names for all the isomers of $C_5H_{11}Cl$.

F. Cycloalkanes

Using the following format for the structure of cyclopentane, give the structures of all isomers of 1,2-dichlorocyclopentane and 1,3-dichlorocyclopentane:

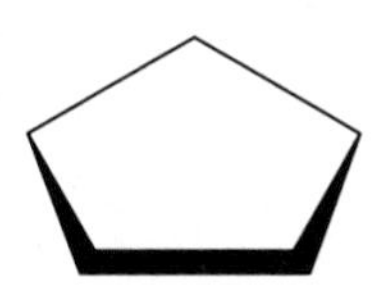

G. Alkenes

Give the structures and names of all isomers of $C_2H_2Cl_2$.

QUESTIONS

1. Write and name all isomers for the alkane C_6H_{14}.

2. Name the following four compounds:

```
                    CH3
                     |
CH3—CH—CH2—CH—CH3
     |
     CH2
     |
     CH3
```

____

```
         Cl
         |
CH3—C—CH2CH3
         |
         CH3
```

____

```
              CH3
               |
CH3—CH2—CH—CH3
```

____

```
                   C2H5
                    |
CH3—CH2—CH—CHCH3
                         |
                         CH(CH3)2
```

____

3. Give the structural formulas for the following.

(a) 2-chloropentane

(b) 3-chloro-3-methylpentane

(c) 2,2,3-trimethylhexane

(d) 2,2-dimethyl-4-ethylhexane

NOTES AND CALCULATIONS

Preparation of Aspirin and Oil of Wintergreen

EXPERIMENT 38

OBJECTIVE

To illustrate the synthesis of organic compounds.

APPARATUS AND CHEMICALS

4 g salicylic acid	125-mL Erlenmeyer flask
6 mL acetic anhydride	600-mL beaker
5 mL methyl alcohol	10-mL graduated cylinder
1 percent $FeCl_3$ solution	large test tube
concentrated H_2SO_4	ring stand
ice	wire gauze
50 mL absolute ethanol	filter paper
small watch glass	funnel

DISCUSSION

Esters are derivatives of organic acids. They can be prepared by the reaction of carboxylic acids with alcohols in the presence of a catalytic amount of mineral acid. The reaction

$$\underset{\text{Acetic acid}}{CH_3C(=O){-}OH} + \underset{\text{Methyl alcohol}}{CH_3OH} \rightleftharpoons \underset{\text{Methyl acetate}}{CH_3C(=O){-}OCH_3} + H_2O$$

is termed *esterification*. Esters usually have pleasant, fruitlike odors and are responsible for the flavors and fragrances of many fruits and flowers. For example, benzyl acetate has been found to be a principal ingredient of jasmine. Some common esters and their odors are octyl acetate, oranges; *n*-pentyl acetate, bananas; and butyl butyrate, pineapples. Generally speaking, the odor of natural products is due to more than one substance. For example, the volatile oil from pineapples contains at least six compounds.

In this experiment, you will prepare two esters of *o*-hydroxybenzoic acid, which is also called salicylic acid. One of the esters is acetylsalicylic acid, or, as it is commonly called, aspirin. It can be prepared by heating salicylic acid and acetic acid for an extended period of time. However, instead of using acetic acid, you will use acetic anhydride, because it is more reactive and will accomplish the esterification more quickly. The reaction is

$$\text{Salicylic acid } (C_6H_4(OH)CO_2H) + \text{Acetic anhydride } ((CH_3CO)_2O) \xrightarrow{H^+} \text{Acetylsalicylic acid } (C_6H_4(OCOCH_3)CO_2H) + CH_3CO_2H$$

Salicylic acid Acetic anhydride Acetylsalicylic acid

Aspirin is one of the oldest and generally most useful drugs known. It is both an analgesic (painkiller) and an antipyretic (reduces fever). Over 30 billion tablets are made in the United States annually. This is more than 200 tablets per person per year!

The second ester you will prepare is methyl salicylate, which is a component of oil of wintergreen. It is prepared by esterification of the carboxylic acid group, —CO_2H, of salicylic acid with methyl alcohol. The reaction is

$$\text{Salicylic acid } (C_6H_4(OH)COOH) + CH_3OH \rightleftharpoons \text{Methyl salicylate } (C_6H_4(OH)COOCH_3) + H_2O$$

Salicylic acid Methyl alcohol Methyl salicylate

It is used quite extensively as a flavoring agent and in rubbing liniments for sore muscles.

Phenols are a class of compounds in which the —OH group is attached to an aromatic ring. Note that this group is present in both salicylic acid and methyl salicylate. Many phenols, but not all, form colored complexes with ferric chloride. The colors range from green through blue and red through violet. Hence, a 1 percent $FeCl_3$ solution is employed as a test for the presence of phenols.

PROCEDURE

A. Synthesis of Aspirin

Place about 3 grams of salicylic acid in a 125 mL Erlenmeyer flask. Record the exact weight of salicylic acid used on the report sheet. Cautiously add 6 mL of acetic anhydride and then 5 drops of concentrated sulfuric acid. *Acetic anhydride and concentrated H_2SO_4 can cause severe burns if they come in contact with skin. If you get any of these reagents on you, immediately wash the area with copious amounts of water.* Thoroughly mix the reagents by swirling the flask. Place the flask in a beaker of water warmed to 80–90°C. Heat the flask and its contents for 20 minutes. Remove the Erlenmeyer flask from the hot water and allow it to cool to room temperature. Add 40 mL of distilled water to the mixture in the Erlenmeyer flask. Thoroughly mix the contents by swirling. Cool the mixture in an ice bath to complete the crystallization. Filter and wash the crystals on the filter with a little ice-cold water. Allow the crystals to drain and then press them between several sheets of paper toweling or filter paper. Allow the crystals to air dry.

Your product is not likely to be pure, and certainly it is *not* suitable for human ingestion. Dissolve a few crystals of salicylic acid and a few crystals of your aspirin in 5 mL of water in separate test tubes. Add a drop of 1 percent ferric chloride solution to each test tube and note the color. Does your aspirin contain any unreacted salicylic acid?

Most solids may be purified by *recrystallization*. This is usually achieved by dissolving the substance in a suitable solvent at the boiling point, filtering the hot solution by gravity to remove any suspended insoluble particles, and letting crystalization proceed as the solution cools.The following is a conventional procedure for recrystallizing aspirin: Dissolve about 6 g in about 20 mL of absolute ethanol (calculate the volume of ethanol needed for your weight of aspirin) in a 125-mL Erlenmeyer flask and warm the alcohol in a water bath (NOT OVER A FLAME, SINCE ETHANOL IS VERY FLAMMABLE). When the aspirin has dissolved, add 50 mL of warm (50°C) distilled water. (If any crystals form at this point, heat the solution over the water bath again until they all dissolve.) Let the solution cool slowly, with the mouth of the flask covered with a watch glass. After crystallization is complete, filter the crystals by gravity, wash them with a little ice-cold distilled water, and air dry. Test the recrystallized material with 1 percent $FeCl_3$ solution for phenolic impurities. Did the recrystallization procedure remove the phenolic impurities? Your asprin is still probably not pure and is *not* suitable for human ingestion.

After your aspirin has dried, obtain and record its weight. Based upon the weight of salicylic acid used and the weight of your product, calculate the percent yield of aspirin. Turn in your product to your laboratory instructor in a sample bottle or test tube labeled with your name or save it for Experiment 39.

B. Synthesis of Methyl Salicylate

Place 1 g of salicylic acid and 5 mL of methyl alcohol in a large test tube. Add 3 drops of concentrated sulfuric acid and then place the test tube in a water bath at 70°C for about 15 min. Note the odor. Add a drop of 1 percent ferric chloride solution to the test tube and note any color change. Would you expect a color change?

REVIEW QUESTIONS

Before beginning this experiment in the laboratory, you should be able to answer the following questions:

1. What are esters? What is their general structure?
2. What is the role of the mineral acids in the esterification process?
3. From what substances can esters be prepared?
4. What is one general physical property of esters?
5. Describe a test for phenols.
6. What is the carboxylic acid group?
7. What is the purpose of recrystallization?

8. 1.02 g of acetylsalicylic acid was obtained from 1 g of salicylic acid by reaction with excess acetic anhydride. Calculate the percent yield of acetylsalicylic acid.
9. Two esters with the empirical formula $C_3H_6O_2$ may be prepared. Write structural formulas for these esters and name them.
10. How would you prepare the two esters in question 9?
11. Esters may be saponified by reaction with a strong base. Predict the products of the reaction.

$$CH_3\overset{\overset{\displaystyle O}{\|}}{C}OCH_3 + NaOH \xrightarrow{H_2O}$$

12. Give the structural formulas for isopropyl acetate and ethyl salicylate.

Name ______________________________ Desk __________

Date ______________ Laboratory Instructor ______________________________

REPORT SHEET FOR EXPERIMENT 38

PREPARATION OF ASPIRIN AND OIL OF WINTERGREEN

A. Synthesis of Aspirin

Weight of salicylic acid __________ g

Weight of *dry* recrystallized aspirin __________ g

Percent yield

Moles of salicylic acid used (mol wt of salicylic acid = 138) __________ mol

Theoretical number of moles of aspirin __________ mol

Theoretical grams of aspirin (mol wt of aspirin = 180) __________ g

$$\text{Percent yield} = \frac{\text{grams aspirin obtained}}{\text{theoretical grams aspirin}} \times 10^2$$

$$= \frac{\_\_\_\_\_\_\_\_\ \text{g}}{\text{g}} \times 10^2$$

__________ % yield

Color of $FeCl_3$ plus salicylic acid ______________________________

Color of $FeCl_3$ plus aspirin ______________________________

Color of $FeCl_3$ plus recrystallized aspirin ______________________________

Did the recrystallization remove phenolic impurities? ______________________________

B. Synthesis of Methyl Salicylate

Odor ______________________________

Color of $FeCl_3$ plus product ______________________________

QUESTIONS

1. Why was *cold* as opposed to warm water used to wash the aspirin that you prepared?

2. Explain why you would or would not expect to observe a color change if $FeCl_3$ were added to the following.
(a) Pure aspirin

(b) Oil of wintergreen

3. Write the structure of the products that you would expect from the following reactions:
(a)

$$C_6H_5CO_2H + CH_3CH_2OH \xrightarrow{H^+}$$

(b) CH_3CH_2OH + acetic anhydride $\xrightarrow{H^+}$

(c) $CH_3CH_2CO_2H + CH_3CH_2CH_2OH \xrightarrow{H^+}$

4. How would you prepare each of the following?
(a) Ethyl acetate (two ways)

(b) Methyl acetate

5. If your experimental yield of aspirin is greater than 100 percent, how could this occur?

EXPERIMENT 39

Analysis of Aspirin

OBJECTIVE

To determine the purity of aspirin by acid-base titrations; to become acquainted with the concept of back-titration analyses.

APPARATUS AND CHEMICALS

aspirin (student preparation or nonbuffered commercial tablets)
0.1 *N* NaOH, standardized
0.1 *N* HCl, standardized
95 percent ethyl alcohol
boiling chips
phenolphthalein solution

50-mL burets (2)
250-mL Erlenmeyer flasks (3)
buret clamp
ring stand
Bunsen burner
analytical balance

DISCUSSION

The aspirin you prepared in Experiment 38 is not likely to be pure. The most likely impurities are acids, either acetic or salicylic. Even most commercial aspirin tablets are not 100 percent acetylsalicylic acid. Most aspirin tablets contain a small amount of "binder," which helps prevent the tablets from crumbling. Even though the binder is chemically inert and was deliberately added by the manufacturer, its presence means that aspirin tablets are not 100 percent acetylsalicylic acid. Moreover, moisture can hydrolyze aspirin; thus aspirin that is not kept dry can decompose. You may be able to detect a vinegarlike odor in aspirin if it has been exposed to moisture for an extended period of time. The hydrolysis product responsible for this odor is, in fact, acetic acid. It is formed in the following way:

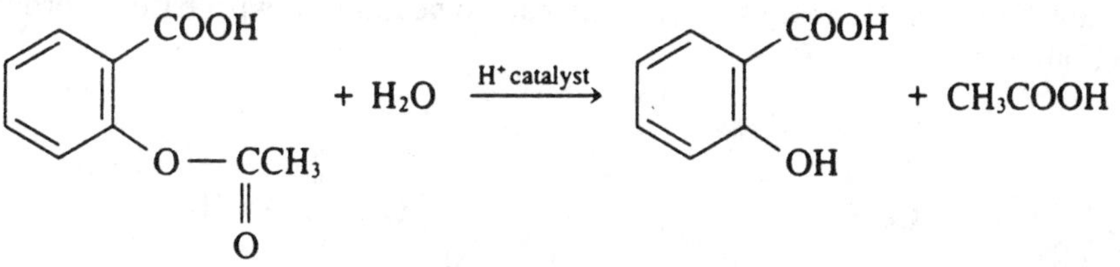

$$C_6H_4(COOH)(O{-}C(=O)CH_3) + H_2O \xrightarrow{H^+\ catalyst} C_6H_4(COOH)(OH) + CH_3COOH$$

In this experiment you will determine the purity of either the aspirin you prepared or commercial tablets. In particular, you will determine the percentage of acetylsalicylic acid in the material you analyze. The basis of the analyses is similar to that of Experiment 19 and involves acid-base titrations.

A *titration* is a process for determining the amount of analyte* present in a solution by the incremental addition of known volumes of a standard solution

*An analyte is the substance being determined in any analytical procedure.

until reaction between the analyte and titrant is judged to be complete. Occasionally, it is convenient or necessary to add an excess of the titrant and then titrate the excess with another reagent. This process is called *back-titration*. In this technique, a measured amount of the reagent, which would normally be the titrant, is added to the sample so that there is a slight excess. After the reaction with the analyte is allowed to go to completion, the amount of excess (unreacted) reagent is determined by titration with another standard solution. Hence, by knowing the number of millimoles (mmol) of reagent taken and measuring the number in excess, we can calculate the number of millimoles of analyte by difference:

$$\text{Millimoles reagent reacted} = \text{total millimoles} - \text{millimoles back-tritated}$$

EXAMPLE 39.1

When nitrogen is analyzed by the Kjeldahl method, all the nitrogen in the sample is converted into NH_3. The NH_3 is distilled into a solution containing excess acid, and the excess acid is titrated with a standard base solution. If the nitrogen from a 1.325-g fertilizer sample is converted into NH_3 and distilled into 50.00 mL of 0.2030 *M* HCl, and if 25.32 mL of 0.1980 *M* KOH is required to back-titrate the excess HCl, how much NH_3 has been liberated from the fertilizer? Calculate the percentage of nitrogen in the fertilizer.

Solution:

$$\text{mmol NH}_3 = \text{mmol HCl} - \text{mmol KOH}$$

$$\begin{aligned}
\text{mmol NH}_3 &= \text{mmol N} \\
&= (50.00 \text{ mL})(0.2030 \text{ mmol/mL}) - (25.32 \text{ mL})(0.1980 \text{ mmol/mL}) \\
&= 10.15 - 5.01 \\
&= 5.14 \text{ mmol}
\end{aligned}$$

$$\begin{aligned}
\text{Weight NH}_3 &= (5.14 \times 10^{-3} \text{ mol})(17.0 \text{ g/mol}) \\
&= 0.087 \text{ g}
\end{aligned}$$

$$\begin{aligned}
\%\text{ N} &= \frac{(5.14 \text{ mmol})(14.00 \text{ mg/mmol})}{1325 \text{ mg}} \times 100 \\
&= 5.43\%
\end{aligned}$$

At low temperature, acetylsalicylic acid can be neutralized with base according to Equation [1]:

$$C_6H_4(O{-}\overset{O}{\overset{\|}{C}}CH_3)(CO_2H) + OH^- \longrightarrow C_6H_4(O{-}\overset{O}{\overset{\|}{C}}{-}CH_3)(CO_2^-) + H_2O \qquad [1]$$

If no acidic impurities were present, you could determine the purity of the aspirin in exactly the same manner that you determined the percent KHP in Experiment 19. (It would be a good idea to review Experiment 19 and especially Example 19.3 before proceeding.) However, if acid impurities are present, titration of the aspirin will neutralize not only the acetylsalicylic acid (Equation

[1]) but the acidic impurities as well. Thus from such a titration the *total* number of milliequivalents of acid present in the aspirin can be calculated by measuring the volume of standard NaOH required to reach the phenolphthalein end point. The total number of milliequivalents (meq) of acid may be calculated according to Equation [2]:

$$\text{Total milliequivalents acid} = \text{milliliters NaOH} \times \text{normality NaOH} \qquad [2]$$

In order to determine the amount of acetylsalicylic acid present in your material, we will take advantage of the fact that this substance will react with additional base reasonably rapidly at elevated temperatures according to Equation [3]:

$$C_6H_4(OCOCH_3)(CO_2^-) + OH^- \xrightarrow{\Delta} C_6H_4(OH)(CO_2^-) + CH_3CO_2^- \qquad [3]$$

This reaction represents what is termed a base-promoted hydrolysis or saponification of esters. The reaction is the reverse of the esterification process. After you have neutralized all acidic material in the aspirin, you will add a known excess amount of milliequivalents of base to cause this reaction to occur. The excess base that is not consumed in the hydrolysis will be determined by a *back-titration* with standard HCl. From these data you can calculate the grams of acetylsalicylic acid in your material. Example 39.2 illustrates such a calculation.

EXAMPLE 39.2

A 0.5130-g sample of aspirin prepared by a student required 27.98 mL of 0.1000 *N* NaOH for neutralization. An additional 42.78 mL of 0.1000 *N* NaOH was added, and the sample was heated to hydrolyze the acetylsalicylic acid. After the reaction mixture cooled, the excess base was back-titrated with 14.29 mL of 0.1056 *N* HCl. How many grams of acetylsalicylic acid are in the sample? What is the percentage of acetylsalicylic acid (or the purity)?

Solution: First recognize that the 27.98 mL of base was used to neutralize all acidic material present in the sample. The total number of milliequivalents of base added for the hydrolysis reaction is

$$\begin{aligned} \text{meq NaOH} &= 42.78 \text{ mL} \times 0.100\ N \\ &= 42.78 \text{ meq} \end{aligned}$$

The milliequivalents of HCl used in the titration corresponds to the number of milliequivalents of base that were *not* consumed in the hydrolysis reaction, Equation [3]:

$$\begin{aligned} \text{meq HCl} &= \text{meq excess NaOH} \\ &= 0.105\ N \times 14.29 \text{ mL} \\ &= 1.509 \text{ meq} \end{aligned}$$

The difference between the milliequivalents of base added for the hydrolyses and those which were not consumed equals the number of milliequivalents of base that brought about hydrolysis. From Equation [3] we see that this is exactly equal to the number of milliequivalents of acetylsalicylic acid:

$$4.278 \text{ meq} - 1.509 \text{ meq} = 2.769 \text{ meq}$$

The molecular weight and equivalent weight of acetylsalicylic acid are the same and equal 180.2. Hence, the number of grams of this acid can be found from Equation [4] in Experiment 18 as follows:

$$\text{Grams} = 2.769 \text{ meq} \times \frac{180.2 \text{ g}}{10^3 \text{ meq}}$$

$$= 0.4989 \text{ g acid}$$

Thus

$$\% \text{ purity} = \frac{0.4989 \text{ g}}{0.5130 \text{ g}} \times 100 = 97.25\%$$

PROCEDURE

Weigh to the nearest milligram about 0.5 g of the aspirin you prepared or commercial nonbuffered tablets into a clean, dry 250-mL Erlenmeyer flask. Prepare two burets, one for acid and the other for base, by rinsing them with standard acid and base as described in Experiment 19. After you fill the burets, record the normalities on you report sheet and also record the initial buret readings. Remember, the buret reading can be estimated to ±0.01 mL.

Add about 25 mL of 95% ethyl alcohol (CAUTION: *Ethyl alcohol is flammable*!) that has been cooled to about 15°C to the flask and swirl the flask to dissolve the aspirin. Add 2 drops of phenolphthalein and rapidly titrate the sample with standard 0.1 *N* NaOH to a faint pink end point. Record the volume of NaOH used. This volume of base corresponds to that which is required to neutralize *all* acids present in your sample, that is, impurities as well as the acetylsalicylic acid.

To hydrolyze the aspirin, you will add additional NaOH to the flask from your buret. This may require your refilling the buret with more standard base. Record the initial buret reading. The amount of base to use for the hydrolysis is determined as follows: Add 15 mL to the volume of base required in the previous titration. Add *about* this volume of NaOH to the Erlenmeyer flask from the buret. Record the buret reading; you will know the exact volume of NaOH used for the hydrolysis from these buret readings. Heat the mixture for 15 min in a bath of boiling water in a 600-mL beaker, as shown in Figure 39.1, to hydrolyze the aspirin. (CAUTION: *Remember that ethyl alcohol is flammable*!) Swirl the flask occasionally. Cool the flask to room temperature with cold tap water or an ice bath. If the solution is not pink, add two more drops of phenolphthalein indicator. Record the initial volume of HCl and back titrate the excess base with the standard HCl until the pink color disappears. Record the volume of HCl used. Repeat this procedure with your other two samples.

From these data calculate the grams of acetylsalicylic acid in your aspirin samples, the percentage of aspirin, the average percentage of aspirin, and the standard deviation.

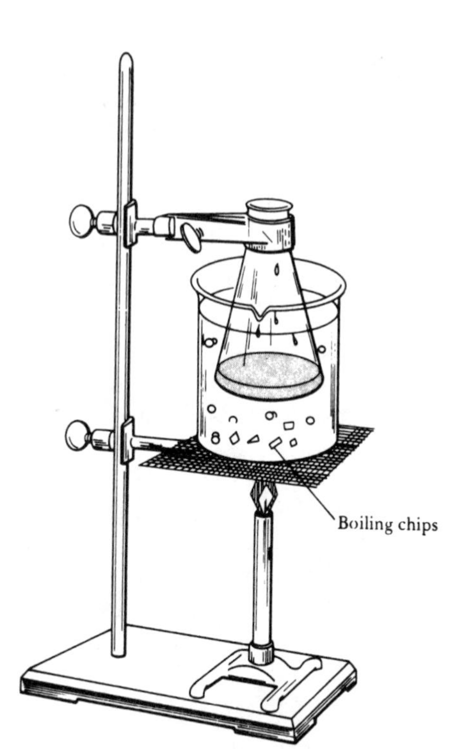

Figure 39.1

REVIEW QUESTIONS

Before beginning this experiment in the laboratory, you should be able to answer the following questions:

1. Define the following: analyte, titration, titrant, standard solution, and standardization.
2. Write balanced equations for the cold neutralization of acetylsalicylic acid. If the mixture is hot, what other reaction may occur?
3. Illustrate the base-promoted hydrolysis of an ester.
4. If 35.3 mL of 0.25 *N* NaOH exactly neutralize 38.5 mL of HCl, what is the normality of the HCl? What is the molarity?
5. How many milliequivalents of NaOH are present in 38.5 mL of 0.125 *N* NaOH? How many grams of NaOH are present?
6. What volume of 0.10 *N* NaOH would be required to exactly neutralize 0.70 g of pure aspirin?
7. Account for the fact that old aspirin may smell like vinegar.
8. What is a buffer?
9. Why might a buffer be added to aspirin?
10. Why would you not use this procedure to analyze buffered aspirin?

NOTES AND CALCULATIONS

Name ____ Desk ____________

Date ____________________ Laboratory Instructor ______________________________

REPORT SHEET FOR EXPERIMENT 39

ANALYSIS OF ASPIRIN

Check one: My preparation ( ); commercial tables ( )

Brand name ____________

Normality of NaOH ____________ *N*

Normality of HCl ____________ *N*

Weight of Aspirin

	Trial 1	Trial 2	Trial 3
Weighing bottle initial weight	____________	____________	____________
Weighing bottle final weight	____________	____________	____________
Weight of aspirin	____________	____________	____________

Volume of NaOH *Required to Neutralize all Acid Material*

	Trial 1	Trial 2	Trial 3
Final reading	____________	____________	____________
Initial reading	____________	____________	____________
Volume of NaOH	____________	____________	____________
Milliequivalents of NaOH	____________	____________	____________

Assuming that the only acid present is acetylsalicylic acid, calculate the grams of acetylsalicylic acid present in the aspirin.

Volume of NaOH *Used in Hydrolysis*

	Trial 1	Trial 2	Trial 3
Final reading	____________	____________	____________
Initial reading	____________	____________	____________
Volume of NaOH	____________	____________	____________
Milliequivalents of NaOH	____________	____________	____________

Volume of HCl *in Back-Titration*

	Trial 1	**Trial 2**	**Trial 3**
Final reading	____________	____________	____________
Initial reading	____________	____________	____________
Volume of HCl	____________	____________	____________
Milliequivalents of HCl	____________	____________	____________

Analysis

	Trial 1	**Trial 2**	**Trial 3**
Grams of acetylsalicylic acid in sample	____________	____________	____________
Percent of aspirin	____________	____________	____________
Average percentage of aspirin	____________	Standard deviation	____________

Show your calculations

From the above data, calculate the number of milliequivalents of acidic impurities in each of your three analyses.

For each of your three analyses, what is the ratio of acetylsalicylic acid to other acid impurities?

Ion-Exchange Resins: Analysis of a Calcium, Magnesium, or Zinc Salt

EXPERIMENT 40

OBJECTIVE

To become acquainted with the technique of ion-exchange chromatography and some of its applications.

APPARATUS AND CHEMICALS

- 50-mL burets (2)
- glass wool
- deionized water
- cation-exchange resin (Dowex 50W-X 8)
- 250-mL Erlenmeyer flasks (3)
- 250-mL beaker
- phenolphthalein indicator solution
- ~0.1 *M* sodium hydroxide (400 mL)
- potassium hydrogen phthalate
- unknown calcium, magnesium, or zinc salt
- buret clamp and ring stand
- polyethylene bottle
- 6 *M* HCl (150 mL)
- pH paper
- 100-mL beaker
- glass rod

DISCUSSION

I. Ion-Exchange Theory

Chromatography is one of the more important and versatile techniques available to separate the components of a mixture. *Chromatography* is a physical means of *separation* in which the components of a mixture become distributed between two different phases. These two phases are commonly called the *stationary phase* and the *mobile phase*. The mobile phase is usually a liquid or a gas, while the stationary phase is usually a solid. If the stationary phase is a polymer which possesses exchangeable ions, such as those illustrated in Figure 40.1, then the chromatographic technique is called *ion-exchange chromatography*. Polymers with exchangeable ionic functional groups have become known as *ion-exchange resins*. These ion-exchange resins are long-chain polymers that contain polar functional groups and have no appreciable water solubility. They come in the form of spherical beads that swell when placed in contact with water, so that the water solution can enter the interior of the bead and come into intimate contact with these functional groups. The polar functional groups are strong electrolytes and are thus completely ionized when in contact with water solutions. Ions of like charge equilibrate between the solution and the resin according to Equations [1] and [2]:

$$R_z^-H^+(s) + M^+(aq) \rightleftharpoons R_z^-M^+(s) + H^+(aq) \quad [1]$$

$$R_z^+OH^-(s) + X^-(aq) \rightleftharpoons R_z^+X^-(s) + OH^-(aq) \quad [2]$$

Figure 40.1

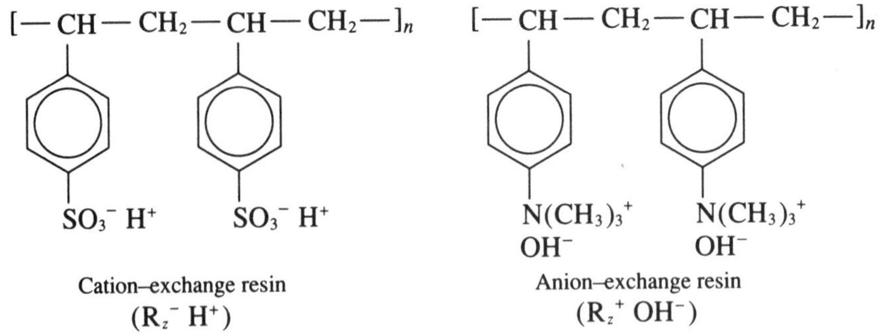

$[-CH-CH_2-CH-CH_2-]_n$ $[-CH-CH_2-CH-CH_2-]_n$

$SO_3^-\ H^+$ $SO_3^-\ H^+$ $N(CH_3)_3^+\ OH^-$ $N(CH_3)_3^+\ OH^-$

Cation–exchange resin
$(R_z^-\ H^+)$

Anion–exchange resin
$(R_z^+\ OH^-)$

where R_z symbolizes the resin. A resin that behaves according to Equation [1] is called a cation-exchange resin, and one that behaves according to Equation [2] is called an anion-exchange resin. In general, the positions of these equilibria are influenced both by the nature and the concentrations of the ions in solution. Ions with a greater positive charge and a smaller size tend to have a greater affinity for the cation-exchange resin than do larger ions with smaller charges. As shown in Equation [1], one H^+ ion is liberated for each M^+ cation. Two H^+ ions would be liberated for each M^{2+} cation, three H^+ ions for each M^{3+} cation, and so forth. Thus the extent of exchange of cations with H^+ on a resin would be expected to decrease in the orders: $Th^{4+} > Al^{3+} > Ca^{2+} > Na^+$ and $Cs^+ > Rb^+ > K^+ > Na^+ > Li^+$. In the latter series there appears to be a discrepancy because we expect the Cs^+ ion to be larger than the Li^+ ion and thus to have a smaller charge density. However, the Li^+ ion is hydrated and is tightly bound to six H_2O molecules, whereas the Cs^+ ion is bound to only four H_2O molecules, making the $Li(H_2O)_6^+$ ion larger than the $Cs(H_2O)_4^+$ ion.

In the same way, when small, highly negative anions come into contact with an anion-exchange resin, they liberate one OH^- for each negative charge. Thus we would expect the position of equilibrium [2] to be influenced by the charge density of the anion also. As a consequence, since anions are not so tightly bound to H_2O molecules as are cations, we expect the extent of exchange of anions in solutions with OH^- ions on the resin to decrease in the orders: $F^- > Cl^- > Br^- > I^-$ and $PO_4^{3-} > SO_4^{2-} > ClO_4^-$, for example. Because ion-exchange resins possess the ability to remove cations and anions from water and replace them with H^+ or OH^- ions, respectively, ion-exchange resins are commonly used to *deionize* water and to soften water. You may have a water softener in your home; if you do, you can see that it is just a large container of cation-exchange resin.

II. Hard Water

The term *hard water* has its origin in the fact that mineral impurities such as Ca^{2+}, Mg^{2+}, and Fe^{3+}, which are frequently present in natural water, react with soap to form gummy precipitates. They also react with CO_3^{2-} and SO_4^{2-} in the water to leave what is called a *boiler scale* ($CaCO_3$, $MgCO_3$, or $CaSO_4$) on walls of the vessels in which hard water is boiled. Ordinary soap is a water-soluble sodium (solid soaps) or potassium (liquid soaps) salt of a long-chain organic acid, RCO_2H, such as sodium stearate, $C_{17}H_{35}CO_2Na$. Sodium stearate reacts with Mg^{2+} or Ca^{2+} ions to form a sticky, insoluble precipitate according to Equation [3].

$$2C_{17}H_{35}CO_2^-(aq) + Ca^{2+}(aq) \rightleftharpoons Ca(C_{17}H_{35}CO_2)_2(s) \qquad [3]$$

This results in waste of the soap and the formation of an undesirable film on items that the soap is used to clean. You may have observed the formation of such a precipitate and the fact that soap does not form much of a foam or lather in hard water.

Suppose, then, that a solution containing calcium chloride, $CaCl_2$, was slowly passed through a bed of cation-exchange resin contained in a column. In this case equilibrium [4] would occur:

$$2R_z^-H^+(s) + Ca^{2+}(aq) + 2Cl^-(aq) \rightleftharpoons [(R_z^-)_2Ca^{2+}](s) + 2H^+(aq) + 2Cl^-(aq) \qquad [4]$$

The *eluate*, or solution, coming out of the column (see Figure 40.2) would contain H^+ and Cl^- ions, and the Ca^{2+} ions would be retained by the column

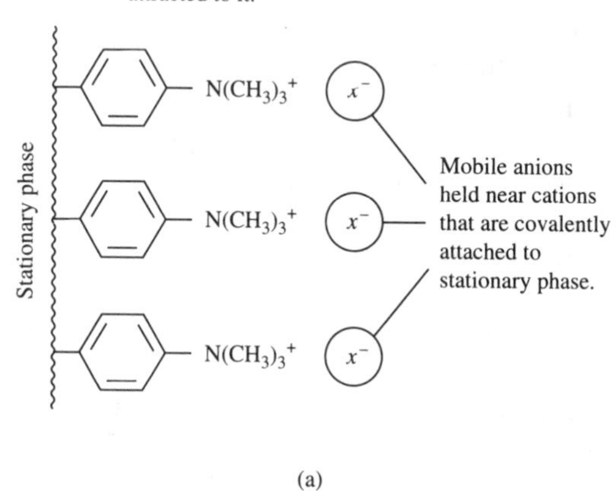

(a)

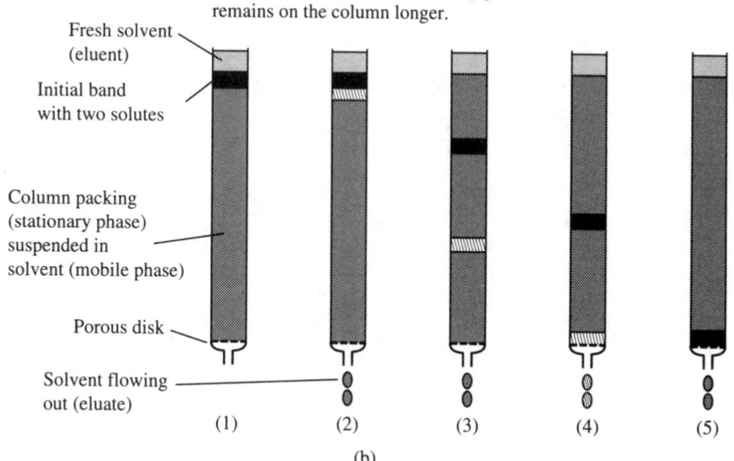

(b)

Figure 40.2

because they have a greater affinity for the resin than do the H^+ ions. If this eluate were then passed through an anion-exchange resin in the OH^- form, then equilibrium [5] would occur on the column:

$$2H^+(aq) + 2Cl^-(aq) + 2R_z^+OH^-(s) \rightleftharpoons 2R_z^+Cl^-(s) + 2H^+(aq) + 2OH^-(aq) \quad [5]$$

Thus, passing a $CaCl_2$ solution successively through a cation- and then an anion-exchange resin would produce a solution containing neither Ca^{2+} nor Cl^- ions in the eluate. Since OH^- ions react with H^+ ions according to Equation [6], the eluate would be pure water:

$$H^+(aq) + OH^-(aq) \longrightarrow H_2O(l) \quad [6]$$

Water treated in this way is said to be *deionized*.

In order to carry out such a process in actuality, it is necessary to have a large volume of resin with respect to the volume of solution, because ion-exchange resins typically have an exchange capacity of only about 5 meq/g on a dry basis. Also, it is necessary to have a long contact time between the solution and the resin so that all the ionic sites on the resin may contact the ions in solution. Thus we generally use a column such as illustrated in Figure 40.2 and slowly pass through it the solution to be exchanged.

Ion-exchange resins are useful not only for separation and purification purposes, but also for analytical purposes. Suppose you passed a solution containing a known mass of a pure soluble calcium salt (which contained an unknown amount of calcium) through a cation-exchange resin. Equilibrium [7] would occur on the column:

$$2R_z^-H^+(s) + Ca^{2+}(aq) \rightleftharpoons [(R_z^-)_2Ca^{2+}](s) + 2H^+(aq) \quad [7]$$

The liberated H^+ in the eluate from the column could be determined by a simple acid-base titration with a standard NaOH solution. Assuming that all the Ca^{2+} was captured by the column, the calcium content of the sample can be easily calculated.

EXAMPLE 40.1

A 0.2940-g sample of a soluble calcium salt was passed through a cation-exchange column in the H^+ form and the H^+ in the eluate required 26.22 mL of 0.1025 *M* NaOH to reach a phenolphthalein end point. Calculate the percentage calcium in the original sample.

Solution: The acid-base reaction is $H^+(aq) + OH^-(aq) \longrightarrow H_2O(l)$.

$$\text{Moles } H^+ = (26.22 \text{ mL})\left(\frac{1 \text{ L}}{1000 \text{ mL}}\right)\left(\frac{0.1025 \text{ mol}}{\text{L}}\right)$$

$$= 2.688 \times 10^{-3} \text{ mol}$$

$$\text{Moles } Ca^{2+} = \tfrac{1}{2} \text{ mol } H^+ \text{ (see Equation [7])}$$

$$= \frac{2.688 \times 10^{-3} \text{ mol}}{2}$$

$$= 1.344 \times 10^{-3} \text{ mol}$$

$$\text{Mass Ca}^{2+} = (1.344 \times 10^{-3}\ \text{mol})(40.08\ \text{g/mol})$$
$$= 53.86 \times 10^{-3}\ \text{g}$$
$$\%\ \text{Ca}^{2+} = \left(\frac{53.86 \times 10^{-3}\ \text{g Ca}^{2+}}{0.2940\ \text{g sample}}\right)(100)$$
$$= 18.31\ \%\ \text{Ca}^{2+}$$

PROCEDURE

Record all data directly onto the report sheets.

In this experiment you are to determine the percent calcium, magnesium, or zinc in an unknown by performing a cation exchange followed by an acid-base titration of the eluate. Your instructor will tell you which cation you are determining. You will first standardize your base solution (NaOH, the titrant).

A. Standardization of NaOH

Place about 400 mL of approximately 0.1 *M* NaOH solution into a polyethylene bottle. Standardize the NaOH solution as described in Experiment 19 by titrating against KHP using phenolphthalein indicator. Remember, the buret reading can be estimated to ±0.01 mL.

B. Preparation of the Ion-Exchange Column

Weigh out about 10 g of the cation ion-exchange resin to the nearest gram. The resin must be in the acid form. In order to ensure that it is, slurry the resin in a 250-mL Erlenmeyer flask with a mixture of 50 mL of 6 *M* HCl and 100 mL of deionized water. Allow the slurry to stand, with occasional swirling, for 15 min. Then decant the acid solution and wash the resin with five 50-mL portions of deionized water, decanting the wash liquid after each washing. Finally add 50 mL of deionized water to the resin.

Using a long wire or glass rod, push a small plug of glass wool to the bottom of a 50-mL buret. Then put 5 mL of deionized water on top of the glass wool in the buret, clamp the buret vertically in a buret clamp on a ring stand, and transfer the resin slurry to the buret, with the aid of more deionized water, if necessary. Drain off excess water occasionally from the bottom of the buret, but never allow the liquid level in the buret to fall below the level of the resin, because this usually causes channels to develop in the resin column and greatly reduces its efficiency. Wash the resin by allowing deionized water to flow through it until the pH of the effluent is the same as the pH of the deionized water (use pH paper to check the pH of the deionized water and the eluate). Then drain water from the buret until the water level is just barely above the top of the resin.

Obtain a sample of a soluble calcium, magnesium, or zinc salt from your laboratory instructor. Weigh out with the aid of weighing paper approximately 0.25 to 0.30 g of this salt to the nearest milligram. Transfer the weighed salt to a 100-mL beaker and dissolve it in about 25 mL of deionized water. Transfer the salt solution to the ion-exchange column, rinse the beaker twice with 5-mL portions of deionized water, and add the rinsings to the ion-exchange column. As the liquid level begins to reach the top of the column, drain liquid (the eluate) from the bottom of the column, and collect the eluate in a 250-mL Erlenmeyer flask. The buret stopcock should be adjusted so that the eluate flow rate is about 2-3 mL/min. Continue to drain solution from the column while periodically adding deionized water to the top of the column until you have collected about 125 mL of solution in the Erlenmeyer flask. At all times maintain the liquid level in the column above the level of the resin bed. When about 125 mL of eluate has been collected, take a small drop of eluate on the end of a glass rod as it emerges from the buret and test it with pH paper to ensure that the pH of the eluate is the same as the pH of the deionized-water wash. Continue to elute the column with deionized water until the two pH values are the same. Close the buret and set the flask and eluate aside and label the flask "Trial 1."

Repeat the above procedure with two more approximately 0.3-g samples of your unknown and label the two Erlenmeyer flasks "Trial 2" and "Trial 3."

Add 2 to 3 drops of phenolphthalein indicator solution to each of the three Erlenmeyer flasks and titrate each of the eluate samples with your standardized NaOH solution to a faint pink end point.

Remove the resin from the buret, regenerate it by slurrying it with 50 mL of 6 *M* HCl, and wash five times with deionized water as described for the preparation of the column. Return the regenerated resin to the resin stock bottle.

From your data, calculate the number of moles of cation in your sample and the percentage of cation in your sample.

REVIEW QUESTIONS

Before beginning this experiment in the laboratory, you should be able to answer the following questions:

1. Describe how an ion-exchange column works.
2. Define stationary and mobile phases.
3. Define eluant and eluate.
4. Why do smaller, highly charged ions have a greater affinity for an ion-exchange column than do larger ions with smaller charges?
5. What are hard water and deionized water?
6. Would you expect Na^+ or Mg^{2+} to be more slowly eluted from an ion-exchange column?
7. A 0.400-g sample of pure $ZnCl \cdot nH_2O$ was placed on an ion-exchange column in the H^+ form. The H^+ released in the eluate required 45.26 mL of 0.1024 *M* NaOH to reach a phenolphthalein end point. Calculate the percent zinc in the sample.
8. Calculate the value of n for the data in question 7.
9. If a 0.153-g sample of NaCl were placed on an ion-exchange column in the H^+ form and the eluate titrated with 0.1321 *M* NaOH, what volume of the NaOH solution would be required to reach a phenolphthalein end point?

Name ____________________ Desk __________

Date __________ Laboratory Instructor ____________________

Unknown cation ____________________ Unknown no. __________

REPORT SHEET FOR EXPERIMENT 40

ION-EXCHANGE RESINS: ANALYSIS OF A CALCIUM, MAGNESIUM, OR ZINC SALT

A. Standardization of NaOH

	Trial 1	Trial 2	Trial 3
Weight of bottle + KHP	______	______	______
Weight of bottle	______	______	______
Weight of KHP used	______	______	______
Final buret reading	______	______	______
Initial buret reading	______	______	______
mL of NaOH Used	______	______	______
Molarity of NaOH	______	______	______
Average molarity	______		
Standard Deviation	______		

(show calculations and standard deviation)

B. Determination of Cation

	Trial 1	Trial 2	Trial 3
Weight of sample + paper	______	______	______
Weight of paper	______	______	______
Weight of sample	______	______	______
Final buret reading	______	______	______
Initial buret reading	______	______	______
mL of NaOH used	______	______	______
Moles H^+	______	______	______
Moles M^{2+}	______	______	______

Mass M^{2+} __________ __________ __________

Percent M^{2+} __________ __________ __________

Average __________ Standard Deviation __________

(Show calculations)

QUESTIONS

1. Assuming that your salt was a hydrated form of a sulfate, $MSO_4 \cdot nH_2O$, calculate n. What values are quoted in handbooks for common hydrates of your cation?

2. A 0.300-g sample of a compound of empirical formula $CrCl_3 \cdot 6H_2O$ was dissolved in water and passed through a cation-exchange column in the H^+ form. The eluate required 33.45 mL of 0.100 M NaOH to reach a phenolphthalein end point. Explain this result in terms of each of the following.
   (a) The charge on Cr in the compound

(b) The structure of the compound, that is, $[Cr(H_2O)_6]^{3+}Cl_3$, $[Cr(H_2O)_5Cl]^{2+}Cl_2$, $[Cr(H_2O)_4Cl_2]^{+}Cl$, or $[Cr(H_2O)_3Cl_3]$

(c) The percentage of Cr in the compound

Appendices

APPENDIX A

Chemical Arithmetic

Elementary mathematics is frequently used in the study of general chemistry. Exponential arithmetic, significant figures, and logarithms are of particular importance and widespread application in these calculations. These are discussed in turn in this appendix.

EXPONENTIAL ARITHMETIC

Many quantities that we measure in chemistry are either very large or very small. Because of this, it becomes convenient to express numbers in scientific notation. Scientific notation is a way of expressing all numbers as a product. The two members of the product are, first, a number between 1 and 10 and, second, the power of 10 that places the decimal point. This second number is called the *exponential term* and is written as 10 with a right-hand superscript (the exponent). The exponent denotes the power of 10, that is, how many times 10 is multiplied by 10. Some examples of the exponential method of expressing numbers are given below:

$$1000 = 1 \times 10^3 \qquad 0.1 = 1 \times 10^{-1}$$

$$100 = 1 \times 10^2 \qquad 0.01 = 1 \times 10^{-2}$$

$$10 = 1 \times 10^1 \qquad 0.001 = 1 \times 10^{-3}$$

$$1 = 1 \times 10^0 \qquad 2386 = 2.386 \times 1000 = 2.386 \times 10^3$$

$$0.123 = 1.23 \times 0.1 = 1.23 \times 10^{-1}$$

As should be evident from the above examples, the power (exponent) of 10 is equal to the number of places the decimal is shifted to give the digit number. The efficacy of using exponential numbers becomes readily apparent when one compares writing 1,230,000,000 with writing 1.23×10^9, or 0.000,000,000,36 with 3.6×10^{-10}.

Once numbers have been expressed as exponentials, the question arises: How does one perform mathematical operations such as addition or multiplication with exponentials? The answers to this question are illustrated in the following examples.

Addition of Numbers in Scientific Notation

Convert all the numbers to the same power of 10 and add the digit terms of the numbers.

EXAMPLE

$5.0 \times 10^{-2} + 3 \times 10^{-3}$

$$\begin{array}{r} 50 \times 10^{-3} \\ +\ \ 3 \times 10^{-3} \\ \hline 53 \times 10^{-3} \end{array} = 5.3 \times 10^{-2} \qquad \text{or} \qquad \begin{array}{r} 5.0 \times 10^{-2} \\ +\ 0.3 \times 10^{-2} \\ \hline 5.3 \times 10^{-2} \end{array}$$

Subtraction of Numbers in Scientific Notation

Convert all the numbers to the same power of 10 and subtract the digit terms of the numbers.

EXAMPLE

$5.0 \times 10^{-6} - 4 \times 10^{-7}$

$$\begin{array}{r} 5.0 \times 10^{-6} \\ -\ 0.4 \times 10^{-6} \\ \hline 4.6 \times 10^{-6} \end{array} \qquad \text{or} \qquad \begin{array}{r} 50 \times 10^{-7} \\ -\ \ \ 4 \times 10^{-7} \\ \hline 46 \times 10^{-7} \end{array} = 4.6 \times 10^{-6}$$

Multiplication of Numbers in Scientific Notation

Multiply the digit terms in the usual way and add the exponents of the exponential terms (that is, $10^a \times 10^b = 10^{a+b}$).

EXAMPLES

$$(4.2 \times 10^{-8})(2 \times 10^{3}) = 8.4 \times 10^{(-8+3)} = 8.4 \times 10^{-5}$$
$$(4.2 \times 10^{-8})(2 \times 10^{-3}) = 8.4 \times 10^{-11}$$
$$(4.2 \times 10^{8})(2 \times 10^{-3}) = 8.4 \times 10^{(8+(-3))} = 8.4 \times 10^{5}$$

Division of Numbers in Scientific Notation

Divide the digit terms of the numerator by the digit term of the denominator and subtract the exponents of the exponential terms (that is, $10^a/10^b = 10^{a-b}$).

EXAMPLES

$$\frac{4.2 \times 10^{-8}}{2 \times 10^{3}} = 2.1 \times 10^{-11}$$
$$\frac{4.2 \times 10^{-8}}{2 \times 10^{-3}} = 2.1 \times 10^{-5}$$
$$\frac{4.2 \times 10^{8}}{2 \times 10^{-3}} = 2.1 \times 10^{(8-(-3))} = 2.1 \times 10^{11}$$

Squaring of Exponentials

Square the digit term in the usual way and multiply the exponent of the exponential term by 2 (that is, $(10^a)^2 = 10^{2a}$).

EXAMPLES

$$(4 \times 10^{-2})^2 = 16 \times 10^{-2\times 2} = 16 \times 10^{-4} = 1.6 \times 10^{-3}$$
$$(5 \times 10^{4})^2 = 25 \times 10^{4\times 2} = 25 \times 10^{8} = 2.5 \times 10^{9}$$
$$(1.2 \times 10^{3})^2 = 1.44 \times 10^{3\times 2} = 1.44 \times 10^{6}$$

Raising Numbers in Scientific Notation to a General Power

Raise the digit term to the power in the usual way and multiply the exponent of the exponential term by the power (that is, $(10^a)^b = 10^{ab}$).

EXAMPLES

$$(2 \times 10^{-2})^3 = 8 \times 10^{-2\times 3} = 8 \times 10^{-6}$$
$$(1 \times 10^{3})^5 = 1 \times 10^{3\times 5} = 1 \times 10^{15}$$

Extraction of Square Roots of Numbers in Scientific Notation

Decrease or increase the exponential term so that the power of 10 is evenly divisible by 2. Extract the square root of the digit term by inspection, by logarithms, or by calculator and divide the exponential term by 2 (that is, $\sqrt{10^a} = 10^{a/2}$).

EXAMPLE

Find the square root of 1.6×10^{-7}.

$$1.6 \times 10^{-7} = 16 \times 10^{-8}$$
$$\sqrt{16 \times 10^{-8}} = \sqrt{16} \times \sqrt{10^{-8}} = 4 \times 10^{-8/2} = 4 \times 10^{-4}$$

Combined Operations

Any combination of the above operations is performed by doing each operation individually and combining the results.

EXAMPLE

Solve the equation $a/(2.5 \times 10^{-3}) = (7.4 \times 10^{8})/(3.9 \times 10^{-6})$.

$$a = \frac{(2.5 \times 10^{-3})(7.4 \times 10^{8})}{(3.9 \times 10^{-6})} = 4.7 \times 10^{11}$$

Because the operations of multiplication and division are commutative, the order of operations is immaterial.

$$a = \frac{(2.5 \times 10^{-3})(7.4 \times 10^{8})}{(3.9 \times 10^{-6})} = \frac{1.85 \times 10^{6}}{3.9 \times 10^{-6}} = 4.7 \times 10^{11}$$

SIGNIFICANT FIGURES

Many operations in the chemistry laboratory involve measurements of some kind. Examples are weighing a compound or measuring the volume of a liquid. It is important to record these data properly so that the number recorded cor-

Figure A.1

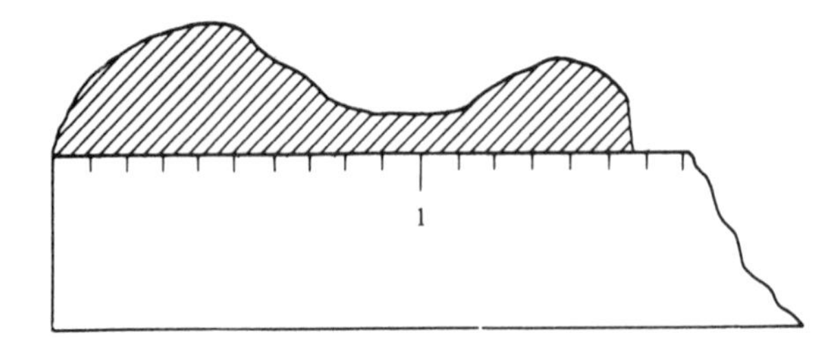

rectly represents the accuracy of the measurement. The following brief discussion is intended to serve as a guide for calculating and reporting numerical results.

Every observed measurement that is made is really an approximation. For example, the length of the object in Figure A.1 is between 1.5 and 1.6 units. Its length is seen to be approximately 1.56 units. There is uncertainty in the last digit, 6; it is estimated. Consider the recorded weight of an object as 2.6 grams. This means that the object was weighed to the nearest tenth (0.1) of a gram and that its exact weight is between 2.55 g and 2.65 g. In recording a result, it is the last digit that represents a degree of uncertainty; for example, in 2.6 g it is the number 6 that represents some degree of uncertainty. We say the number 2.6 g contains two significant figures, the numbers 2 and 6 being the significant figures. If the recorded weight of the object were $2.63\underline{3}$ g, there would be four significant figures (2, 6, 3, and 3), and this would mean that the object was weighed to the nearest thousandth (0.001 g) of a gram. Thus it is the last (underscored) 3 which has been estimated. *Significant figures* refer to those digits we know with certainty plus the first doubtful or estimated digit.

Zeros

A zero may or may not represent a significant figure. The number 5.00 represents three significant figures, the zeros being significant; this result implies that the measurement was made to the nearest one-hundredth part. Similarly, the zeros in 5.034, 7.206, 9.310, 10.20, and 2.001 are all regarded as significant figures. Each of the five preceding numbers contains four significant figures.

It is a common error to omit the zeros that indicate the reliability of a number. An object that weighs 2 g when weighed to the nearest 0.01 g on a triple-beam balance should be recorded as 2.00 g, *not* 2 g.

When the zero is used to locate a decimal point, as in 0.023, the zero is *not* a significant figure. There are only two significant figures in each of the following: 0.027, 0.00078, and 0.0010. These numbers could be written 2.7×10^{-2}, 7.8×10^{-4}, and 1.0×10^{-3}.

Exact Numbers

Some numbers by their very nature are exact numbers. When you say you are holding three test tubes in your hand, you mean exactly three test tubes, not two or four. Thus if this figure 3 is used in a calculation, you may regard it as containing as many significant figures as desired: 3.0000. . . .

Another example of exact numbers is a number used to relate quantities within the same system of units:

$$1 \text{ ft} = 12 \text{ in.}$$

$$1 \text{ cm} = 10 \text{ mm}$$

This quality of exactness follows from the definition of the equality. Note, however, that the relationship between units in two different systems is not exact. For example,

$$1 \text{ kg} = 2.205 \text{ lb}$$

One kilogram does not exactly equal 2.205 lb, because the two systems have been defined independently of one another. Thus 1 kg is approximately equal to 2.205 lb.

Rounding off Numbers

A number may be rounded off to the desired number of significant figures by dropping one or more digits at the end of the number. The following rules should be observed when rounding off a number:

1. When the first digit dropped is less than 5, the last digit retained should remain unchanged. Examples: 17.373 becomes 17.37; 1.5324 becomes 1.53 if rounded off to three significant figures and 1.5 if rounded off to two significant figures.
2. When the first digit dropped is equal to or greater than 5, the last digit retained is increased by 1. Examples: 19.765 becomes 19.77; 8.1525 becomes 8.153 if rounded off to four significant figures and 8.2 if rounded off to two significant figures.

Significant Figures and Calculations

In any calculation in which experimental results are used, the final result should contain only as many significant figures as are justified by the experiment. Thus the least precise measurement dictates the number of significant figures that should be present in the final answer.

1. Addition and subtraction: In addition and subtraction retain only as many decimal places in the result as there are in that component with the least number of decimal places. For example,

$$\begin{array}{r} 21.1 \\ 2.035 \\ \underline{6.12} \\ 29.255 \end{array} \text{ becomes } 29.3$$

2. Multiplication and division: In multiplication and division the answer should contain only as many significant figures as are contained in the factor with the least number of significant figures. For example,

$$21.71 \times 0.029 \times 89.2 = 56.159428$$

The number 0.029 contains only two significant figures; therefore, according to the rule above, the answer should be rounded off to contain two significant figures: 56.

Problems

1. How many significant figures are present in each of the following numbers: (a) 454 g; (b) 22.01 cm; (c) 18.00 mL; (d) 0.020 g?
2. Add each of the following: (a) 311 cm + 10.1 cm + 1.21 cm; (b) 18.00 mL + 1.2 mL + 0.71 mL; (c) 3.286 ft + 7.01 ft + 0.001 ft.
3. Multiply each of the following: (a) 3.70×1.11; (b) 3.70×2.2; (c) 3.70×0.022.
4. Divide each of the following: (a) $\frac{98.98}{4.90}$; (b) $\frac{75.24}{1.1}$; (c) $\frac{37.1}{2.5312}$

Answer: 1. (a) 3; (b) 4; (c) 4; (d) 2 2. (a) 322 cm; (b) 19.9 mL; (c) 10.30 ft. 3. (a) 4.11; (b) 8.1 (c) 0.081 4. (a) 20.2; (b) 68; (c) 14.7

THE USE OF LOGARITHMS AND EXPONENTIAL NUMBERS

The *common logarithm* of a number is the power to which the number 10 must be raised to equal that number. For example, the logarithm of 100 is 2 because the number 10 must be raised to the second power to be equal to 100, that is, $\log_{10} 100 = 2$, or $10^2 = 100$. Additional examples are given in Table A.1.

Table A.1 Some Examples of Common Logarithms

Number	Number Expressed Exponentially	Logarithm
10,000	10^4	4
1,000	10^3	3
10	10^1	1
1	10^0	0
0.1	10^{-1}	−1
0.01	10^{-2}	−2
0.001	10^{-3}	−3
0.0001	10^{-4}	−4

What is the logarithm of 60? Because 60 lies between 10 and 100, which have logarithms of 1 and 2, respectively, the logarithm of 60 must lie between 1 and 2. The logarithm of 60 is 1.78, that is, $60 = 10^{1.78}$, or $\log_{10} 60 = 1.78$.

Every logarithm is made up of two parts, called the *characteristic* and the *mantissa*. The characteristic is that part of the logarithm which lies to the left of the decimal point; thus the characteristic of the logarithm of 60 is 1. The mantissa is that part of the logarithm which lies to the right of the decimal point; thus the mantissa of the logarithm of 60 is .78. The characteristic of the logarithm of a number greater than 1 is 1 less than the number of digits to the left of the decimal point of the number, as shown by the following table:

Number	Characteristic	Number	Characteristic
60	1	2.340	0
600	2	23.40	1
6,000	3	234.0	2
52,840	4	2340.0	3

The mantissa of the logarithm of a number is found in Table A.2. Its value is independent of the position of the decimal point. Thus the logarithms of 2.340, 23.40, 234.0, and 2340.0 all have the same mantissa, so the logarithms of these numbers are 0.3692, 1.3692, 2.3692, and 3.3692, respectively.

The meaning of the mantissa and characteristic can be better understood from a consideration of their relationship to exponential numbers. For example, 2340 may be written as 2.340×10^3. The logarithm of (2.340×10^3) = the logarithm of 2.340 plus the logarithm of 10^3. Recall that $(10^a) \times (10^b) = 10^{a+b}$. To find the logarithm of a product, $\log 10^{a+b} = a + b = \log 10^a + \log 10^b$. The logarithm of 2.340 is 0.3692 and the logarithm of 10^3 is 3. Thus the logarithm of 2340 = 3 + 0.3692 = 3.3692.

The logarithm of a number less than 1 has a negative value, and a convenient method of obtaining the logarithm of such a number is given below.

EXAMPLE

Obtain the logarithm of 0.00234. When expressed exponentially, $0.00234 = 2.34 \times 10^{-3}$. The logarithm of 2.34×10^{-3} = the logarithm of 2.34 plus the logarithm of 10^{-3}. The logarithm of 2.34 is 0.369, and the logarithm of 10^{-3} is −3. Thus the logarithm of 0.00234 = 0.369 + (−3) = 2.631.

To multiply two numbers, we add the logarithms of the numbers.

EXAMPLE

Find 412 × 353.

	Logarithm of 412	= 2.615
+	Logarithm of 353	= 2.548
	Logarithm of product	= 5.163

The number whose logarithm is 5.163 is called the *antilogarithm* of 5.163, and it is 1.45×10^5, or 145,000. Thus 412 × 353 = 145,000, or $(4.12 \times 10^2)(3.53 \times 10^2) = 14.5 \times 10^4 = 1.45 \times 10^5$.

To divide two numbers we subtract the logarithms of the numbers.

EXAMPLE

Find 412/353.

	Logarithm of 412	= 2.615
−	Logarithm of 353	= 2.548
	Logarithm of quotient	= 0.067

Table A.2 Four-place Common Logarithms

Natural numbers	0	1	2	3	4	5	6	7	8	9	Proportional Parts								
											1	2	3	4	5	6	7	8	9
10	0000	0043	0086	0128	0170	0212	0253	0294	0334	0374	4	8	12	17	21	25	29	33	37
11	0414	0453	0492	0531	0569	0607	0645	0682	0719	0755	4	8	11	15	19	23	26	30	34
12	0792	0828	0864	0899	0934	0969	1004	1038	1072	1106	3	7	10	14	17	21	24	28	31
13	1139	1173	1206	1239	1271	1303	1335	1367	1399	1430	3	6	10	13	16	19	23	26	29
14	1461	1492	1523	1553	1584	1614	1644	1673	1703	1732	3	6	9	12	15	18	21	24	27
15	1761	1790	1818	1847	1875	1903	1931	1959	1987	2014	3	6	8	11	14	17	20	22	25
16	2041	2068	2095	2122	2148	2175	2201	2227	2253	2279	3	5	8	11	13	16	18	21	24
17	2304	2330	2355	2380	2405	2430	2455	2480	2504	2529	2	5	7	10	12	15	17	20	22
18	2553	2577	2601	2625	2648	2672	2695	2718	2742	2765	2	5	7	9	12	14	16	19	21
19	2788	2810	2833	2856	2878	2900	2923	2945	2967	2989	2	4	7	9	11	13	16	18	20
20	3010	3032	3054	3075	3096	3118	3139	3160	3181	3201	2	4	6	8	11	13	15	17	19
21	3222	3243	3263	3284	3304	3324	3345	3365	3385	3404	2	4	6	8	10	12	14	16	18
22	3424	3444	3464	3483	3502	3522	3541	3560	3579	3598	2	4	6	8	10	12	14	15	17
23	3617	3636	3655	3674	3692	3711	3729	3747	3766	3784	2	4	6	7	9	11	13	15	17
24	3802	3820	3838	3856	3874	3892	3909	3927	3945	3962	2	4	5	7	9	11	12	14	16
25	3979	3997	4014	4031	4048	4065	4082	4099	4116	4133	2	3	5	7	9	10	12	14	15
26	4150	4166	4183	4200	4216	4232	4249	4265	4281	4298	2	3	5	7	8	10	11	13	15
27	4314	4330	4346	4362	4378	4393	4409	4425	4440	4456	2	3	5	6	8	9	11	13	14
28	4472	4487	4502	4518	4533	4548	4564	4579	4594	4609	2	3	5	6	8	9	11	12	14
29	4624	4639	4654	4669	4683	4698	4713	4728	4742	4757	1	3	4	6	7	9	10	12	13
30	4771	4786	4800	4814	4829	4843	4857	4871	4886	4900	1	3	4	6	7	9	10	11	13
31	4914	4928	4942	4955	4969	4983	4997	5011	5024	5038	1	3	4	6	7	8	10	11	12
32	5051	5065	5079	5092	5105	5119	5132	5145	5159	5172	1	3	4	5	7	8	9	11	12
33	5185	5198	5211	5224	5237	5250	5263	5276	5289	5302	1	3	4	5	6	8	9	10	12
34	5315	5328	5340	5353	5366	5378	5391	5403	5416	5428	1	3	4	5	6	8	9	10	11
35	5441	5453	5465	5478	5490	5502	5514	5527	5539	5551	1	2	4	5	6	7	9	10	11
36	5563	5575	5587	5599	5611	5623	5635	5647	5658	5670	1	2	4	5	6	7	8	10	11
37	5682	5694	5705	5717	5729	5740	5752	5763	5775	5786	1	2	3	5	6	7	8	9	10
38	5798	5809	5821	5832	5843	5855	5866	5877	5888	5899	1	2	3	5	6	7	8	9	10
39	5911	5922	5933	5944	5955	5966	5977	5988	5999	6010	1	2	3	4	5	7	8	9	10
40	6021	6031	6042	6053	6064	6075	6085	6096	6107	6117	1	2	3	4	5	6	8	9	10
41	6128	6138	6149	6160	6170	6180	6191	6201	6212	6222	1	2	3	4	5	6	7	8	9
42	6232	6243	6253	6263	6274	6284	6294	6304	6314	6325	1	2	3	4	5	6	7	8	9
43	6335	6345	6355	6365	6375	6385	6395	6405	6415	6425	1	2	3	4	5	6	7	8	9
44	6435	6444	6454	6464	6474	6484	6493	6503	6513	6522	1	2	3	4	5	6	7	8	9
45	6532	6542	6551	6561	6571	6580	6590	6599	6609	6618	1	2	3	4	5	6	7	8	9
46	6628	6637	6646	6656	6665	6675	6684	6693	6702	6712	1	2	3	4	5	6	7	7	8
47	6721	6730	6739	6749	6758	6767	6776	6785	6794	6803	1	2	3	4	5	5	6	7	8
48	6812	6821	6830	6839	6848	6857	6866	6875	6884	6893	1	2	3	4	4	5	6	7	8
49	6902	6911	6920	6928	6937	6946	6955	6964	6972	6981	1	2	3	4	4	5	6	7	8
50	6990	6998	7007	7016	7024	7033	7042	7050	7059	7067	1	2	3	3	4	5	6	7	8
51	7076	7084	7093	7101	7110	7118	7126	7135	7143	7152	1	2	3	3	4	5	6	7	8
52	7160	7168	7177	7185	7193	7202	7210	7218	7226	7235	1	2	2	3	4	5	6	7	7
53	7243	7251	7259	7267	7275	7284	7292	7300	7308	7316	1	2	2	3	4	5	6	6	7

Natural numbers	0	1	2	3	4	5	6	7	8	9	Proportional Parts								
											1	2	3	4	5	6	7	8	9
55	7404	7412	7419	7427	7435	7443	7451	7459	7466	7474	1	2	2	3	4	5	5	6	7
56	7482	7490	7497	7505	7513	7520	7528	7536	7543	7551	1	2	2	3	4	5	5	6	7
57	7559	7566	7574	7582	7589	7597	7604	7612	7619	7627	1	2	2	3	4	5	5	6	7
58	7634	7642	7649	7657	7664	7672	7679	7686	7694	7701	1	1	2	3	4	4	5	6	7
59	7709	7716	7723	7731	7738	7745	7752	7760	7767	7774	1	1	2	3	4	4	5	6	7
60	7782	7789	7796	7803	7810	7818	7825	7832	7839	7846	1	1	2	3	4	4	5	6	6
61	7853	7860	7868	7875	7882	7889	7896	7903	7910	7917	1	1	2	3	4	4	5	6	6
62	7924	7931	7938	7945	7952	7959	7966	7973	7980	7987	1	1	2	3	3	4	5	6	6
63	7993	8000	8007	8014	8021	8028	8035	8041	8048	8055	1	1	2	3	3	4	5	5	6
64	8062	8069	8075	8082	8089	8096	8102	8109	8116	8122	1	1	2	3	3	4	5	5	6
65	8129	8136	8142	8149	8156	8162	8169	8176	8182	8189	1	1	2	3	3	4	5	5	6
66	8195	8202	8209	8215	8222	8228	8235	8241	8248	8254	1	1	2	3	3	4	5	5	6
67	8261	8267	8274	8280	8287	8293	8299	8306	8312	8319	1	1	2	3	3	4	5	5	6
68	8325	8331	8338	8344	8351	8357	8363	8370	8376	8382	1	1	2	3	3	4	4	5	6
69	8388	8395	8401	8407	8414	8420	8426	8432	8439	8445	1	1	2	2	3	4	4	5	6
70	8451	8457	8463	8470	8476	8482	8488	8494	8500	8506	1	1	2	2	3	4	4	5	5
71	8513	8519	8525	8531	8537	8543	8549	8555	8561	8567	1	1	2	2	3	4	4	5	5
72	8573	8579	8585	8591	8597	8603	8609	8615	8621	8627	1	1	2	2	3	4	4	5	5
73	8633	8639	8645	8651	8657	8663	8669	8675	8681	8686	1	1	2	2	3	4	4	5	5
74	8692	8698	8704	8710	8716	8722	8727	8733	8739	8745	1	1	2	2	3	4	4	5	5
75	8751	8756	8762	8768	8774	8779	8785	8791	8797	8802	1	1	2	2	3	3	4	5	5
76	8808	8814	8820	8825	8831	8837	8842	8848	8854	8859	1	1	2	2	3	3	4	5	5
77	8865	8871	8876	8882	8887	8893	8899	8904	8910	8915	1	1	2	2	3	3	4	4	5
78	8921	8927	8932	8938	8943	8949	8954	8960	8965	8971	1	1	2	2	3	3	4	4	5
79	8976	8982	8987	8993	8998	9004	9009	9015	9020	9026	1	1	2	2	3	3	4	4	5
80	9031	9036	9042	9047	9053	9058	9063	9069	9074	9079	1	1	2	2	3	3	4	4	5
81	9085	9090	9096	9101	9106	9112	9117	9122	9128	9133	1	1	2	2	3	3	4	4	5
82	9138	9143	9149	9154	9159	9165	9170	9175	9180	9186	1	1	2	2	3	3	4	4	5
83	9191	9196	9201	9206	9212	9217	9222	9227	9232	9238	1	1	2	2	3	3	4	4	5
84	9243	9248	9253	9258	9263	9269	9274	9279	9284	9289	1	1	2	2	3	3	4	4	5
85	9294	9299	9304	9309	9315	9320	9325	9330	9335	9340	1	1	2	2	3	3	4	4	5
86	9345	9350	9355	9360	9365	9370	9375	9380	9385	9390	1	1	2	2	3	3	4	4	5
87	9395	9400	9405	9410	9415	9420	9425	9430	9435	9440	0	1	1	2	2	3	3	4	4
88	9445	9450	9455	9460	9465	9469	9474	9479	9484	9489	0	1	1	2	2	3	3	4	4
89	9494	9499	9504	9509	9513	9518	9523	9528	9533	9538	0	1	1	2	2	3	3	4	4
90	9542	9547	9552	9557	9562	9566	9571	9576	9581	9586	0	1	1	2	2	3	3	4	4
91	9590	9595	9600	9605	9609	9614	9619	9624	9628	9633	0	1	1	2	2	3	3	4	4
92	9638	9643	9647	9652	9657	9661	9666	9671	9675	9680	0	1	1	2	2	3	3	4	4
93	9685	9689	9694	9699	9703	9708	9713	9717	9722	9727	0	1	1	2	2	3	3	4	4
94	9731	9736	9741	9745	9750	9754	9759	9763	9768	9773	0	1	1	2	2	3	3	4	4
95	9777	9782	9786	9791	9795	9800	9805	9809	9814	9818	0	1	1	2	2	3	3	4	4
96	9823	9827	9832	9836	9841	9845	9850	9854	9859	9863	0	1	1	2	2	3	3	4	4
97	9868	9872	9877	9881	9886	9890	9894	9899	9903	9908	0	1	1	2	2	3	3	4	4
98	9912	9917	9921	9926	9930	9934	9939	9943	9948	9952	0	1	1	2	2	3	3	4	4

The antilogarithm of 0.0671 is 1.17. Thus

$$\frac{412}{353} = 1.17 \quad \text{or} \quad \frac{4.12 \times 10^2}{3.53 \times 10^2} = 1.17$$

Combined operations are performed in precisely the same manner.

Significant Figures and Common Logarithms

For the common logarithm of a measured quantity, the number of digits after the decimal point (the number of digits in the mantissa) equals the number of significant figures in the original number. For example, if 23.5 is a measured quantity (three significant figures) then log 23.5 = 1.371 (three significant figures in the mantissa). The characteristic, 1, just places the decimal point and is not a significant figure.

EXAMPLE

Find $\frac{(353)(295)}{(412)}$

$$\begin{array}{rll} & \text{Logarithm of } 353 & = 2.548 \\ + & \text{Logarithm of } 295 & = 2.470 \\ \hline & \text{Logarithm of } 353 \times 295 & = 5.018 \\ - & \text{Logarithm of } 412 & = 2.615 \\ \hline & \text{Logarithm of quotient} & = 2.403 \end{array}$$

The antilogarithm of 2.403 is 253. Thus (353)(296)/(412) = 253.

The extraction of roots of numbers by means of logarithms is a simple procedure.

EXAMPLE

Find $\sqrt[3]{7235} = (7235)^{1/3}$

$$\begin{aligned} \text{Logarithm of } 7235 &= 3.8594 \\ \tfrac{1}{3} \text{ logarithm of } 7235 &= 1.2865 \\ \text{Antilogarithm } 1.2865 &= 1.934 \times 10^1 = 19.34 \end{aligned}$$

Thus $19.34 = (7235)^{1/3}$.

Powers are found in the same fashion.

EXAMPLE

$(353)^3 = ?$

$$\begin{aligned} \text{Logarithm } 353 &= 2.548 \\ 3 \text{ logarithm } 353 &= 7.644 \\ \text{Antilogarithm } 7.644 &= 4.40 \times 10^7 \end{aligned}$$

Thus $(3.53 \times 10^2)^3 = 4.40 \times 10^7$.

Finding the antilogarithm of a negative logarithm is best illustrated by example.

EXAMPLE

Find the antilogarithm of -7.1594. First convert the mantissa to a positive value, since there are no tables of negative logarithms. This is done as follows:

$$-7.1594 = -8 + 0.8406$$

Then find the antilogarithm of this logarithm as follows:

$$\begin{aligned}\text{Antilog } -7.1594 &= \text{antilog } (-8) \times \text{antilog } (0.8406) \\ &= 10^{-8} \times 6.930\end{aligned}$$

Thus antilog of $-7.1594 = 6.930 \times 10^{-8}$.

After a little practice you should find that logarithms simplify the mathematical operations involving very large or very small numbers. Operations involving roots and powers are most easily performed using logarithms. Some basic rules to remember are the following:

$$10^a \times 10^b = 10^{(a+b)} \qquad \log (10^a \times 10^b) = \log 10^a + \log 10^b = a + b$$

$$\frac{10^a}{10^b} = 10^{(a-b)} \qquad \log (10^a/10^b) = \log 10^a - \log 10^b = a - b$$

$$(10^a)^b = 10^{ab} \qquad \log (10^a)^b = b \log 10^a = ba$$

$$\sqrt{10^a} = (10^a)^{1/2} = 10^{a/2} \qquad \log (10^a)^{1/2} = \tfrac{1}{2} \log 10^a = a/2$$

Graphical Interpretation of Data; Calibration Curves and Least-Squares Analysis

APPENDIX

B

GRAPHICAL INTERPRETATION OF DATA

Relationships between two or more variables can easily be visualized when the data are plotted or graphed. In such a graph the horizontal axis (x-axis) represents the experimentally varied variable, called the *independent variable*. The vertical axis (y-axis) represents the *dependent variable,* which responds to a change in the independent variable. For example, consider the data in Table B.1 for the mass of mercury as a function of volume. In this case volume is the independent variable and mass the dependent variable. A graph of these data is shown in Figure B.1. Clearly, the relationship between mass and volume of mercury is linear. The equation for a linear relationship is of the form $y = ax + b$, where a is the slope of the line ($\Delta y/\Delta x$) and b is the intercept with the y-axis. That is, b is the value of y when $x = 0$. Therefore, the equation for the data in Table B.1 is

Table B.1 Relationship Between Mass and Volume of Mercury

Volume (mL)	Mass (g)
2.00	27.0
2.50	33.75
3.00	40.5
3.50	47.25
4.00	54.0
4.50	60.75

$$\text{Mass} = (a)\text{volume} + b$$

where $a = 13.5$ g/mL and $b = 0$.

or

$$\text{Mass} = \left(\frac{13.5\ \text{g}}{\text{mL}}\right)\text{volume} + 0$$

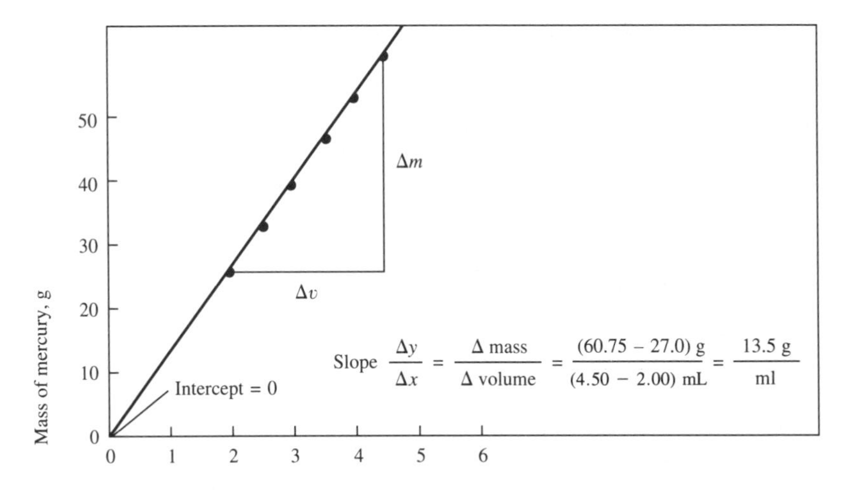

Figure B.1

In this case, the slope, 13.5 g/mL, has the units of mass per unit volume, which is density. In practice, experimental data have uncertainties, and all points, therefore, usually do not lie precisely on the line. In these cases, one draws the best straight line or calculates the best straight line from a least squares analysis and then determines the slope and intercept of this line, as above.

CALIBRATION CURVES: LEAST-SQUARES ANALYSIS

Many analytical methods require a calibration step in which standards containing known amounts of analyte (x) are analyzed in the same way as the unknown. The experimental result (y) is then plotted versus x to give a calibration curve such as that shown in Figure B.2. These plots often are straight lines or linear relations. Seldom, however, do the experimental data fall exactly on the line, usually because of the experimental errors. The experimenter is then obligated to draw the "best" straight line through the experimental points. Statistical methods are available that allow for an objective determination of such a line and also for an estimation of the uncertainties associated with this line. Statisticians call this technique *linear-regression analyses*. We will use the simplest of these techniques, called the *method of least squares*.

We assume that a linear relation of the form

$$y = ax + b$$

does in fact exist.

Method

The line generated by a least-squares method is the one which minimizes the squares of the individual vertical displacements, or *residuals* (Figure B.2), from that line. In addition to providing the best fit of the experimental points to the line, the method calculates the slope a and intercept b of the line.

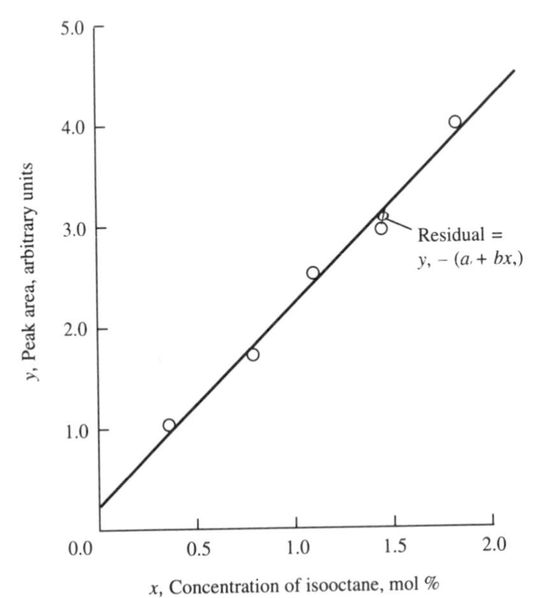

Figure B.2

We define:

$$S_{xx} = \Sigma(x_i - \bar{x})^2 = \Sigma x_i^2 - \frac{(\Sigma x_i)^2}{n} \qquad [1]$$

$$S_{yy} = \Sigma(y_i - \bar{y})^2 = \Sigma y_i^2 - \frac{(\Sigma y_i)^2}{n} \qquad [2]$$

$$S_{xy} = \Sigma(x_i - \bar{x})(y_i - \bar{y}) = \Sigma x_i y_i - \frac{\Sigma x_i \Sigma yi}{n} \qquad [3]$$

where x_i and y_i are individual pairs of values for x and y defining each of the points to be plotted. The quantity n is the number of points, and $\bar{x}$ and $\bar{y}$ are the average values of x and y, that is,

$$\bar{x} = \frac{\Sigma x_i}{n} \qquad \bar{y} = \frac{\Sigma y_i}{n}$$

Note that S_{xx} and S_{yy} are simply the sums of the squares of the deviations from the means for individual values of x and y.

From S_{xx}, S_{yy}, and S_{xy} we may calculate:

1. The slope of the line, a:

$$a = \frac{S_{xy}}{S_{xx}}$$

2. The intercept of the line, b:

$$b = \bar{y} - a\bar{x}$$

3. The standard deviation about the regression, S_r, which is based upon deviations of the individual points from the line:

$$S_r = \sqrt{\frac{S_{yy} - a^2 S_{xx}}{n - 2}}$$

EXAMPLE B.1

Given the following data, carry out a least-squares analysis.

x_i	y_i	x_i^2	y_i^2	$x_i y_i$
0.352	1.09	0.12390	1.1881	0.38368
0.803	1.78	0.64481	3.1684	1.42934
1.08	2.60	1.16640	6.7600	2.80800
1.38	3.03	1.90140	9.1809	4.18140
1.75	4.01	3.06250	16.0801	7.01750
5.365	12.51	6.90201	36.3775	15.81942

Solution:

$$S_{xx} = \Sigma x_i^2 - \frac{(\Sigma x_i)^2}{n} = 6.90201 - \frac{(5.365)^2}{5} = 1.145365$$

$$S_{yy} = \Sigma y_i^2 - \frac{(\Sigma y_i)^2}{n} = 36.3755 - \frac{(12.51)^2}{5} = 5.07748$$

$$S_{xy} = \Sigma x_i y_i - \frac{\Sigma x_i \Sigma y_i}{n} = 15.81992 - \frac{(5.365)(12.51)}{5} = 2.39669$$

$$a = \frac{2.39669}{1.145365} = 2.0925 = 2.09$$

$$b = \bar{y} - a\bar{x} = \frac{12.51}{5} - 2.09\frac{(5.365)}{5} = 0.257$$

Thus the best linear equation is

$$y = 2.09x + 0.257$$

and

$$S_r = \sqrt{\frac{S_{yy} - b^2 S_{xx}}{n - 2}} = \sqrt{\frac{5.07748 - (2.0925)^2(1.145365)}{5 - 2}}$$

$$= \pm 0.144 = \pm 0.14$$

Appendix C List of the Elements with Their Symbols and Atomic Masses

	Symbol	Atomic number	Atomic mass		Symbol	Atomic number	Atomic mass		Symbol	Atomic number	Atomic mass
Actinium	Ac	89	227.0278	Hafnium	Hf	72	178.49	Praseodymium	Pr	59	140.9077
Aluminum	Al	13	26.98154	Hahnium	Ha	105	(262)[b]	Promethium	Pm	61	(145)
Americium	Am	95	(243)[a]	Helium	He	2	4.00260	Protactinium	Pa	91	231.0359
Antimony	Sb	51	121.75	Holmium	Ho	67	164.9304	Radium	Ra	88	226.0254
Argon	Ar	18	39.948	Hydrogen	H	1	1.0079	Radon	Rn	86	(222)
Arsenic	As	33	74.9216	Indium	In	49	114.82	Rhenium	Re	75	186.207
Astatine	At	85	(210)	Iodine	I	53	126.9045	Rhodium	Rh	45	102.9055
Barium	Ba	56	137.33	Iridium	Ir	77	192.22	Rubidium	Rb	37	85.4678
Berkelium	Bk	97	(247)	Iron	Fe	26	55.847	Ruthenium	Ru	44	101.07
Beryllium	Be	4	9.01218	Krypton	Kr	36	83.80	Rutherfordium	Rf	104	(261)[b]
Bismuth	Bi	83	208.9804	Lanthanum	La	57	138.9055	Samarium	Sm	62	150.36
Boron	B	5	10.81	Lawrencium	Lr	103	(260)	Scandium	Sc	21	44.9554
Bromine	Br	35	79.904	Lead	Pb	82	207.2	Selenium	Se	34	78.96
Cadmium	Cd	48	112.41	Lithium	Li	3	6.941	Silicon	Si	14	28.0855
Calcium	Ca	20	40.08	Lutetium	Lu	71	174.967	Silver	Ag	47	107.8682
Californium	Cf	98	(251)	Magnesium	Mg	12	24.305	Sodium	Na	11	22.98977
Carbon	C	6	12.011	Manganese	Mn	25	54.9380	Srontium	Sr	38	87.62
Cerium	Ce	58	140.12	Mendelevium	Md	101	(258)	Sulfur	S	16	32.06
Cesium	Cs	55	132.9054	Mercury	Hg	80	200.59	Tantalum	Ta	73	180.9479
Chlorine	Cl	17	35.453	Molybdenum	Mo	42	95.94	Technetium	Tc	43	(98)
Chromium	Cr	24	51.996	Neodymium	Nd	60	144.24	Tellurium	Te	52	127.60
Cobalt	Co	27	58.9332	Neon	Ne	10	20.179	Terbium	Tb	65	158.9254
Copper	Cu	29	63.546	Neptunium	Np	93	237.0482	Thallium	Tl	81	204.383
Curium	Cm	96	(247)	Nickel	Ni	28	58.69	Thorium	Th	90	232.0381
Dysprosium	Dy	66	162.50	Niobium	Nb	41	92.9064	Thulium	Tm	69	168.9342
Einsteinium	Es	99	(254)	Nitrogen	N	7	14.0067	Tin	Sn	50	118.69
Erbium	Er	68	167.26	Nobelium	No	102	(259)	Titanium	Ti	22	47.88
Europium	Eu	63	151.96	Osmium	Os	76	190.2	Tungsten	W	74	183.85
Fermium	Fm	100	(257)	Oxygen	O	8	15.9994	Unnilhexium	Unh	106	(263)[b]
Fluorine	F	9	18.998403	Palladium	Pd	46	106.4	Uranium	U	92	238.0289
Francium	Fr	87	(223)	Phosphorus	P	15	30.97376	Vanadium	V	23	50.9415
Gadolinium	Gd	64	157.25	Platinum	Pt	78	195.08	Xenon	Xe	54	131.29
Gallium	Ga	31	69.72	Plutonium	Pu	94	(244)	Ytterbium	Yb	70	173.04
Germanium	Ge	32	72.59	Polonium	Po	84	(209)	Yttrium	Y	39	88.9059
Gold	Au	79	196.9665	Potassium	K	19	39.0983	Zinc	Zn	30	65.38
								Zirconium	Zr	40	91.22

[a] Approximate values for radioactive elements are listed in parentheses.
[b] The official name and symbol has not been agreed to.

Appendix D Periodic Table of the Elements

1A	2A	3B	4B	5B	6B	7B	8B			1B	2B	3A	4A	5A	6A	7A	8A
1 **H**																	2 **He**
3 **Li**	4 **Be**											5 **B**	6 **C**	7 **N**	8 **O**	9 **F**	10 **Ne**
11 **Na**	12 **Mg**											13 **Al**	14 **Si**	15 **P**	16 **S**	17 **Cl**	18 **Ar**
19 **K**	20 **Ca**	21 **Sc**	22 **Ti**	23 **V**	24 **Cr**	25 **Mn**	26 **Fe**	27 **Co**	28 **Ni**	29 **Cu**	30 **Zn**	31 **Ga**	32 **Ge**	33 **As**	34 **Se**	35 **Br**	36 **Kr**
37 **Rb**	38 **Sr**	39 **Y**	40 **Zr**	41 **Nb**	42 **Mo**	43 **Tc**	44 **Ru**	45 **Rh**	46 **Pd**	47 **Ag**	48 **Cd**	49 **In**	50 **Sn**	51 **Sb**	52 **Te**	53 **I**	54 **Xe**
55 **Cs**	56 **Ba**	57 **La**	72 **Hf**	73 **Ta**	74 **W**	75 **Re**	76 **Os**	77 **Ir**	78 **Pt**	79 **Au**	80 **Hg**	81 **Tl**	82 **Pb**	83 **Bi**	84 **Po**	85 **At**	86 **Rn**
87 **Fr**	88 **Ra**	89 **Ac**	104 **Rf**	105 **Ha**	[106]	[107]	[108]	[109]									

58 **Ce**	59 **Pr**	60 **Nd**	61 **Pm**	62 **Sm**	63 **Eu**	64 **Gd**	65 **Tb**	66 **Dy**	67 **Ho**	68 **Er**	69 **Tm**	70 **Yb**	71 **Lu**
90 **Th**	91 **Pa**	92 **U**	93 **Np**	94 **Pu**	95 **Am**	96 **Cm**	97 **Bk**	98 **Cf**	99 **Es**	100 **Fm**	101 **Md**	102 **No**	103 **Lw**

Metals

Semimetals

Nonmetals

Summary of Solubility Properties of Ions and Solids

APPENDIX

	Cl^-	SO_4^{2-}	CO_3^{2-} PO_4^{3-}	CrO_4^{2-}	OH^- O^{2-}	H_2S pH = 0.5	S^{2-}, pH = 9
Li^+, Na^+, K^+, NH_4^+	S	S	S	S	S	S	S
Ba^{2+}	S	I	A	A	S⁻	S	S
Ca^{2+}	S	S⁻	A	S	S⁻	S	S
Mg^{2+}	S	S	A	S	A	S	S
Fe^{3+}	S	S	A	A	A	S	A
Cr^{3+}	S	S	A	A	A	S	A
Al^{3+}	S	S	A, B	A, B	A, B	S	A, B
Ni^{2+}	S	S	A, N	A, N	A, N	S	A⁺, O⁺
Co^{2+}	S	S	A	A	A	S	A⁺, O⁺
Zn^{2+}	S	S	A, B, N	A, B, N	A, B, N	S	A
Mn^{2+}	S	S	A	A	A	S	A
Cu^{2+}	S	S	A, N	A, N	A, N	O	O
Cd^{2+}	S	S	A, N	A, N	A, N	A⁺, O	A⁺, O
Bi^{3+}	A	A	A	A	A	O	O
Hg^{2+}	S	S	A	A	A	O⁺, C	O⁺, C
Sn^{2+}, Sn^{4+}	A, B	A, B	A, B	A, B	A, B	A⁺, C	A⁺, C
Sb^{3+}	A, B	A, B	A, B	A, B	A, B	A⁺, C	A⁺, C
Ag^+	A⁺, N	S⁻, N	A, N	A, N	A, N	O	O
Pb^{2+}	HW, B, A⁺	B	A, B	B	A, B	O	O
Hg_2^{2+}	O⁺	S⁻, A	A	A	A	O⁺	O⁺

Key: S, soluble in water.
A, soluble in acid (6 *M* HCl or other nonprecipitating, nonoxidizing acid).
B, soluble 6 *M* NaOH.
O, soluble in hot 6 *M* HNO_3.
N, soluble in 6 *M* NH_3.
I, insoluble in any common reagent.
S⁻, slightly soluble in water.
A⁺, soluble in 12 *M* HCl.
O⁺, soluble in aqua regia.
C, soluble in 6 *M* NaOH containing excess S^{2-}.
HW, soluble in hot water.

Example: For Cd^{2+} and OH^- the entry is A, N. This means that $Cd(OH)_2(s)$, the product obtained when solutions containing Cd^{2+} and OH^- are mixed, will dissolve to the extent of at least 0.1 mol/L when treated with 6 *M* HCl or 6 *M* NH_3. Since 6 *M* HNO_3, 12 *M* HCl, and aqua regia are at least as strongly acidic as 6 *M* HCl, $Cd(OH)_2(s)$ would also be soluble in those reagents.

Solubility Rules

Water-soluble salts	
Na^+, K^+, NH_4^+	All sodium, potassium, and ammonium slats are soluble.
NO_3^-, ClO_3^-, $C_2H_3O_2^-$	All nitrates, chlorates, and acetates are soluble.
Cl^-	All chlorides are soluble except AgCl, Hg_2Cl_2, and $PbCl_2$.[a]
Br^-	All bromides are soluble except AgBr, Hg_2Br_2, $PbBr_2$,[a] and $HgBr_2$.[a]
I^-	All iodides are soluble except AgI, Hg_2I_2, PbI_2, and HgI_2.
SO_4^{2-}	All sulfates are soluble except $CaSO_4$,[a] $SrSO_4$, $BaSO_4$, Hg_2SO_4, $PbSO_4$, and Ag_2SO_4.
Water-insoluble salts	
CO_3^{2-}, SO_3^{2-}, PO_4^{3-}, CrO_4^{2-}	All carbonates, sulfites, phosphates, and chromates are insoluble except those of alkali metals and NH_4^+.
OH^-	All hydroxides are insoluble except those of alkali metals and $Ca(OH)_2$,[a] $Sr(OH)_2$,[a] and $Ba(OH)_2$.
S^{2-}	All sulfides are insoluble except those of the alkali metals, alkaline earths, and NH_4^+.

[a] Slightly soluble

Solubility–Product Constants for Compounds at 25°C

APPENDIX

Name	Formula	K_{sp}
Barium carbonate	$BaCO_3$	5.1×10^{-9}
Barium chromate	$BaCrO_4$	1.2×10^{-10}
Barium fluoride	BaF_2	1.0×10^{-6}
Barium hydroxide	$Ba(OH)_2$	5×10^{-3}
Barium oxalate	BaC_2O_4	1.6×10^{-7}
Barium phosphate	$Ba_3(PO_4)_2$	3.4×10^{-23}
Barium sulfate	$BaSO_4$	1.1×10^{-10}
Cadmium carbonate	$CdCO_3$	5.2×10^{-12}
Cadmium hydroxide	$Cd(OH)_2$	2.5×10^{-14}
Cadmium sulfide	CdS	8.0×10^{-27}
Calcium carbonate	$CaCO_3$	2.8×10^{-9}
Calcium chromate	$CaCrO_4$	7.1×10^{-4}
Calcium fluoride	CaF_2	3.9×10^{-11}
Calcium hydroxide	$Ca(OH)_2$	5.5×10^{-6}
Calcium phosphate	$Ca_3(PO_4)_2$	2.0×10^{-29}
Calcium sulfate	$CaSO_4$	9.1×10^{-6}
Cerium(III) fluoride	CeF_3	8×10^{-16}
Chromium(III) fluoride	CrF_3	6.6×10^{-11}
Chromium(III) hydroxide	$Cr(OH)_3$	6.3×10^{-31}
Cobalt(II) carbonate	$CoCO_3$	1.4×10^{-13}
Cobalt(II) hydroxide	$Co(OH)_2$	1.6×10^{-15}
Cobalt(III) hydroxide	$Co(OH)_3$	1.6×10^{-44}
α-Cobalt(II) sulfide[a]	CoS	4.0×10^{-21}
Copper(I) bromide	$CuBr$	5.3×10^{-9}
Copper(I) chloride	$CuCl$	1.2×10^{-6}
Copper(I) sulfide	Cu_2S	2.5×10^{-48}
Copper(II) carbonate	$CuCO_3$	1.4×10^{-10}
Copper(II) chromate	$CuCrO_4$	3.6×10^{-6}
Copper(II) hydroxide	$Cu(OH)_2$	2.2×10^{-20}
Copper(II) phosphate	$Cu_3(PO_4)_2$	1.3×10^{-37}
Copper(II) sulfide	CuS	6.3×10^{-36}
Gold(I) chloride	$AuCl$	2.0×10^{-13}
Gold(III) chloride	$AuCl_3$	3.2×10^{-25}
Iron(II) carbonate	$FeCO_3$	3.2×10^{-11}
Iron(II) hydroxide	$Fe(OH)_2$	8.0×10^{-16}
Iron(II) sulfide	FeS	6.3×10^{-18}
Iron(III) hydroxide	$Fe(OH)_3$	4×10^{-38}
Lanthanum fluoride	LaF_3	7×10^{-17}
Lanthanum iodate	$La(IO_3)_3$	6.1×10^{-12}
Lead carbonate	$PbCO_3$	7.4×10^{-14}
Lead chloride	$PbCl_2$	1.6×10^{-5}

[a] Some substances exist in more than one crystalline form; the prefix indicates the particular form for which K_{sp} is listed.

Name	Formula	K_{sp}
Lead chromate	$PbCrO_4$	2.8×10^{-13}
Lead fluoride	PbF_2	2.7×10^{-8}
Lead hydroxide	$Pb(OH)_2$	1.2×10^{-15}
Lead sulfate	$PbSO_4$	1.6×10^{-8}
Lead sulfide	PbS	8.0×10^{-28}
Magnesium hydroxide	$Mg(OH)_2$	1.8×10^{-11}
Magnesium oxalate	MgC_2O_4	8.6×10^{-5}
Manganese carbonate	$MnCO_3$	1.8×10^{-11}
Manganese hydroxide	$Mn(OH)_2$	1.9×10^{-13}
Manganese(II) sulfide	MnS	1.0×10^{-13}
Mercury(I) chloride	Hg_2Cl_2	1.3×10^{-18}
Mercury(I) oxalate	$Hg_2C_2O_4$	2.0×10^{-13}
Mercury(I) sulfide	Hg_2S	1.0×10^{-47}
Mercury(II) hydroxide	$Hg(OH)_2$	3.0×10^{-26}
Mercury(II) sulfide	HgS	4×10^{-53}
Nickel carbonate	$NiCO_3$	6.6×10^{-9}
Nickel hydroxide	$Ni(OH)_2$	1.6×10^{-14}
Nickel oxalate	NiC_2O_4	4×10^{-10}
α-Nickel sulfide[a]	NiS	3.2×10^{-19}
Silver arsenate	Ag_3AsO_4	1.0×10^{-22}
Silver bromide	$AgBr$	5.0×10^{-13}
Silver carbonate	Ag_2CO_3	8.1×10^{-12}
Silver chloride	$AgCl$	1.8×10^{-10}
Silver chromate	Ag_2CrO_4	1.1×10^{-12}
Silver cyanide	$AgCN$	1.2×10^{-16}
Silver iodide	AgI	8.3×10^{-17}
Silver sulfate	Ag_2SO_4	1.4×10^{-5}
Silver sulfide	Ag_2S	6.3×10^{-50}
Strontium carbonate	$SrCO_3$	1.1×10^{-10}
Tin(II) hydroxide	$Sn(OH)_2$	1.4×10^{-28}
Tin(II) sulfide	SnS	1.0×10^{-25}
Zinc carbonate	$ZnCO_3$	1.4×10^{-11}
Zinc hydroxide	$Zn(OH)_2$	1.2×10^{-17}
Zinc oxalate	ZnC_2O_4	2.7×10^{-8}
α-Zinc sulfide[a]	ZnS	1.1×10^{-21}

[a] Some substances exist in more than one crystalline form; the prefix indicates the particular form for which K_{sp} is listed.

APPENDIX

Dissociation Constants for Acids at 25°C

Name	Formula	K_{a1}	K_{a2}	K_{a3}
Acetic	$HC_2H_3O_2$	1.8×10^{-5}		
Arsenic	H_3AsO_4	5.6×10^{-3}	1.0×10^{-7}	3.0×10^{-12}
Arsenous	H_3AsO_3	6×10^{-10}	1.6×10^{-12}	
Ascorbic	$HC_6H_7O_6$	8.0×10^{-5}		
Benzoic	$HC_7H_5O_2$	6.5×10^{-5}		
Boric	H_3BO_3	5.8×10^{-10}		
Carbonic	H_2CO_3	4.3×10^{-7}	5.6×10^{-11}	
Chloroacetic	$HC_2H_2O_2Cl$	1.4×10^{-3}		
Citric	$H_3C_6H_5O_7$	3.5×10^{-4}		
Cyanic	HCNO	7.4×10^{-4}	$1.7 \times {}^{-5}$	4.0×10^{-7}
Formic	$HCNO_2$	1.8×10^{-4}		
Hydroazoic	HN_3	1.9×10^{-5}		
Hydrocyanic	HCN	4.9×10^{-10}		
Hydrofluoric	HF	6.8×10^{-4}		
Hydrogen chromate ion	$HCrO_4^-$	3.0×10^{-7}		
Hydrogen peroxide	H_2O_2	2.4×10^{-12}		
Hydrogen selenate ion	$HSeO_4^-$	2.2×10^{-2}		
Hydrogen sulfate ion	HSO_4	1.2×10^{-2}		
Hydrogen sulfide	H_2S	5.7×10^{-8}	1.3×10^{-13}	
Hypobromous	HBrO	2×10^{-9}		
Hypochlorous	HClO	3.0×10^{-8}		
Hypoiodus	HIO	2×10^{-11}		
Iodic	HIO_3	1.7×10^{-1}		
Lactic	$HC_3H_5O_3$	1.4×10^{-4}		
Malonic	$H_2C_3H_2O_4$	1.5×10^{-3}	2.0×10^{-6}	
Nitrous	HNO_2	4.5×10^{-4}		
Oxalic	$H_2C_2O_4$	5.9×10^{-2}	6.4×10^{-5}	
Paraperiodic	H_5IO_6	2.8×10^{-2}	5.3×10^{-9}	
Phenol	HC_6H_5O	1.3×10^{-10}		
Phosphoric	H_3PO_4	7.5×10^{-3}	6.2×10^{-8}	4.2×10^{-13}
Propionic	$HC_3H_5O_2$	1.3×10^{-5}		
Pyrophosphoric	$H_4P_2O_7$	3.0×10^{-2}	4.4×10^{-3}	
Selenous	H_2SeO_3	2.3×10^{-3}	5.3×10^{-9}	
Sulfuric	H_2SO_4	strong acid	1.2×10^{-2}	
Sulfurous	H_2SO_3	1.7×10^{-2}	6.4×10^{-8}	
Tartaric	$H_2C_4H_4O_6$	1.0×10^{-3}	4.6×10^{-5}	

APPENDIX

H Dissociation Constants for Bases

Name	Formula	K_b
Ammonia	NH_3	1.8×10^{-5}
Aniline	$C_6H_5NH_2$	4.3×10^{-10}
Dimethylamine	$(CH_3)_2NH$	5.4×10^{-4}
Ethylamine	$C_2H_5NH_2$	6.4×10^{-4}
Hydrazine	H_2NNH_2	1.3×10^{-6}
Hydroxylamine	$HONH_2$	1.1×10^{-8}
Methylamine	CH_3NH_2	4.4×10^{-4}
Pyridine	C_5H_5N	1.7×10^{-9}
Trymethylamine	$(CH_3)_3N$	6.4×10^{-5}

Selected Standard Reduction Potentials (25°C)

APPENDIX

I

Half-reaction	$E°$ (V)
$Ag^+(aq) + e^- \rightleftharpoons Ag(s)$	+0.799
$AgBr(s) + e^- \rightleftharpoons Ag(s) + Br^-(aq)$	+0.095
$AgCl(s) + e^- \rightleftharpoons Ag(s) + Cl^-(aq)$	+0.222
$Ag(CN)_2^-(aq) + e^- \rightleftharpoons Ag(s) + 2CN^-(aq)$	−0.31
$Ag_2CrO_4(s) + 2e^- \rightleftharpoons 2Ag(s) + CrO_4^{2-}(aq)$	+0.446
$AgI(s) + e^- \rightleftharpoons Ag(s) + I^-(aq)$	−0.151
$Ag(S_2O_3)_2^{3-} + e^- \rightleftharpoons Ag(s) + 2S_2O_3^{2-}(aq)$	+0.01
$Al^{3+}(aq) + 3e^- \rightleftharpoons Al(s)$	−1.66
$H_2AsO_4(aq) + 2H^+(aq) + 2e^- \rightleftharpoons H_3AsO_3(aq) + H_2O(l)$	+0.559
$Ba^{2+}(aq) + 2e^- \rightleftharpoons Ba(s)$	−2.90
$BiO^+(aq) + 2H^+(aq) + 3e^- \rightleftharpoons Bi(s) + H_2O(l)$	+0.32
$Br_2(l) + 2e^- \rightleftharpoons 2Br^-(aq)$	+1.065
$BrO_3^-(aq) + 6H + (aq) + 5e^- \rightleftharpoons \frac{1}{2}Br_2(l) + 3H_2O(l)$	+1.52
$Ca^{2+}(aq) + 2e^- \rightleftharpoons Ca(s)$	−2.87
$2CO_2(g) + 2H+(aq) + 2e^- \rightleftharpoons H_2C_2O_4(aq)$	−0.49
$Cd^{2+}(aq) + 2e^- \rightleftharpoons Cd(s)$	−0.403
$Ce^{4+}(aq) + e^- \rightleftharpoons Ce^{3+}(aq)$	+1.61
$Cl_2(g) + 2e^- \rightleftharpoons 2Cl^-(aq)$	+1.359
$HClO(aq) + H^+(aq) + e^- \rightleftharpoons \frac{1}{2}Cl_2(g) + H_2O(l)$	+1.63
$ClO^-(aq) + H_2O(l) + 2e^- \rightleftharpoons Cl^-(aq) + 2OH^-(aq)$	+0.89
$ClO_3^-(aq) + 6H^+(aq) + 5e^- \rightleftharpoons \frac{1}{2}Cl_2(g) + 3H_2O(l)$	+1.47
$Co^{2+}(aq) + 2e^- \rightleftharpoons Co(s)$	−0.277
$Co^{3+}(aq) + e^- \rightleftharpoons Co^{2+}(aq)$	+1.842
$Cr^{3+}(aq) + 3e^- \rightleftharpoons Cr(s)$	−0.74
$Cr^{3+}(aq) + e^- \rightleftharpoons Cr^{2+}(aq)$	−0.41
$Cr_2O_7^{2-}(aq) + 14H^+(aq)\ 6e^- \rightleftharpoons 2Cr^{3+}(aq) + 7H_2O(l)$	+1.33
$CrO_4^{2-}(aq) + 4H_2O(l) + 3e^- \rightleftharpoons Cr(OH)_3(s) + 5OH^-(aq)$	−0.13
$Cu^2(aq) + 2e^- \rightleftharpoons Cu(s)$	+0.337
$Cu^{2+}(aq) + e^- \rightleftharpoons Cu^+(aq)$	+0.153
$Cu^+(aq) + e^- \rightleftharpoons Cu(s)$	+0.521
$CuI(s) + e^- \rightleftharpoons Cu(s) + I^-(aq)$	−0.185
$F_2(g) + 2e^- \rightleftharpoons 2F^-(aq)$	+2.87
$Fe^{2+}(aq) + 2e^- \rightleftharpoons Fe(s)$	−0.440
$Fe^{3+}(aq) + e^- \rightleftharpoons Fe^{2+}(aq)$	+0.771
$Fe(CN)_6^{3-}(aq) + e^- \rightleftharpoons Fe(CN)_6^{4-}(aq)$	+0.36
$2H^+(aq) + 2e^- \rightleftharpoons H_2(g)$	0.000
$2H_2O(l) + 2e^- \rightleftharpoons H_2(g) + 2OH^-(aq)$	−0.83
$HO_2^-(aq) + H_2O(l) + 2e^- \rightleftharpoons 3OH^-(aq)$	+0.88
$H_2O_2(aq) + 2H^+(aq) + 2e^- \rightleftharpoons 2H_2O(l)$	+1.776
$Hg_2^{2+}(aq) + 2e^- \rightleftharpoons 2Hg(l)$	+0.789

Half-reaction	$E°$ (V)
$2Hg^{2+}(aq) + 2e^- \rightleftharpoons Hg_2^{2+}(aq)$	+0.920
$Hg^{2+}(aq) + 2e^- \rightleftharpoons Hg(l)$	+0.854
$I_2(s) + 2e^- \rightleftharpoons 2I^-(aq)$	+0.536
$IO_3^-(aq) + 6H^+(aq) + 5e^- \rightleftharpoons \frac{1}{2}I_2(s) + 3H_2O(l)$	+1.195
$K^+(aq) + e^- \rightleftharpoons K(s)$	−2.925
$Li^+(aq) + e^- \rightleftharpoons Li(s)$	−3.05
$Mg^{2+}(aq) + 2e^- \rightleftharpoons Mg(s)$	−2.37
$Mn^{2+}(aq) + 2e^- \rightleftharpoons Mn(s)$	−1.18
$MnO_2(s) + 4H^+(aq) + 2e^- \rightleftharpoons Mn^{2+}(aq) + 2H_2O(l)$	+1.23
$MnO_4^-(aq) + 8H^+(aq) + 5e^- \rightleftharpoons Mn^{2+} + 4H_2O(l)$	+1.51
$MnO_4^-(aq) + 2H_2O(l) + 3e^- \rightleftharpoons MnO_2 + 4OH^-(aq)$	+0.59
$HNO_2(aq) + H^+(aq) + e^- \rightleftharpoons NO(g) + H_2O(l)$	+1.00
$N_2(g) + 4H_2O(l) + 4e^- \rightleftharpoons 4OH^-(aq) + N_2H_4(aq)$	−1.16
$N_2(g) + 5H^+(aq) - 4e^- \rightleftharpoons N_2H_5^+(aq)$	−0.23
$NO_3^-(aq) + 4H^+(aq) + 3e^- \rightleftharpoons NO(g) + 2H_2O(l)$	+0.96
$Na^+(aq) + e^- \rightleftharpoons Na(s)$	−2.71
$Ni^{2+}(aq) + 2e^- \rightleftharpoons Ni(s)$	−0.28
$O_2(g) + 4H^+(aq) + 4e^- \rightleftharpoons 2H_2O(l)$	+1.23
$O_2(g) + 2H_2O(l) + 4e^- \rightleftharpoons 4OH^-(aq)$	+0.40
$O_2(g) + 2H^+(aq) + 2e^- \rightleftharpoons H_2O_2(aq)$	+0.68
$O_3(g) + 2H^+(aq) + 2e^- \rightleftharpoons O_2(g) + H_2O(l)$	+2.07
$Pb^{2+}(aq) + 2e^- \rightleftharpoons Pb(s)$	−0.126
$PbO_2(s) + HSO_4^-(aq) + 3H^+(aq) + 2e^- \rightleftharpoons PbSO_4(s) + 2H_2O(l)$	+1.685
$PbSO_4(s) + H^+(aq) + 2e^- \rightleftharpoons Pb(s) + HSO_4^-(aq)$	−0.356
$PtCl_4^{2-}(aq) + 2e^- \rightleftharpoons Pt(s) + 4Cl^-(aq)$	+0.73
$S(s) + 2H^+(aq) + 2e^- \rightleftharpoons H_2S(g)$	+0.141
$H_2SO_3(aq) + 4H^+(aq) + 4e^- \rightleftharpoons S(s) + 3H_2O(l)$	+0.45
$HSO_4^-(aq) + 3H^+(aq) + 2e^- \rightleftharpoons H_2SO_3(aq) + H_2O(l)$	+0.17
$Sn^{2+}(aq) + 2e^- \rightleftharpoons Sn(s)$	−0.136
$Sn^{4+}(aq) + 2e^- \rightleftharpoons Sn^{2+}(aq)$	+0.154
$VO_2^+(aq) + 2H^+(aq) + e^- \rightleftharpoons VO^{2+}(aq) + H_2O(l)$	+1.00
$Zn^{2+}(aq) + 2e^- \rightleftharpoons Zn(s)$	−0.763

APPENDIX

Qualitative-Analysis Techniques

J

I. MIXING SOLUTIONS AND PRECIPITATION

When one solution is added to another in a small test tube, it is important that the two be thoroughly mixed. Mixing can be accomplished by using a *clean* stirring rod, or it can be achieved by holding the test tube at the top in one hand and stroking or "tickling" it with the fingers of the other hand, as shown in Figure J.1. When precipitation reagents are added to solutions in the test tube and it is believed that precipitation is complete, centrifuge the sample. Always balance the centrifuge by placing a test tube filled with water to about the same level as your sample test tube directly *across* the centrifuge head from your sample. It usually requires only about 30 s of centrifugation for the precipitate to settle to the bottom of the test tube.

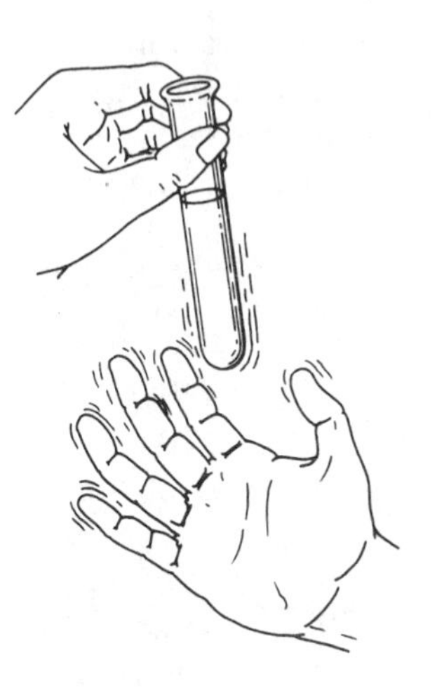

Figure J.1

II. DECANTATION AND WASHING OF PRECIPITATES

The liquid above the precipitate is the *supernatant liquid,* or the *decantate*. The best way to remove this liquid without disturbing the precipitate is to withdraw it by means of a capillary pipet, as shown in Figure J.2. We loosely refer to this operation as *decantation*. Because the preciptiate separated from the supernatant liquid by this technique will be wet with the decantate, it is necessary to *wash* the precipitate free of contaminating ions. Washing is usually accomplished by adding about 10 drops of water to the precipitate, stirring with a stirring rod, and repeating the centrifuging and decanting.

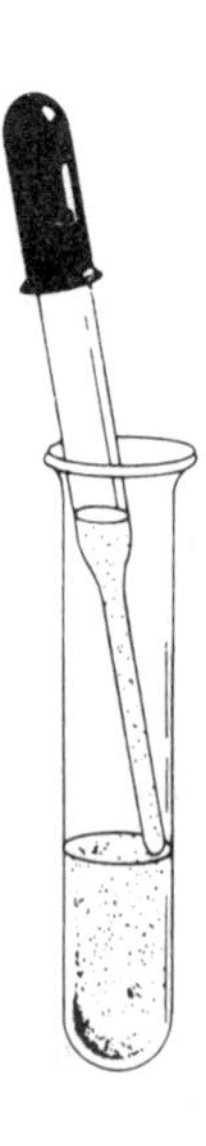

Figure J.2 Use of a dropper (also called a capillary pipet) to withdraw liquid from above a solid.

III. TESTING ACIDITY

Instructions sometimes require making a solution acidic or basic to litmus by adding acid or base. Always be sure that the solution is thoroughly mixed after adding the acid or base; then, by means of a clean stirring rod, remove a drop of the solution and apply it to litmus paper. Do not dip the litmus paper directly into the solution. Remember, just because you have added acid (or base) to a solution does not ensure that it is acidic (or alkaline).

IV. HEATING SOLUTIONS IN SMALL TEST TUBES

The safest way to heat solutions in small test tubes is by means of a water bath, as described in Experiment 1 and shown in Figure J.3. A pipe cleaner wrapped around the test tube serves as a convenient handle for placing the tube into or removing it from the bath.

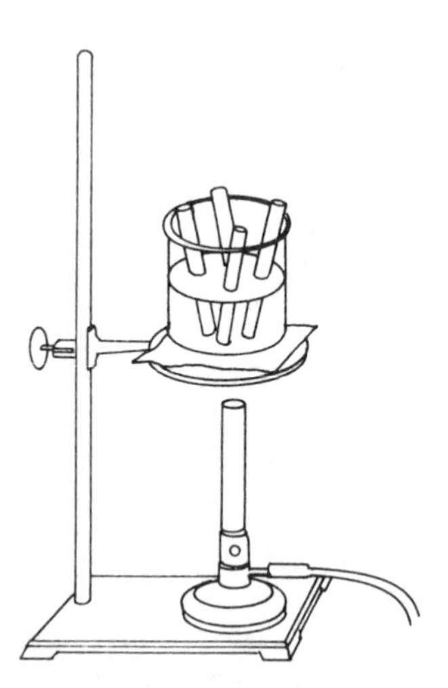

Figure J.3 Water bath for heating solutions in test tubes.

APPENDIX

Answers to Selected Review Questions

EXPERIMENT 1

1. The basic units of measurement in the SI system are mass, kilogram; temperature, °C; volume, liter; length, meter; energy, joule.
2. (1.25 mL)(1000 mL/L) = 1250 mL = 1250 cm^3
4. Convection currents will tend to buoy the object and thus lead to an inaccurate weight. Moreover, hot objects can damage the balance.
5. Thermometers and volumetric glassware are designed to be accurate, but their manufacture is subject to human error, and they should be calibrated.
6. Precision is a measure of the internal consistency of a replicate set of data.
7. Density is mass per unit volume. The determination of density requires a measurement of both mass and volume and so could not be done with a single unitary measurement.

9.

$$\text{Mean} = \frac{10.4 + 10.1 + 10.2}{3} = 10.2$$

$$\text{Average dev. from mean} = \frac{0.2 + 0.1 + 0.0}{3} = 0.1$$

10.

$$d = \frac{m}{V}; \quad V = \frac{m}{d} = \frac{23.2 \text{ g}}{1.3 \text{ g/mL}} = 18 \text{ mL}$$

EXPERIMENT 2

1. Melting points, boiling points, solubility properties, color and densities are five physical properties.
3. Methylene chloride, m.p. −97°C is a liquid, and diphenyl, m.p. 70°C, is a solid at room temperature because room temperature is normally about 20°C.
4. The density of benzophenone could be determined in water because benzophenone is not soluble in water.

6. Thermomenter, pipets, and other pieces of laboratory equipment are mass produced and subject to human error. Consequently, they should always be calibrated.
7. Bromoform is not miscible with water but is miscible with cyclohexane.
11. The liquid is ether, ethyl propyl.

EXPERIMENT 3

1. A mixture differs from an impure substance only in the relative amounts of materials. If the substance contains primarily one component, it is called an impure substance. If, on the other hand, it contains several components in similar amounts, it is called a mixture. In reality all impure substances are mixtures.
2. Sublimation is the process by which a substance changes physical states from the solid to the gaseous and back to the solid state without passing through the liquid state.
3. Filtration differs from decantation in that the liquid phase passes through a semipermeable substance such as a filter, whereas in decantation the liquid phase is separated from the solid phase by carefully pouring off the liquid. Decantation does not involve the use of another substance to achieve the separation.
4. A hot object creates a buoyancy effect by radiating energy in the form of heat and will appear to have a reduced mass. In addition the hot object may cause damage to the delicate balance and so should be allowed to cool to room temperature before it is weighed.
5. All of the original sample should be recovered; thus the sum of the weights of the components, NH_4Cl,SiO_2, and NaCl, should precisely equal the total weight of the sample as no matter is being converted into energy in this experiment.
7. Since $ZnCl_2$ is water soluble and cyclohexane is not and cyclohexane is soluble in cyclohexane and $ZnCl_2$ is not, water will extract $ZnCl_2$ and cyclohexane will extract cyclohexane.

EXPERIMENT 4

1. Before a chemical equation can be written, you must know the reactants and products of the reaction.
2. A color change or formation of a solid or a gas are indicative of a chemical reaction.
3. NO_2 is brown in color and has a pungent odor, while NO is colorless and has a slightly pleasant odor.
4. Metathesis reactions are atom-or group-transfer reactions. For example, $AgNO_3 + NaCl \rightarrow AgCl + NaNO_3$ is a metathesis reaction in which chloride transfer occurs.
5. A precipitate is an insoluble substance that separates from a homogeneous chemical reaction as a solid.
6. $2KBrO_3(s) \rightarrow 2KBr(s) + 3O_2(g)$; $ZnCl_2(aq) + 2AgNO_3(aq) \rightarrow Zn(NO_3)_2(aq) + 2AgCl(s)$

EXPERIMENT 5

1. A compound is composed of a definite number of whole atoms in a fixed proportion.
2. We use relative atomic weights because the actual weights of atoms are exceedingly small and are awkward numbers to manipulate.
3. Formula weights are used for ionic compounds that do not exist as discrete molecular entities. Molecular weights are used for covalent compounds that do exist as discrete molecular entities.
4. The formula weight of $BaCO_3$ is:

$$\begin{array}{ccc} \text{B} & \text{C} & 0 \\ 137.3 + & 12.0 + & 3(16.0) = 197.3 \text{ amu} \end{array}$$

$$\%\text{Ba} = \frac{137.3 \text{ amu}}{197.3 \text{ amu}} \times 100 = 69.6\%$$

$$\%\text{C} = \frac{12.0 \text{ amu}}{197.3 \text{ amu}} \times 100 = 6.1\%$$

$$\%\text{O} = \frac{48.0 \text{ amu}}{197.3 \text{ amu}} \times 100 = 24.3\%$$

5. Assuming 100 g of compound, this would contain 65.95 g of Ba and 34.05 g of Cl. Thus we have

$$\text{Moles Ba} = (65.95 \text{ g})\left(\frac{1 \text{ mol Ba}}{137.3 \text{ g}}\right) = 0.480 \text{ mol}$$

$$\text{Moles Cl} = (34.05 \text{ g})\left(\frac{1 \text{ mol Cl}}{35.45 \text{ g}}\right) = 0.960 \text{ mol}$$

or $Ba_{0.480}Cl_{0.960}$, which implies that the smallest whole-number combining ratio of the constituent elements (that is, the empirical formula) is $BaCl_2$.

6. The law of definite proportions states simply that compounds form from elements in definite proportions by weight such that all samples of the same compound will contain the same ratios of the constituent elements.
7. Molecular formulas are always identical to or multiples of the empirical formulas because the empirical formulas represent the smallest whole-number combining ratios and the molecular formulas the actual numbers of the constituent elements in chemical compounds.

EXPERIMENT 6

1. Numerous examples of redox and metathesis reactions could be cited. Two are $CuO + H_2 \rightarrow Cu + H_2O$ (redox) and $AgNO_3 + NaI \rightarrow AgI + NaNO_3$ (metathesis).
2. Reactions will proceed to completion whenever one of the products is

physically removed from the reaction medium. This often occurs when a gas or a solid is formed.

3. Percent yield = (actual yield/theoretical yield) (100).
4. Materials can be separated one from another by distillation, sublimation, filtration, decantation, sedimentation, chromatography, and extraction.
5. Exothermic reactions are usually accompanied by a temperature increase, while endothermic reactions are usually accompanied by a temperature decrease.
6. The balanced chemical equation is $Cu + 4HNO_3 \rightarrow Cu(NO_3)_2 + 2H_2O + 2NO_2$. From this equation we have moles $Cu(NO_3)_2$ = moles Cu and

$$\text{Moles Cu} = \frac{0.93 \text{ g}}{63.5 \text{ g/mol}} = 0.146 \text{ mol}$$

The theoretical yield is $Cu(NO_3)_2 = (0.146 \text{ mol})(188 \text{ g/mol}) = 2.75 \text{ g}$

$$\% \text{ yield} = \frac{1.65 \text{ g} \times 100}{2.75 \text{ g}} = 60\%$$

7. The maximum percent yield in any reaction is 100%.

EXPERIMENT 7

1. Many chemicals are hazardous, and the haphazard mixing of chemicals could lead to the production of species with very dangerous properties.
2. Addition of strong base to a solution containing NH_4^+ will release gaseous NH_3, which will cause a piece of moist, red litmus held above the reaction solution to turn blue.
3. Addition of an acid to a solution containing CO_3^{2-} will release gaseous CO_2, which will react with $Ba(OH)_2$ to precipitate $BaCO_3$.
4. Solid chloride salts will release gaseous HCl when heated with concentrated H_2SO_4, and solutions of chloride salts will precipitate AgCl if treated with $AgNO_3$ solution.
5. Solutions of sulfate salts will precipitate $BaSO_4$ when treated with $BaCl_2$.
6. Solutions of iodide salts will react with Cl_2 to liberate I_2, which will appear brown in H_2O and purple in mineral oil.
7. $2LiCl(s) + H_2SO_4(aq) \rightarrow 2HCl(g) + Li_2SO_4(aq)$;
$NH_4^+(aq) + OH^-(aq) \rightarrow NH_3(g) + H_2O(l)$;
$AgNO_3(aq) + I^-(aq) \rightarrow AgI(s) + NO_3^-(aq)$;
$NaHCO_3(s) + H^+(aq) \rightarrow CO_2(g) + H_2O(l) + Na^+(aq)$.

EXPERIMENT 8

1. Gravimetric analyses involve a weighing as the determining measurement, while volumetric analyses involve a volume measurement as the determining measurement.

2. Stoichiometry is the mole ratio of atoms in a compound or compounds in a chemical reaction and refers to the amounts of substances involved in reactions.
3. Silver chloride is photosensitive and reacts with light to produce silver metal and chlorine gas, which will lead to a low result if the silver chloride is not protected from light.
4. Indeterminate errors are just that, indeterminant. They cannot be ascertained or eliminated, but rather occur in a random fashion and so are accounted for with statistics.
5. Standard deviation measures precision.
6. If the filter paper were opened in this way, the solution and precipitate would simply pass through, and no filtration would result.
7. Since photodecomposition liberates Cl_2 from AgCl according to the equation $2AgCl \xrightarrow{hv} 2Ag + Cl_2$, the Ag will weigh less than the AgCl and your results will be low.

EXPERIMENT 9

1. A coordinate covalent, or dative, bond is formed by the union of a Lewis acid with a Lewis base. It is a two-electron covalent bond wherein the base donates both electrons in the bond. In the formation of a covalent bond, one electron comes from each species. For example:

$H_3N:$	+	BF_3	$\longrightarrow$	$H_3N:BF_3$
Lewis base		Lewis acid		coordinate covalent bond
		H. + .H	$\longrightarrow$	H:H
				covalent bond

2.

$$\text{Moles } Co^{2+} = \text{moles } [Co(C_5H_5N)_4(NCS)_2]$$

$$\text{Grams } Co^{2+} = (\text{moles } Co^{2+})(\text{g-at. wt Co})$$

Hence,

$$\text{Grams Co in sample} = \frac{(\text{g-at.wt Co})(\text{g}[Co(C_5H_5N)_4(NCS)_2]}{(\text{g-mol wt}[Co(C_5H_5N)_4(NCS)_2])}$$

$$= \frac{(58.93)(\text{g}[Co(C_5H_5N)_4(NCS)_2])}{(491.29)}$$

$$= (0.1199)(\text{g}[Co(C_5H_5N)_4(NCS)_2])$$

$$\%\ Co = \frac{(\text{g Co in sample})(100)}{\text{g-weight sample}}$$

3. Chelation is the process of forming metal-ligand bonds that result in a cyclic structure; the structure usually contains five or six atoms in a ring in which a central metallic ion is held in a coodination complex. The word is derived from the Greek word *chēlē,* meaning claw.

4. A Lewis acid is an electron-pair acceptor, and a Lewis base is an electron-pair donor, as illustrated in the answer to question **1**.

5.
$$\text{Average \% Cu} = \frac{9.15 + 9.06 + 9.22}{3} = 9.14$$

Standard deviation

$$= \sqrt{\frac{(9.15 - 9.14)^2 + (9.14 - 9.06)^2 + (9.14 - 9.22)^2}{2}}$$

$$= \sqrt{\frac{(0.01)^2 + (0.08)^2 + (0.08)^2}{2}}$$

$$= \sqrt{\frac{0.001 + 0.0064 + (0.0064)}{2}}$$

$$= \sqrt{\frac{0.0129}{2}} = \sqrt{0.0065} = 0.01$$

EXPERIMENT 10

1. Chromatography is yet another technique that permits the separation of substances.
2. An R_f value is the distance that a substance moves on a chromatographic support relative to the distance that the solvent moves on the same support under the same conditions. It aids in the qualitative identification of substances.
3. The reactions are, for nickel, $Ni^{2+} + 2DMG^- \rightarrow Ni(DMG)_2$, where DMG^- is the monoanion of dimethylglyoxime; and for copper, $Cu^{2+} + 4NH_3 \rightarrow [Cu(NH_3)_4]^{2+}$.
4. The solvent moves along the chromatographic support because of capillary action and dipole-dipole attraction.

10. The petri dish (or beaker) should be covered so as to maintain equilibrium between the solvents in the liquid and gaseous phases. If the vessel were not covered, volatile liquid would escape.

EXPERIMENT 11

1. Red, orange, yellow, green, blue, indigo, violet.
3. $\lambda = c/\nu = 3 \times 10^8 \text{ msec}^{-1}/76 \text{ sec}^{-1} = 3.9 \times 10^6\text{m}$
4. $E = h\nu = 6.63 \times 10^{-34} \text{ J} \cdot \text{sec} \times 76 \text{ sec}^{-1} = 5.04 \times 10^{-32} \text{ J}$
5. Green light has the higher frequency and greater energy because it has the shorter wavelength.

EXPERIMENT 12

1. The volume of an ideal gas increases as the temperature increases at constant pressure.
2. The volume of an ideal gas decreases as the pressure increases at constant temperature.
3. The volume of an ideal gas increases as the number of molecules increases at constant temperature and pressure.
4. $PV = nRT$; when the units of R are L-atm/mol-K, P is in atm, V in L and T in K.
5. Combining the ideal-gas law and the definition of moles, we have

$$PV = nRT \text{ and moles} = \frac{\text{weight}}{\mu\text{W}}$$

$$n = \frac{PV}{RT} = \frac{\text{weight}}{\mu\text{W}} \quad \text{or} \quad \mu\text{W} = \frac{(\text{weight})(RT)}{(PV)}$$

$$\mu\text{W} = \frac{(\text{weight})(RT)}{(PV)}$$

6.

$$\mu\text{W} = \frac{(0.75\text{ g})\left(\dfrac{0.082\text{ L-atm}}{\text{mol-K}}\right)(300\text{ K})}{\left(\dfrac{700\text{ mm}}{760\text{ mm/atm}}\right)(0.300\text{ L})}$$

$$= 67\text{ g/mol}$$

7. STP is 273 K and 1 atm, whence

$$\text{V} = (250\text{ mL})\left(\frac{273\text{ K}}{300\text{ K}}\right)\left(\frac{650\text{ mm}}{760\text{ mm}}\right)$$

$$= 210\text{ mL}$$

EXPERIMENT 13

1. Gases obey the ideal gas law at high temperatures and low pressures.
2. From $PV = nRT$, $R = PV/nT$ and STP is 1 atm and 0°C, or 273 K, whence

$$R = \frac{(1\text{ atm})(22.4\text{ L})}{(1\text{ mol})(273\text{ K})} = 0.082\frac{\text{L-atm}}{\text{mole-K}}$$

3. Equalizing the water levels equalizes the pressures and ensures that the total pressure in the bottle is atmospheric and does not contain a contribution from the pressure due to the height of the water column.
4. Since gaseous and liquid water are in dynamic equilibrium, there will al-

ways be some water vapor above a sample of liquid water. Since the vapor pressure of water is reasonably high at ambient temperature, it makes a significant contribution to the total pressure.

5. An error analysis allows you to judge the reliability of your data and gives an indication of the potential sources of error.
6. The ideal-gas law assumes that there are no forces of attraction between the individual gaseous molecules. Whenever this isn't so, real molecules will not obey the ideal-gas law. This would be expected to occur at very high pressures and at very low temperatures where molecules are so close to one another that they necessarily interact.
7. $$PV = nRT$$

$$P = \frac{nRT}{V} = \frac{\left(\frac{0.05\ \text{g}}{2\ \text{g/mol}}\right)\left(\frac{0.082\ \text{L-atm}}{\text{mol-K}}\right)(298\ \text{K})}{0.100\ \text{L}}$$

$$= 6.10\ \text{atm}$$

assuming no gas in the void volume. Clearly this does present a problem, as H_2 gas is extremely explosive. Because lead-storage batteries produce H_2, sealing them would be dangerous.

EXPERIMENT 14

1. Most metals have high thermal and electrical conductivities, high luster, malleability, and ductility, while nonmetals usually have low thermal and electrical conductivities, low luster, low malleability, and low ductility. In addition, relative to nonmetals, metals have low ionization potentials and low electron affinities.
2. Ionization energy is the energy required to remove an electron from an atom in the gaseous state.
3. Electron affinity is the energy produced when an electron is added to a species and is the inverse of the ionization energy in a physical sense, but the values are not just of opposite sign because you are considering slightly different processes; in each case, they differ by one electron.
4. An oxidation must always be accompanied by a reduction because the species being oxidized must transfer an electron to some other species that is reduced. The electron cannot just be given up to free space.
5. By systematically observing the displacement reactions among metals and their cations, it is possible to determine the relative oxidation potentials of the metals. The metal with the lower oxidation potential will reduce a cation of a metal with a higher oxidation potential.
6. $2Mg + O_2 \rightarrow 2MgO$; $Zn + 2HCl \rightarrow ZnCl_2 + H_2$; $Zn + Cu^{2+} \rightarrow Zn^{2+} + Cu$.

EXPERIMENT 15

1. Hydrogen is produced from H_2O at the cathode. Hydrogen thus undergoes a change in oxidation state from +1 in H_2O to O in H_2, and this is a reduction reaction.
2. A coulomb is the quantity of electrical charge passing a point in a circuit in 1 s when the current is 1 ampere.
3. In general, reduction occurs at the cathode and oxidation at the anode in an electrochemical cell regardless of whether the cell is producing energy (a galvanic cell) or consuming energy (an electrolytic cell).
4. The water column exerts a pressure that is directly proportional to the height of the water column, just as atmospheric pressure and altitude are related. The higher the altitude above sea level, the less the atmosphere and the lower the atmospheric pressure.
5. The sulfuric acid is present to increase the rate of the reaction by increasing the ability of the solution to conduct an electrical current. According to Faraday's law, the more current that passes through this solution, the larger the number of species that will react. The more ions that are present in solution, the greater will be the conductivity of the solution and the greater the current flow. Sulfuric acid is a strong electrolyte and is almost completely dissociated in solution to produce the ions H_3O^+ and SO_4^{2-}. It does not directly enter into the electrochemical reaction.

EXPERIMENT 16

1. A faraday is the charge on 1 mol of electrons and equals 96,500 coulombs. A salt bridge is an ion-containing conducting medium used to physically and chemically separate two half-cells in an electrochemical cell. An anode is an electrode at which oxidation occurs, and a cathode is an electrode at which reduction occurs in an electrochemical cell. A voltaic cell is an electrochemical cell that produces electrical energy by means of a spontaneous redox reaction. An electrolytic cell is one that requires energy in order to bring about a redox reaction.
2. The cell $Ag|Ag^+||Cu^{2+}|Cu$ represents the reaction $2Ag + Cu^{2+} \rightarrow 2Ag^+ + Cu$.
3. $E° = 0.34 - 0.08\text{ V} = 0.26\text{ V}$
4. The cell $Zn|Zn^{2+}(0.10\ M)||Cu^{2+}(0.40\ M)|Cu$ represents the reaction $Zn + Cu^{2+} \rightarrow Zn^{2+} + Cu$, for which

   $$E = E° - (0.059/2)\log([Zn^{2+}]/[Cu^{2+}]),$$

   whence

   $$E = 1.100\text{ V} - 0.0295\log([0.10]/[0.40]) = 1.100 + 0.0178 = 1.178\text{ V}.$$
5. $E° = -1.18 + 2.90 = 1.72$ V. To calculate K_{eq}, we use the equation $E = E° - (0.059/n)\log K_{eq}$. Because at equilibrium E = O, we have

$$E^\circ = \left(\frac{0.059}{n}\right)\log K_{eq}$$

$$\log K_{eq} = \frac{nE^\circ}{0.059} = \frac{(2)(1.72)}{0.059} = 58.3050$$

$$K_{eq} = \text{antilog}(58.3050) = 2.02 \times 10^{58}$$

The free energy for the given cell is

$$\Delta G^\circ = -n\mathcal{F}E^\circ = -(2 \text{ mol } e-)\left(\frac{96{,}500 \text{ J}}{V\text{-mol } e-}\right)(1.72 \text{ V})$$

$$= -331{,}960 \text{ J} = -331.96 \text{ kJ}$$

6. Most spontaneous reactions (ΔG negative) are exothermic (ΔH negative). Because voltaic cells have spontaneous reactions, we would expect ΔH to be negative for most voltaic cells.

EXPERIMENT 17

1. Ionic bonding results from transfer of electrons from one species to another to form a chemical bond, while covalent bonding results from sharing electron density (equally or unequally) between the two partners forming the chemical bond. Ionic bonding will result from the union of atoms having low ionization potentials with atoms of high electron affinity. Covalent bonding will result from the union of atoms with similar or identical ionization potentials and electron affinities.

2. An anhydride is literally a compound without water. Thus, $H_2SO_4 - H_2O = SO_3$; SO_3 is the anhydride of H_2SO_4.

3. According to the Brønsted-Lowry definition, an acid is a proton donor and a base is a proton acceptor.

4. The autodissociation of water is $2H_2O \rightleftarrows H_3O^+ + OH^-$, for which $K_{eq} = [H_3O^+][OH^-]/[H_2O]^2$ and $K_w = [H_3O^+][OH^-]$. Since the molar concentration of water changes so little in this reaction, it is essentially a constant and is incorporated in K_w. $K_w = 55.6\ K_{eq}$ for dilute aqueous solutions.

5. The concept of pH is introduced to avoid the manipulation of very small numbers, because the concentrations of H_3O^+ are usually very small in aqueous solution.

6. The following table illustrates when aqueous solutions are acidic, basic, or neutral in terms of $[H^+]$, $[OH^-]$, and pH:

	$[H^+]$	$[OH^-]$	pH
Acidic	$> 10^{-7}\ M$	$< 10^{-7}\ M$	< 7
Basic	$< 10^{-7}\ M$	$> 10^{-7}\ M$	> 7
Neutral	$= 10^{-7}\ M$	$= 10^{-7}\ M$	$= 7$

7. pH = $-\log[H^+] = -\log(10^{-4}) = 4$;
$[OH^-] = 10^{-14}/[H^+] = 10^{-14}/10^{-4} = 10^{-10}$ *M*.
8. $Li_2O + H_2O \rightarrow 2LiOH$; $N_2O_5 + H_2O \rightarrow 2HNO_3$.

EXPERIMENT 18

1. Solute is is the lesser component and solvent the greater component in a solution.
2. Three colligative properties are boiling point, freezing point, and vapor pressure. They are called colligative properties because they are related to the number and energy of collisions between particles and not to what the particles are.
3. A volatile substance has a high vapor pressure, and a nonvolatile substance has a low vapor pressure at room temperature. Obviously, volatility is a relative term and depends upon temperature and pressure. Volatility increases with increasing temperature and decreases with increasing pressure.
6. Supercooling involves the lowering of the temperature of a substance below its normal freezing point without the solidification of the substance. Supercooling can be minimized by cooling slowly with rapid stirring.
7. Mol of benzene is 6.50 g/(78 g/mol) = 0.083 mol. Consequently, a solution of 6.5 g of benzene in 160 g of $CHCl_3$ is 0.083 mol/0.160 kg or 0.52 molal. The freezing-point lowering is thus $\Delta T = K_{fp}m$, or ΔT = (4.68°C/m)(0.52 *m*) = 2.43°C, and the freezing point is −63.5°C − 2.43°C = −65.9°C.
8. ΔT = 80.6°C − 75.4°C = 5.2°C; $m = \Delta T/K_{fp}$ = 5.2°C/(6.9°C/*m*) = 0.753 *m*; *m* = moles solute/1000 g solvent, whence 1.00 g/12.5 g = *x* g/1000 g, x = 80 g, and 80 g = 0.753 mol, or MW = 80 g/0.753 mol = 106.2 g/mol.

EXPERIMENT 19

1. Standardization is the process of determining the concentration of a solution. It is usually achieved by titrating the solution to be standardized against a known amount of a primary standard substance according to a known reaction.
2. Titration is the technique of accurately measuring the volume of a solution that is required to react with a known amount of another reagent.
3. Normality is the number of gram equivalent weights of solute in a liter of solution. Molarity is the number of moles of solute in a liter of solution. Normality (*N*) = equivalents solute/liter solution; molarity (*M*) = moles solute/liter solution; *N* = (*M*)(equivalents/moles).
4. You weigh by difference because this is in general more accurate than weighing directly.
5. An equivalence point is the point in a titration where the numbers of equivalents of the two reactants are precisely equal. An end point is the

point in a titration where some indicator (such as a dye or electrode) undergoes a discernible change. Ideally, one hopes that these two points coincide, but in practice they differ slightly. It becomes of primary importance to minimize the difference if accuracy is desired.

6. Parallax is the apparent displacement or the difference in apparent direction of an object as seen from two different points not on a straight line with the object. It should be avoided because it introduces an error in the measurement.
7. Carbon dioxide should be removed from the water because it is an acid anhydride and reacts with water to produce carbonic acid according to the reaction $CO_2 + H_2O \rightleftarrows H_2CO_3$. Its presence would lead to an erroneous determination of the amount of a particular acid or base in solution. It is particularly detrimental to sodium hydroxide solutions as these absorb carbon dioxide to produce sodium carbonate.
8.

$$\text{Normality} = \frac{\text{equivalents solute}}{\text{liter solution}}$$

$$= \frac{(1.89\text{ g})(2\text{ equiv/mol})}{(126\text{ g/mol})(0.100\text{ L})} = 0.300N$$

EXPERIMENT 20

1. (a) Molecular: $2HNO_3(aq) + BaCO_3(s) \rightarrow Ba(NO_3)_2(aq) + CO_2(g) + H_2O(l)$; ionic: $2H^+(aq) + 2NO_3^-(aq) + BaCO_3(s) \rightarrow Ba^{2+}(aq) + 2NO_3^-(aq) + CO_2(g) + H_2O(l)$; net ionic: $BaCO_3(s) + 2H^+(aq) \rightarrow Ba^{2+}(aq) + H_2O(l) + CO_2(g)$
   (b) As above; insoluble product $PbCl_2(s)$ is formed; net ionic: $Pb^{2+}(aq) + 2Cl^-(aq) \rightarrow PbCl_2(s)$
   (c) net ionic: $HC_2H_3O_2(aq) + OH^-(aq) \rightarrow H_2O(l) + C_2H_3O_2^-(aq)$.
2. The following are not water soluble: $CuCO_3$, ZnS.
5. $BaCl_2$, $AgNO_3$, HCl and HNO_3 are strong electrolytes.
6. HF and NH_3 are weak electrolytes.

EXPERIMENT 21

1. According to the Brønsted-Lowry definition, an acid is a proton donor and a base is a proton acceptor.
2. A weak acid dissociates in aqueous solution according to the equilibrium $HA + H_2O \rightleftarrows H_3O^+ + A^-$, for which the equilibrium constant is $K_{eq} = [H_3O^+][A^-]/[HA][H_2O]$, and the dissociation constant is $K_a = [H_3O^+][A^-]/[HA]$ or $K_a = [H_2O]K_{eq}$.
3. The pH at the equivalence point in an acid-base titration depends upon the nature of the species present. For the titration of a strong acid and a strong base, the pH will be 7 because a salt that does not hydrolyze will be formed (for example, NaOH + HCl). For the titration of a strong acid with a weak base (for example, HCl + NH_4OH), the pH will be less

than 7; and for the titration of a weak acid with a strong base (e.g., HOAc + NaOH), the pH will be greater than 7. This is so because the salt formed in each case (NH_4Cl and NaOAc) will undergo hydrolysis reactions with water.

4. $pK_a = -\log K_a = -\log(3.6 \times 10^{-6}) = 6 - \log 3.6 = 6 - 0.56 = 5.44$.
5. Two electrodes are necessary for an electrical measurement because some current flow must occur and this requires both a donor and an acceptor for the electrons.
6. A buffer solution is a solution that is resistant to a pH change. It always contains a weak electrolyte and normally is composed of two species, such as a weak acid and one of its salts or a weak base and one of its salts. Two specific examples are acetic acid plus sodium acetate and ammonium hydroxide plus ammonium chloride. An example of a single-component buffer solution is disodium phosphate, Na_2HPO_4.
7. At one-half equivalence point $[HA] = [A^-]$. Because $HA \rightleftarrows H^+ + A^-$ and $K_a = [H^+][A^-]/[HA]$, at one-half equivalence point, $K_a = [H^+]$. Therefore, $pK_a = pH$, so $pK_a = 5.32$, $K_a = \text{antilog}(-5.32) = \text{antilog}(0.68) \times 10^{-6} = 4.8 \times 10^{-6}$.

EXPERIMENT 22

1. A polyprotic acid is one that possesses more than one ionizable proton. For example, H_2SO_4 is a diprotic acid and H_3PO_4 a triprotic acid because they possess two and three ionizable protons, respectively.
3. In an acid-base reaction, $H_2C_2O_4$ has two equivalents per mole because it possesses two potentially ionizable protons. As a result, a 6.2 *N* solution is 3.1 *M* because normality is two times molarity.
4. There are (0.05 L)(0.3 equiv/L) = 1.5×10^{-2} equivalents of H_3O^+ present in 50 mL of 0.3 *N* H_2SO_4. Since there are 2 equivalents per mole for H_2SO_4, there are 7.5×10^{-3} mol of H_2SO_4 and 1.5×10^{-2} mol of H_3O^+ present in this solution.
5. A pH meter must be standardized because it measures relative potentials and thus relative pH. It is necessary to know to what the measurement is relative; consequently, a standard must be measured and the meter set to the known value for this standard.
6. See the answer to question 7, Experiment 21, for the method: $pK_1 = 3.52$; $K_1 = 3.02 \times 10^{-4}$; $pK_2 = 6.31$; $K_2 = 4.88 \times 10^{-7}$.

EXPERIMENT 23

1. According to the Brønsted definitions, an acid is a proton donor and a base is a proton acceptor.
2. Cu^{2+}, Zn^{2+} and SO_3^{2-} will undergo hydrolysis.
3. For example, $Cu^{2+}(aq) + 2H_2O(l) \rightleftarrows Cu(OH)_2(s) + 2H^+(aq)$.
4. $K_b = 1.0 \times 10^{-14}/2.1 \times 10^{-9} = 4.8 \times 10^{-6}$

5. The conjugate base is HPO_4^{2-} and the conjugate acid is H_3PO_4.
6. For example, $CaSO_4$ may be made from $Ca(OH)_2$ and H_2SO_4.

EXPERIMENT 24

1. $CaF_2 \rightleftharpoons Ca^{2+} + 2F^-$; $K_{sp} = [Ca^{2+}][F^-]^2$

2.
$$\begin{aligned}\text{Moles } Ag^+ &= \left(\frac{5 \text{ mL}}{1000 \text{ mL/L}}\right)(0.004 \text{ mol/L}) \\ &= 2 \times 10^{-5} \text{ mol} \\ \text{Moles } CrO_4^{2-} &= \left(\frac{5 \text{ mL}}{1000 \text{ mL/L}}\right)(0.0024 \text{ mol/L}) \\ &= 1.20 \times 10^{-5} \text{ mol}\end{aligned}$$

3. $2AgNO_3 + K_2CrO_4 \rightleftharpoons 2KNO_3 + Ag_2CrO_4$. The stoichiometry requires 2 mol of $AgNO_3$ for each mole of K_2CrO_4.

$$\begin{aligned}\text{Millimoles } AgNO_3 &= (10 \text{ mL})(0.004 \text{ millimol/mL}) \\ &= 0.04 \text{ millimol} \\ \text{Millimoles } K_2CrO_4 &= (10 \text{ mL})(0.0024 \text{ millimol/mL}) \\ &= 0.024 \text{ millimol}\end{aligned}$$

and 0.024 millimol K_2CrO_4 would require 0.048 millimol $AgNO_3$. Hence, K_2CrO_4 is in excess.

4. $[Ba^{2+}] = (10 \text{ mL}/20 \text{ mL})(1 \times 10^{-4} \text{ mol/L}) = 5 \times 10^{-5}$ mol/L; $[CrO_4^{2-}] = (10 \text{ mL}/20 \text{ mL})(1 \times 10^{-4} \text{ mol/L}) = 5 \times 10^{-5}$ mol/L. Hence, $K_{sp} = [Ba^{2+}][CrO_4^{2-}] = 1.09 \times 10^{-5}$ and ion product = $(5 \times 10^{-5})(5 \times 10^{-5}) = 2.5 \times 10^{-9}$. Since the ion product is less than K_{sp}, no precipitation would occur.
9. A sparingly soluble salt will precipitate when the ion product exceeds K_{sp}.

EXPERIMENT 25

1. An exothermic reaction is a reaction that produces heat; its ΔH will be less than zero, or negative. An endothermic reaction is a reaction that absorbs heat; its ΔH will be greater than zero, or positive.
2. (425 mL)(1.0 g/mL)(50.0°C − 10.0°C)(4.18 J/g − °C) = 71,060 J = 71.1 kJ.
3. The heat capacity of a substance is the amount of energy, usually in the form of heat, necessary to raise the temperature of a specified amount of the substance (usually 1 g) by 1°C.
4. ΔT = 38.8°C − 23.3°C = 15.5°C; heat required = (80.0 g)(4.18 J/g · °C)(15.5°C) = 5180 J.

EXPERIMENT 26

1. The factors that influence the rate of a chemical reaction are temperature, pressure, concentration, particle size, catalysts, and the nature of the species undergoing the reaction.
2. The general form of the rate law for the reaction A + B → is: Rate = $k[A]^x[B]^y$.
3. A reaction that obeys the rate law of the form rate = $k[A]^2[B]^3$ is second order in A and third order in B. This reaction is fifth order overall.
4. The chemical reactions involved in this experiment are $S_2O_8^{2-} + 2I^- \rightarrow I_2 + 2SO_4^{2-}$, $I_2 + 2S_2O_3^{2-} \rightarrow 2I + S_4O_6^{2-}$, and I_2 + starch → starch · I_2(blue), for which the rate of disappearance of $S_2O_3^{2-}$ is $k[S_2O_8^{2-}]^x[I^-]^y$ and the rate of appearance of blue color equals the rate of formation of the starch-iodine complex, which is proportional to the rate of appearance of iodine. Thus the rate of appearance of iodine is $k[S_2O_8^{2-}]^x[I^-]^y$ and the rate of appearance of blue color is $k[I_2]^x[\text{starch}]^y \alpha k[S_2O_8^{2-}]^x[I^-]^y$.
5. Rate = $k[A]^2[B]^2$.
6. $$\text{Rate} = \left(\frac{2 \times 10^{-4}\ \text{mol}}{0.050\ \text{L}}\right)\left(\frac{1}{188\ \text{sec}}\right)$$
$$= 2.1 \times 10^{-5}\ \frac{\text{mol}}{\text{L-sec}}$$
7. From a knowledge of the rate of chemical reactions, chemists can determine how long to work in order to obtain the desired products. Clearly, this is of extreme practical value in the synthesis of new compounds.

EXPERIMENT 27

1. See answer 1 to Experiment 26.
2. (a) A reaction with a rate of law of the form rate = $k[A]^2[B]$ is third order.
   (b) If the concentrations of both A and B are doubled, the rate will increase by a factor of 8.
   (c) k is known as the specific rate constant.
   (d) changing the concentrations of the reactants may change the rate but does not change the specific rate constant. The rate will be quadrupled.
5. Rate = $k[A]^2[B]^2$.

EXPERIMENT 28

I CATIONS

1. Ammonium, NH_4^+; silver, Ag^+; ferric or iron(III), Fe^{3+}; aluminum, Al^{3+}; chromic or chromium(III), Cr^{3+}; calcium, Ca^{2+}; magnesium, Mg^{2+}; nickel or nickel(II), Ni^{2+}; zinc, Zn^{2+}; and sodium, Na^+.

2. Sometimes ions behave very similarly. For example, Ba^{2+} and Pb^{2+} both form yellow precipitates with K_2CrO_4. Hence, other additional confirmatory tests would be required to distinguish between these ions. For example, Pb^{2+} forms a white precipitate with HCl while Ba^{2+} does not.
3. Only three of the ions are colored as follows: Fe^{3+}, rust to yellow; Cr^{3+}, blue-green; Ni^{2+}, green.
4. Because only AgCl precipitates on the addition of HCl to a solution of the ten cations, all chlorides of these ions except AgCl must be soluble.
5. Based on the group separation chart (Figure 28.1) Ag^+ can be separated from all of the other ions by the addition of HCl. Silver forms insoluble AgCl.
6. Examination of the group separation chart shows that in the presence of the buffer NH_3—NH_4Cl, Cr^{3+} precipitates as $Cr(OH)_3$ while Mg^{2+} remains in solution.
7. Addition of chloride ions to a solution containing Al^{+3} and Ag^+ would precipitate AgCl, while Al^{3+} would remain in solution. HCl would be a good source of chloride ions, and the H^+ would help retard hydrolysis of Al^{3+}.
8. $NH_4^+ + OH^- \rightleftarrows NH_3 + H_2O$; $AgCl + 2NH_3 \rightleftarrows [Ag(NH_3)_2]^+ + Cl^-$.

II ANIONS

1. Sulfate, SO_4^{2-}; nitrate, NO_3^-; carbonate, CO_3^{2-}; chloride, Cl^-; bromide, Br^-; and iodide, I^-.
2. See Table 28.1: behavior of anions with concentrated sulfuric acid.
3. No. Because HCl is pungent and forms on the addition of H_2SO_4 to chloride salts, it would be impossible to tell that a second gas that is odorless is also present. Hence you would not be aware of the presence of CO_3^{2-} in the mixture.

EXPERIMENT 29

CATIONS

Part A

1. Ag^+, Hg_2^{2+} and Pb^{2+}
2. $PbCl_2$.
3. AgCl.
4. (a) $BaCl_2$ is water soluble and AgCl is not.
   (b) HCl forms a white precipitate with $AgNO_3$ while HNO_3 does not form a precipitate with $AgNO_3$.
8. With the use of a small pipet.

Part B

1. Pb^{2+}, Cu^{2+}, Bi^{3+}, Sn^{4+}
2. CuS does not dissolve in aqueous $(NH_4)_2S$ while SnS_2 does.
3. Aqueous NH_3 precipitates $Bi(OH)_3$ and $Cu(NH_3)_4^{2+}$ remains in solution.

Part C

1. Fe^{3+}, Ni^{2+}, Mn^{2+}, Al^{3+}.
2. Cu^{2+} (blue), Fe^{3+} (yellow to reddish brown), Al^{3+} (colorless), Ni^{2+} (green).
3. $Fe(OH)_3$ (red-brown), MnS (salmon), $Al(OH)_3$ (white) $Ni(OH)_2$ (green).
4. Aqueous NaOH precipitates $Fe(OH)_3$ while $Al(OH)_4^-$ remains in solution.

7. (a) HCl or H_2SO_4.
   (b) excess NaOH.

Part D

1. Ba^{2+}, Ca^{2+}, NH_4^+, Na^+.
2. $BaCrO_4$.
3. $BaSO_4$ (white), $BaCrO_4$ (yellow), CaC_2O_4 (white).
4. (a) Ba^{2+} (green).
   (b) Ca^{2+} (orange-red).
   (c) Na^+ (yellow).

Part E

ANIONS

1. Sulfate, SO_4^{2-}; nitrate, NO_3^-; carbonate, CO_3^{2-}; chromate, CrO_4^{2-}; phosphate, PO_4^{3-}; chloride, Cl^-; bromide, Br^-; sulfide, S^{2-}; sulfite, SO_3^{2-}; and iodide, I^-.
2. See Table 29.1: behavior of anions with concentrated sulfuric acid.
3. No. Because HCl is pungent and forms on the addition of H_2SO_4 to chloride salts, it would be impossible to tell that a second gas that is odorless is also present; Hence, you would not be aware of the presence of CO_3^{2-} in the mixture.
4. (a) SO_4^{2-}.
   (b) CO_3^{2-}, PO_4^{3-}, S^{2-}, or OH^-.
   (c) Cl^-.
5. SO_4^{2-}, CO_3^{2-}, OH^-, CrO_4^{2-}, PO_4^{3-}, Cl^-, Br^-, and I^-.
6. CO_3^{2-}.

EXPERIMENT 30

1. The concentration, the cell pathlength, wavelength and solvent.
2. Absorbance $A = \log I_0/I$ and percent transmittance $T = (I/I_0)100$, so $T/100 = I/I_0$ or $100/T = I_0/I$ and $\log(100/T) = \log(I_0/I) = A$.
3. A spectrophotometer must possess a light source, a monochromator, a sample cell, a detector, and a meter.
4. The Beer-Lambert law is $A = abc$, where A is the absorbance, a is the absorptivitiy or extinction coefficient, b is the solution path length, and c is the molar concentration of absorbing species.
5. Not all substances precisely obey the Beer-Lambert law over all concentration ranges. A calibration curve will provide the relationship between concentration and absorbance under conditions that are similar to or identical with those used for the analysis.
6. Hydroxylamine is used to reduce Fe^{3+} to Fe^{2+}.
7. Since $\% \ T = (I/I_0)(100)$, $\log \ \% \ T = \log(I/I_0) + \log 100$, or $\log(I_0/I) = 2 - \log \ \% \ T$. $A = 2 - \log \ \% \ T = 2 - \log 100 = 2 - 2 = 0$.

EXPERIMENT 31

1. The volumes of 1×10^{-3} M solutions required to prepare 50 mL of solutions with the cited concentrations are

Concentration (M) $\times 10^{-5}$:	2	5	10	20	50	75
Volume of 10^{-3} M solution required (mL):	1	2.5	5	10	25	37.5

2. The species $(NH_4)_3PO \cdot NH_4VO_3 \cdot Mo_{16}O_{48}$ is thought to be the light absorbing species in this experiment.
3. The reaction $(NH_4)_3PO_4 + NH_4VO_3 + 16MoO_3 \rightarrow (NH_4)_4PVMo_{16}O_{55}$ is thought to be the reaction that forms the heteropoly acid that absorbs the light in this experiment.
6. A calibration curve is constructed by measuring the absorbances of solutions with varying known concentrations. It is constructed in order to ascertain the relationship between absorbance and concentration of the absorbing species.
7. If the measured absorbance of the unknown is greater than that of any of the calibration solutions, you should measure the absorbance of a more dilute solution. If the measured absorbance of the unknown is less than that of any of the calibration solutions, you should measure the absorbance of a more concentrated solution.
9. 1 mg FeSCN/L $= 1 \times 10^{-3}$ g FeSCN/(113.93 g/mol)L $= 8.78 \times 10^{-6}$ M if $T = 0.295$ then $A = 2 - \text{Log } 29.5 = 2 - 1.47 = 0.53$ and

from $A = abc$, $a = A/bc = 0.53/[1.00 \text{ cm}][8.78 \times 10^{-6} M] = 6 \times 10^4$ L/mol-cm.

EXPERIMENT 32

1. Molarity is defined as the number of moles of solute per liter of solution, and normality is defined as the number of equivalents of solute per liter of solution. For oxidation-reduction reactions, the number of equivalents per mole is determined by the number of electrons transferred per mole of reactant in the reaction. Normality is equal to molarity times the number of equivalents per mole.
2. From the relations equiv $Na_2S_2O_3$ = equiv KIO_3; equiv $Na_2S_2O_3 = (V)(N)$; and

$$\text{Equivalents } KIO_3 = \frac{\text{g-wt } KIO_3}{35.67 \text{ g/equiv}}$$

one obtains

$$N_{Na_2S_2O_3} = \frac{(\text{g-wt } KIO_3)}{(35.67 \text{ g/equiv})(V\, Na_2S_2O_3)}$$

so that

$$N_{Na_2S_2O_3} = \frac{(0.1309 \text{ g})}{(35.67 \text{ g/equiv})(0.02375 \text{ L})} = 0.1545\, N$$

3. Chloroform is an antibacterial agent and as such kills bacteria. Because the bacteria are killed, they do not consume oxygen by their growth, so the water sample is preserved by the addition of chloroform.
4. In the reaction $Mn(SO_4)_2 + 2KI \rightleftarrows MnSO_4 + K_2SO_4 + I_2$, manganese is reduced (its oxidation state changes from +4 to +2) and iodide is oxidized (its oxidation state changes from −1 to 0).
5. The blue color returns because oxygen in the air oxidizes iodide to iodine according to the reaction $4I^- + O_2 + 4H^+ \rightleftarrows 2I_2 + 2H_2O$.
6. $(1 \times 10^{-4}$ mol/L)(32 g/mol)(1000 mg/g) = 3.2 mg/L = 3.2 ppm.

EXPERIMENT 33

3. A coordination compound is a compound formed by the reaction of a Lewis acid with a Lewis base; it contains one or more coordinate covalent bonds. An example is the compound $H_3N{:}BF_3$.
4. Geometric isomers are isomers that have the same empirical and molecular formulas but differ in their spatial arrangements of the constituent atoms.

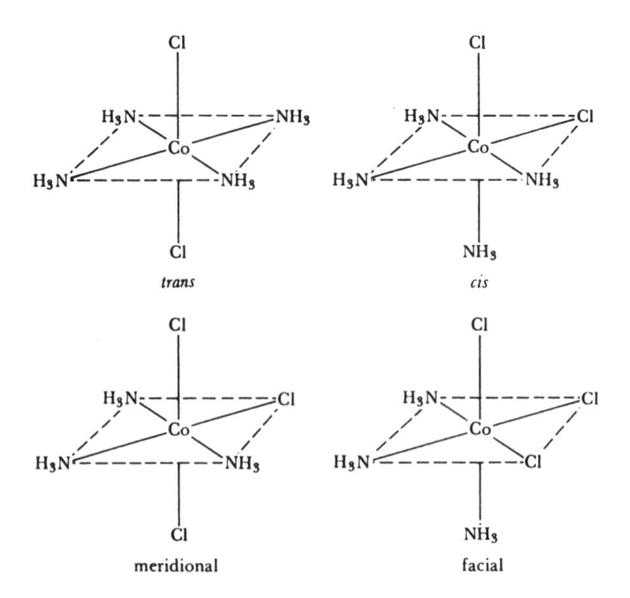

5. The compounds $[Co(NH_3)_4Cl_2]$ and $[Co(NH_3)_3Cl_3]$ can each exist in two geometric isomeric forms, which are shown above.
6. The chlorine atoms are in equivalent environments in *cis*- and *trans*-$[Co(NH_3)_4Cl_2]$ and facial $[Co(NH_3)_3Cl_3]$, but there are two different chlorine environments in meridional $[Co(NH_3)_3Cl_3]$.
7. Dichroism is the property according to which the colors of a crystal are different when the crystal is viewed in the direction of two different axes.
8. Triturate means to cause a semisolid to become a solid by crushing or grinding.
9. You must find the answer to question 9 in the library.
10. You must find the answer to question 10 in the library.
11. It works by forming a soluble iron(II) oxalate complex, $[Fe(C_2O_4)_3]^{4-}$, by reacting with iron oxide (rust).

EXPERIMENT 34

1. (a) $2NaHCO_3(s) \xrightarrow{\Delta} Na_2CO_3(s) + H_2O(g) + CO_2(g)$
   (b) $CaCO_3(s) \xrightarrow{\Delta} CaO(s) + CO_2(g)$
   (c) $CaO(s) + H_2O(l) \rightarrow Ca^{2+}(aq) + 2OH^-(aq)$
   (d) $NH_3(g) + H_2O(l) \rightleftarrows NH_4^+(aq) + OH^-(aq)$
   (e) $2NH_4^+(aq) + CaO(s) \rightarrow Ca^{2+}(aq) + H_2O(l) + 2NH_3(g)$
   (f) $H^+(aq) + HCO_3^-(aq) \rightarrow H_2O(l) + CO_2(g)$
2. Henry's law that states the solubility of a gas in a liquid is directly proportional to the partial pressure of the gas above the solution.
5. The common ion affect of Cl^- would decrease the solubility of NaCl.

EXPERIMENT 35

2. Given that equivalents oxalate = equivalents permanganate and that for $Na_2C_2O_4$ there are 2 equiv/mol so that the equivalent weight is 67 g/equiv, one can calculate the normality:

$$N_{KMnO_4} = \left(\frac{\text{g-wt } Na_2C_2O_4}{67 \text{ g/equiv}}\right)\left(\frac{I}{V \text{ KMnO}_4}\right)$$

$$= \frac{(0.5468 \text{ g})}{(67.00 \text{ g/equiv})(0.03543 \text{ L})} = 0.2303\ N$$

and since there are 5 equiv/mol of permanganate, $M = N/5 = 0.2303/5 = 0.04607\ M$.

3. The average is 15.67 and the standard deviation is 0.10.
4. The solution slowly decolorizes after the equivalence point has been reached because permanganate undergoes photochemical reduction and because it is slowly reduced by water.
5. In the permanganate-oxalate reaction, permanganate undergoes a five-electron change. Consequently, there are 5 equiv/mol of permanganate, and a 1 M solution is 5 N.
6. The permanganate is filtered in order to remove insoluble MnO_2. Because rubber reduces MnO_4^- to MnO_2, the solution should not contact rubber or other organic material.
7. $16H^+ + 2MnO_4^- + 5C_2O_4^{2-} \rightarrow 10CO_2 + 2Mn^{2+} + 8H_2O$.

$$\text{Moles } MnO_4^- = \left(\frac{2}{5}\right) \text{moles } C_2O_4^{2-}$$

$$= \left(\frac{2}{5}\right)\frac{(0.56 \text{ g})\left(\dfrac{2 \text{ mol } C_2O_4^{2-}}{\text{mole } K_2[Cu(C_2O_4)_2] \cdot 2H_2O}\right)}{(353.54 \text{ g/mole } K_2[Cu(C_2O_4)_2] \cdot 2H_2O}$$

$$= 1.267 \times 10^{-3} \text{ mol} = (V_{KMnO_4})(M_{KMnO_4})$$

$$V_{KMnO_4} = \frac{1.267 \times 10^{-3} \text{ mol}}{0.1 \text{ mol/L}}$$

$$= 1.267 \times 10^{-2} \text{ L} = 12.67 \text{ mL}$$

EXPERIMENT 36

2. $6Fe^{2+} + Cr_2O_7^{2-} + 14H^+ \rightarrow 6Fe^{3+} + 2Cr^{3+} + 7H_2O; Sn^{2+} + 2Fe^{3+} \rightleftarrows 2Fe^{2+} + Sn^{4+}$.
3. The equivalent weight of $K_2Cr_2O_7$ in this reaction is the formula weight divided by 6, that is, $294.2/6 = 49.03$. The equivalent weight of an oxidizing or reducing agent is that weight of the substance which gains or loses Avogadro's number of electrons. Therefore, $N = 3.42/(49.03 \times 0.250) = 0.279\ N$.

4. A suitable redox indicator should (a) change colors upon oxidation or reduction, (b) have an oxidization potential somewhere in between those of the oxidizing and reducing agents, and (c) be water soluble.
5. $SnCl_2$ is used to reduce iron(III) to iron(II) prior to titration with dichromate. The $HgCl_2$ is used to oxidize excess $SnCl_2$. Dichromate is capable of oxidizing $SnCl_2$; hence, $SnCl_2$ must be removed. This is achieved by $HgCl_2$ because Hg_2Cl_2 is insoluble.
6. If the solutions contain elemental mercury, it will be oxidized by $Cr_2O_7^{2-}$, and your results will be too high.

EXPERIMENT 37

1. Molecular formulas indicate the composition of molecules but do not indicate how the atoms are arranged; therefore, they do not distinguish between isomers. Structural formulas indicate how the atoms are arranged in molecules as well as the molecular composition. The molecular formula $C_2H_4Cl_2$ may refer to either isomer:

```
      H   H                 H   H
      |   |                 |   |
  H—C—C—H    or    H—C—C—Cl
      |   |                 |   |
      Cl  Cl                H   Cl
       (a)                   (b)
```

2. Condensed structural formulas are a convenient, simple way of writing formulas and are a "shorthand version" of structural formulas. For example, molecules (*a*) and (*b*) above may be written: (*a*) CH_2ClCH_2Cl and (*b*) CH_3CHCl_2.
3. Isomers are compounds with the same molecular formula but different structural formulas; for example, (*a*) and (*b*) above are isomers.
4. Because atoms are arranged differently in isomers, the molecules have different properties. For example, ethyl alcohol, CH_3CH_2OH, has an —OH group and in some ways resembles H—OH. Dimethyl ether, CH_3OCH_3, is an isomer of ethyl alcohol and does not contain the —OH group. Hence, it differs from ethyl alcohol.
5.

```
      H   H
      |   |
  H—C—C—H    or    CH3CH3
      |   |
      H   H
        Ethane

      H   H   H
      |   |   |
  H—C—C—C—H    or    CH3CH2CH3
      |   |   |
      H   H   H
          Propane
```

EXPERIMENT 38

1. Esters are derivatives of carboxylic acids and have the general formula

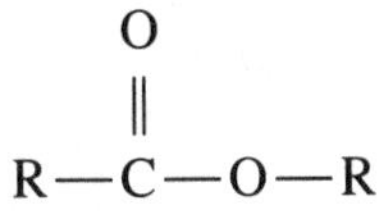

$$R-\overset{\overset{\Large O}{\|}}{C}-O-R$$

2. The mineral acid functions as a catalyst. It lets you accomplish the esterification in a shorter time.
3. Esters can be prepared from carboxylic acids and alcohols. They can also be made from acid anhydrides and alcohols.
4. Esters generally have pleasant, fruitlike odors.
5. Phenols form colored complexes with ferric chloride. The colors may range from green through violet. Hence $FeCl_3$ can be used to test for the presence of phenols.
6. The carboxylic acid group is

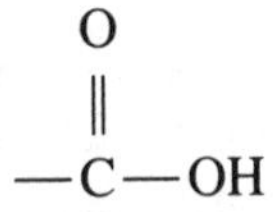

$$-\overset{\overset{\Large O}{\|}}{C}-OH$$

7. Recrystallization is a technique that is often utilized to purify a solid compound by separating impurities that are soluble in the solvent that is used for the recrystallization.
8.

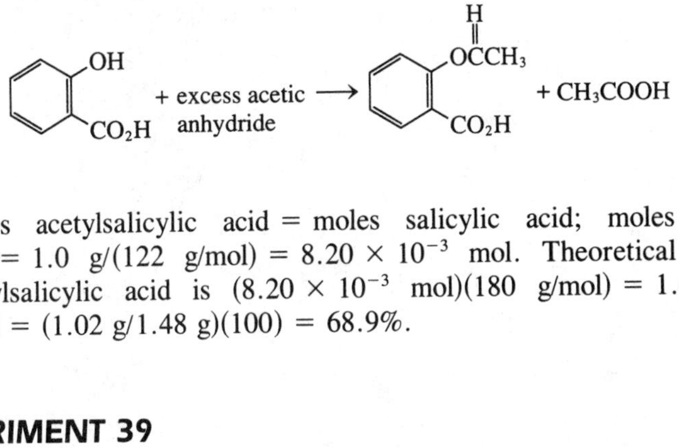

$$C_6H_4(OH)(CO_2H) + \text{excess acetic anhydride} \longrightarrow C_6H_4(O\overset{\overset{H}{\|}}{C}CH_3)(CO_2H) + CH_3COOH$$

Moles acetylsalicylic acid = moles salicylic acid; moles salicylic acid = 1.0 g/(122 g/mol) = 8.20×10^{-3} mol. Theoretical yield of acetylsalicylic acid is (8.20×10^{-3} mol)(180 g/mol) = 1.48 g; % yield = (1.02 g/1.48 g)(100) = 68.9%.

EXPERIMENT 39

1. An analyte is the substance being determined in an analysis; a standard solution is one whose concentration is accurately known; titration is the operation of adding a standard solution to another solution until the chemical reaction between the two solutes is complete; standardization is the process of accurately determining the concentration of a solution.
2. (a)

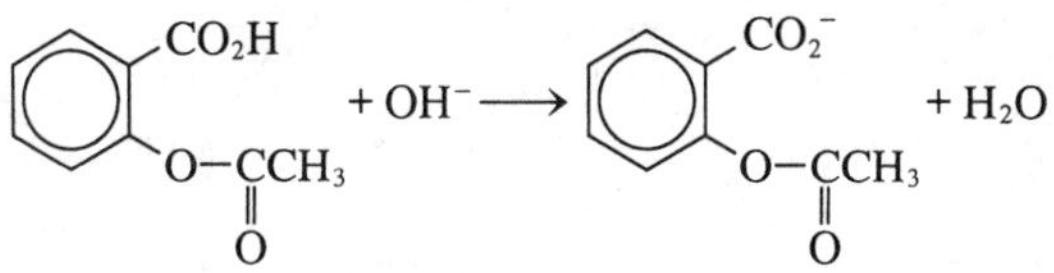

$$C_6H_4(CO_2H)(O-\underset{\underset{\Large O}{\|}}{C}CH_3) + OH^- \longrightarrow C_6H_4(CO_2^-)(O-\underset{\underset{\Large O}{\|}}{C}CH_3) + H_2O$$

(b)

$$C_6H_4(CO_2H)(O\text{-}\overset{\overset{\large O}{\|}}{C}CH_3) + 2OH^- \longrightarrow C_6H_4(CO_2^-)(OH) + CH_3CO_2^- + H_2O$$

3.

$$R\text{—}\overset{\overset{\large O}{\|}}{C}\text{—}OCH_3 + OH^- \longrightarrow R\text{—}\overset{\overset{\large O}{\|}}{C}\text{—}O^- + CH_3OH$$

4. 35.3 mL × 0.25N = 38.5 mL × N; N = (35.3 mL × 0.25 N)/38.5 mL = 0.229 N. Because there is one equivalent per mol, molarity equals normality
5. mL × N = meq; 38.5 mL × 0.125N = 4.81 meq; (40.0 g/eq)(1 eq/1000 meq)(4.81 meq) = 0.193 g.

EXPERIMENT 40

4. The strength of interaction between an ion and an ion-exchange resin increases with an increase in the charge-to-size ratio of the ion because these are electrostatic interactions.
6. Mg^{2+} should be more slowly eluted because it has the larger charge-to-size ratio.

Vapor Pressure of Water at Various Temperatures

APPENDIX **L**

Temperature (°C)	Pressure (mm Hg)	Temperature (°C)	Pressure (mm Hg)
0	4.6	26	25.2
1	4.9	27	26.7
2	5.3	28	28.3
3	5.7	29	30.0
4	6.1	30	31.8
5	6.5	31	33.7
6	7.0	32	35.7
7	7.5	33	37.7
8	8.0	34	39.9
9	8.6	35	42.2
10	9.2	40	55.3
11	9.8	45	71.9
12	10.5	50	92.5
13	11.2	55	118.0
14	12.0	60	149.4
15	12.8	65	187.5
16	13.6	70	233.7
17	14.5	75	289.1
18	15.5	80	355.1
19	16.5	85	433.6
20	17.5	90	525.8
21	18.7	95	633.9
22	19.8	97	682.1
23	21.1	99	733.2
24	22.4	100	760.0
25	23.8	101	787.6

APPENDIX

M Names, Formulas, and Charges of Common Polyatomic Ions

Formula and charge	Name	Formula and charge	Name
NH_4^+	Ammonium	CrO_4^{2-}	Chromate
CH_3COO^-	Acetate	$Cr_2O_7^{2-}$	Dichromate
NO_3^-	Nitrate	SiO_3^{2-}	Silicate
NO_2^-	Nitrite	PO_4^{3-}	Phosphate
OH^-	Hydroxide	HPO_4^{2-}	Hydrogen phosphate
ClO^-	Hypochlorite	$H_2PO_4^-$	Dihydrogen phosphate
ClO_2^-	Chlorite	PO_3^{3-}	Phosphite
ClO_3^-	Chlorate	HPO_3^{2-}	Hydrogen phosphite
ClO_4^-	Perchlorate	H_2PO_3	Dihydrogen phosphite
BrO^-	Hypobromite	IO^-	Hypoiodite
BrO_2^-	Bromite	IO_2^-	Iodite
BrO_3^-	Bromate	IO_3^-	Iodate
BrO_4^-	Perbromate	IO_4^-	Periodate
MnO_4^-	Permanganate	AsO_4^{3-}	Arsenate
MnO_4^{2-}	Manganate	AsO_3^{3-}	Arsenite
CO_3^{2-}	Carbonate	N_3^-	Azide
HCO_3^-	Hydrogen carbonate	CN^-	Cyanide
SO_4^{2-}	Sulfate	SCN^-	Thiocyanate
SO_3^{2-}	Sulfite	OCN^-	Cyanate
$S_2O_3^{2-}$	Thiosulfate	O^{2-}	Oxide
O_2^{2-}	Peroxide	HS^-	Bisulfide or hydrogen sulfide

APPENDIX

N Some Formula Weights

Formula	Weight
AgBr	187.78
AgCl	143.32
Ag_2CrO_4	331.73
AgI	234.77
$AgNO_3$	169.87
Al_2O_3	101.96
$Al_2(SO_4)_3$	342.14
B_2O_3	69.62
$BaCO_3$	197.35
$BaCl_2$	208.25
$BaCrO_4$	253.33
$Ba(OH)_2$	171.36
$BaSO_4$	233.40
CO_2	44.01
$CaCO_3$	100.09
CaO	56.08
$CaSO_4$	136.14
$[Co(C_5H_5N)_4(NCS)_2]$	491.05
CuO	79.54
Cu_2O	143.08
$CuSO_4$	159.60
$[Cu(C_5H_5N)_2(NCS)_2]$	337.54
$Fe(NH_4)_2(SO_4)_2 \cdot 6H_2O$	392.14
FeO	71.85
Fe_2O_3	159.69
Fe_3O_4	231.54
HBr	80.92
HC_2H_3O (acetic)	60.05
$HC_7H_5O_2$ (benzoic)	122.12
HCl	36.46
$HClO_4$	100.46
$H_2C_2O_4 \cdot 2H_2O$	126.07
HNO_3	63.01
H_2O	18.015
H_2O_2	34.01
H_3PO_4	98.00
H_2S	34.08
H_2SO_3	82.08
H_2SO_4	98.08
HgO	216.59
Hg_2Cl_2	472.09
$HgCl_2$	271.50
KBr	119.01
KCl	74.56
$KClO_3$	122.55
K_2CrO_4	194.20
$K_2Cr_2O_7$	294.19
$KHC_8H_4O_4$ (phthalate)	204.23
K_2HPO_4	174.18
KH_2PO_4	136.09
$KHSO_4$	136.17
KI	166.01
KIO_3	214.00
KIO_4	230.00
$KMnO_4$	158.04
KNO_3	101.11
KOH	56.11
K_2SO_4	174.27
$MgNH_4PO_4$	137.35
MgO	40.31
$MgSO_4$	120.37
MnO_2	86.94
Mn_2O_3	157.88
NaBr	102.90
$NaC_2H_3O_2$	82.03
NaCl	58.44
NaCN	49.01
Na_2CO_3	105.99
Na_2O_2	77.98
NaOH	40.00
NaSCN	81.07
Na_2SO_4	142.04
$Na_2S_2O_3 \cdot 5H_2O$	248.18
NH_3	17.03
NH_4Cl	53.49
NH_4NO_3	80.04
$(NH_4)_2SO_4$	132.14
$[Ni(C_5H_5N)_4(NCS)_2]$	490.83
$PbCrO_4$	323.18
PbO	223.19
PbO_2	239.19
$PbSO_4$	303.25
P_2O_5	141.94
SiO_2	60.08
$SnCl_2$	189.60
SnO_2	150.69
SO_2	64.06
SO_3	80.06
$ZnCl_z$	136.27
$[Zn(C_5H_5N)_2(NCS)_2]$	339.37

APPENDIX

Basic SI Units, Some Derived SI Units, and Conversion Factors

	SI unit	Conversion factors
Length	Meter (m)	1 m = 100 centimeters (cm)
		= 1.0936 yards (yd)
		1 cm = 0.3937 inch (in.)
		1 in. = 2.54 cm
		= 0.0254 m
		1 angstrom (Å) = 10^{-10} m
Mass	Kilogram (kg)	1 kg = 1000 grams (g)
		= 2.205 pounds (lb)
		1 lb = 453.6 grams (g)
		1 atomic mass unit (amu) = 1.66053×10^{-24}
Time	Second (s)	1 day (d) = 86,400 s
		1 hour (hr) = 3600 s
		1 minute (min) = 60 s
Electric current	Ampere (A)	
Temperature	Kelvin (K)	O K = −273.15° Celsius (C)
		= −459.67° Fahrenheit (F)
		°F = (9/5)°C + 32°
		°C = (5/9)(°F − 32°)
		K = °C + 273.15°
Luminous intensity	Candela (cd)	
Amount of substance	Mole (mol)	
Volume (derived)	Cubic meter (m^3)	1 liter (L) = 10^{-3} m^3
		= 1.057 quarts (qt)
		1 $in.^3$ = 16.4 cm^3
Force (derived)	Newton	1 dyne (dyn) = 10^{-5} N
	(N = m-kg/s^2)	
Pressure (derived)	Pascal	1 atmosphere (atm) = 101,325 Pa
	(Pa = N/m^2)	= 760 mm Hg
		= 14.70 lb/$in.^2$
		= 1.013×10^6 dyn/cm^2
Energy (derived)	Joule	1 calorie (cal) = 4.184 J
	(J = N-m)	1 electron volt (eV) = 1.6022×10^{-19} J
		1 erg = 6.2420×10^{11} eV
		1 J = 10^7 ergs